VOLUME ONE HUNDRED AND THIRTY SIX

INTERNATIONAL REVIEW OF NEUROBIOLOGY

The Role of Neuropeptides in Addiction and Disorders of Excessive Consumption

INTERNATIONAL REVIEW OF NEUROBIOLOGY

VOLUME 136

SERIES EDITOR

PATRICIA JANAK
Janak Lab, Dunning Hall
Baltimore, MD, USA

PETER JENNER
Division of Pharmacology and Therapeutics
GKT School of Biomedical Sciences
King's College, London, UK

EDITORIAL BOARD

VOLUME ONE HUNDRED AND THIRTY SIX

INTERNATIONAL REVIEW OF NEUROBIOLOGY

The Role of Neuropeptides in Addiction and Disorders of Excessive Consumption

Edited by

TODD E. THIELE
University of North Carolina at Chapel Hill, Chapel Hill, NC, United States

ELSEVIER

ACADEMIC PRESS
An imprint of Elsevier

Academic Press is an imprint of Elsevier
50 Hampshire Street, 5th Floor, Cambridge, MA 02139, United States
525 B Street, Suite 1800, San Diego, CA 92101-4495, United States
The Boulevard, Langford Lane, Kidlington, Oxford OX5 1GB, United Kingdom
125 London Wall, London, EC2Y 5AS, United Kingdom

First edition 2017

Notices
Knowledge and best practice in this field are constantly changing. As new research and experience broaden our understanding, changes in research methods, professional practices, or medical treatment may become necessary.

Practitioners and researchers must always rely on their own experience and knowledge in evaluating and using any information, methods, compounds, or experiments described herein. In using such information or methods they should be mindful of their own safety and the safety of others, including parties for whom they have a professional responsibility.

To the fullest extent of the law, neither the Publisher nor the authors, contributors, or editors, assume any liability for any injury and/or damage to persons or property as a matter of products liability, negligence or otherwise, or from any use or operation of any methods, products, instructions, or ideas contained in the material herein.

ISBN: 978-0-12-812473-4
ISSN: 0074-7742

For information on all Academic Press publications
visit our website at https://www.elsevier.com/books-and-journals

Publisher: Zoe Kruze
Acquisition Editor: Sam Mahfoudh
Editorial Project Manager: Andrea Gallego Ortiz
Production Project Manager: Vignesh Tamil
Cover Designer: Victoria Pearson

Typeset by SPi Global, India

CONTENTS

CONTRIBUTORS

Jessica R. Barson
Drexel University College of Medicine, Philadelphia, PA, United States

Mehdi Farokhnia
Section on Clinical Psychoneuroendocrinology and Neuropsychopharmacology, National Institute on Alcohol Abuse and Alcoholism and National Institute on Drug Abuse, National Institutes of Health, Bethesda, MD, United States

William J. Giardino
Stanford University, Stanford, CA, United States

Markus Heilig
Center for Social and Affective Neuroscience, Linkoping University, Linkoping, Sweden

Katherine M. Holleran
Wake Forest School of Medicine, Winston-Salem, NC, United States

Sara R. Jones
Wake Forest School of Medicine, Winston-Salem, NC, United States

Anushree Karkhanis
Wake Forest School of Medicine, Winston-Salem, NC, United States

Dean Kirson
The Scripps Research Institute, La Jolla, CA, United States

Lorenzo Leggio
Section on Clinical Psychoneuroendocrinology and Neuropsychopharmacology, National Institute on Alcohol Abuse and Alcoholism and National Institute on Drug Abuse, National Institutes of Health, Bethesda, MD; Center for Alcohol and Addiction Studies, Brown University, Providence, RI, United States

Sarah F. Leibowitz
The Rockefeller University, New York, NY, United States

Montserrat Navarro
University of North Carolina, Chapel Hill, NC, United States

Cort A. Pedersen
The University of North Carolina at Chapel Hill, Chapel Hill, NC, United States

Marisa Roberto
The Scripps Research Institute, La Jolla, CA, United States

Stacey L. Robinson
University of North Carolina at Chapel Hill, Chapel Hill, NC, United States

Andrey E. Ryabinin
Oregon Health & Science University, Portland, OR, United States

Jesse R. Schank
University of Georgia, Athens, GA, United States

Samantha R. Spierling
The Scripps Research Institute, La Jolla, CA, United States

Todd E. Thiele
University of North Carolina at Chapel Hill, Chapel Hill, NC, United States

Brendan J. Tunstall
Neurobiology of Addiction Section, Intramural Research Program, National Institute on Drug Abuse, National Institutes of Health, Baltimore, MD, United States

Leandro F. Vendruscolo
Neurobiology of Addiction Section, Intramural Research Program, National Institute on Drug Abuse, National Institutes of Health, Baltimore, MD, United States

Lia J. Zallar
Section on Clinical Psychoneuroendocrinology and Neuropsychopharmacology, National Institute on Alcohol Abuse and Alcoholism and National Institute on Drug Abuse, National Institutes of Health, Bethesda; Neurobiology of Addiction Section, Intramural Research Program, National Institute on Drug Abuse, National Institutes of Health, Baltimore, MD, United States

Eric P. Zorrilla
The Scripps Research Institute, La Jolla, CA, United States

PREFACE

Centrally acting neuropeptides modulate a diverse array of neurobiological functions, including consummatory and ingestive behaviors that are critical for survival. Many neuropeptides modulate the intake of both natural rewards (i.e., those necessary for survival such as food and water) and the consumption of drugs and alcohol. When the actions of these neuropeptides are imbalanced, the result can translate into excessive consumption that threatens the well-being of the organism. Interestingly, in many cases, repeated consumption of food, drugs, and alcohol can promote alterations in neuropeptide systems, which in turn drive increased consumption and further neuropeptide imbalance. This vicious cycle is thought to underlie the transition to drug and alcohol dependence, as well as eating disorders. Thus, targeting neuropeptide systems as therapeutic treatments for disorders of excessive consumption is a promising approach.

The goal of this volume is to highlight some of the most well-studied neuropeptide systems that modulate consumption of drugs, alcohol, and, in some cases, food. I would like to take this opportunity to thank the authors that have taken the time to provide outstanding chapters for this book. After a brief introduction in Chapter 1, Marisa Roberto, Samantha Spierling, Dean Kirson, and Eric Zorrilla in Chapter 2 describe the role of elevated corticotropin-releasing factor (CRF) signaling in modulating negative emotional states that emerge with drug and alcohol dependence, which in turn drives excessive intake through negative reinforcement, a process referred to as the "dark side" of addiction. The role of CRF in compulsive eating disorders is also discussed. In Chapter 3, Anushree Karkhanis, Katherine Holleran, and Sara Jones describe the role of the dynorphin/kappa opioid receptor (KOR) system in modulating the "dark side" of addiction. Like CRF, dynorphin/KOR signaling is upregulated with drug and alcohol dependence which produces a negative emotional state that motivates excessive drug and alcohol intake via negative reinforcement. The dynorphin/KOR system modulates excessive consumption, in part, by blunting dopamine signaling in the brain. In Chapter 4, Lia Zallar, Mehdi Farokhnia, Brendan Tunstall, Leandro Vendruscolo, and Lorenzo Leggio describe the role of the gut peptide, ghrelin, in modulating drug and alcohol intake. Ghrelin acts centrally and has been implicated in both feeding and drug and alcohol use in both preclinical and clinical studies. In Chapter 5,

Montserrat Navarro describes the role of the melanocortin (MC) system in modulating drug and alcohol consumption. For many years the role of the MC system has been studied with respect to food intake and obesity, and the work by Navarro highlights the role of this system in modulating excessive alcohol intake. Given the overlapping control of feeding and alcohol intake, therapeutic drugs aimed at the MC may be useful for treating both obesity and alcohol use disorders. In Chapter 6, Jesse Schank and Markus Heilig describe the role of substance P (SP) and the associated neurokinin-1 receptor (NK1R) in treating alcohol use disorders. They compare SP and the NK1R to the CRF system and provide evidence that NK1R antagonists may have therapeutic value for treating addiction. In Chapter 7, Stacey Robinson and myself overview evidence for a role for neuropeptide Y (NPY) in modulating drug and alcohol use disorders. Repeated drug and alcohol use is associated with reduced NPY signaling in brain regions that modulate emotion which triggers a negative emotional state and increased drug and alcohol intake via negative reinforcement. Thus, like many of the neuropeptides described earlier, therapeutic targets aimed at the NPY system may have value for treating the "dark side" of addiction. In Chapter 8, Jessica Barson and Sarah Leibowitz discuss the role of the hypothalamic neuropeptide, orexin/hypocretin (OX). They describe how OX signaling in the hypothalamus is involved in modulating intake of natural rewards that maintain homeostasis, and how OX signaling outside of the hypothalamus can modulate the intake of alcohol and drugs of abuse. Interestingly, drugs and alcohol stimulate OX signaling which they suggest promotes abuse through a positive feedback loop. In Chapter 9, Cort Pedersen describes the role of oxytocin in modulating drug and alcohol use. Clinical trial evidence suggests that intranasal oxytocin blunts withdrawal responses and reduces alcohol consumption in heavy drinkers. It is hypothesized that oxytocin may reduce alcohol intake by reversing tolerance. Finally, in Chapter 10, Andrey Ryabinin and William Giardino describe the contribution of the urocortin system in alcohol use disorders. Interestingly, urocortin neuropeptides act on CRF receptors and thus are part of the CRF system. They overview evidence, indicating that urocortin 1 modulates excessive alcohol intake.

Once again, I am very grateful to my colleagues for all the time and effort they dedicated to this volume. It is my hope that this volume will impress the reader with the significant role that neuropeptides play in modulating disorders of excessive consumption.

Todd E. Thiele
University of North Carolina at Chapel Hill

CHAPTER ONE

Neuropeptides and Addiction: An Introduction

Todd E. Thiele[1]
University of North Carolina at Chapel Hill, Chapel Hill, NC, United States
[1]Corresponding author: e-mail address: thiele@unc.edu

Abstract

Neuropeptides are short sequences of amino acids that are coexpressed with neurotransmitters and which are widely expressed throughout the central nervous system. There is a large database of data pointing to critical roles for neuropeptides in modulating neurobiological responses to alcohol and drugs of abuse. Continued alcohol and drug use promote allostatic alterations in neuropeptide systems, and these changes contribute to excessive and uncontrolled intake that emerges with dependence. The neuropeptides that are reviewed in this chapter represent some of the most well-studied targets in the current drug and alcohol abuse literature. The goal of this chapter is to convey the significant roles that neuropeptides play in neurobiological responses to alcohol and drugs of abuse, and reinforce the idea that targeting neuropeptides and their receptors continue to be attractive avenues for treating drug and alcohol use disorders, as well as eating disorders.

The goal of this chapter is to provide the reader with an appreciation for the critical role that central neuropeptides play in modulating disorders of excessive consumption, including drug and alcohol abuse and eating disorders. Neuropeptides are short sequences of amino acids, typically defined as having 50 base pairs or less (with large sequences referred to as proteins). Neuropeptides are widely distributed throughout the central nervous system and are cosecreted from neurons with small-molecule neurotransmitters, including glutamate and gamma-aminobutyric acid. Neuropeptides can function as neuromodulators, influencing the synaptic transmission of cosecreted small-molecule neurotransmitters, as well as neurotransmitters. Generally, neuropeptides act on G protein-coupled receptors that are expressed pre- and postsynaptically, often depending on receptor subtype (for a recent review on neuropeptides, see van den Pol, 2012). There are numerous neuropeptides expressed throughout the brain with unique expression and projection patterns. In many cases, specific neuropeptides appear to be involved

International Review of Neurobiology, Volume 136
ISSN 0074-7742
http://dx.doi.org/10.1016/bs.irn.2017.07.001

with diverse sets of neurobiological functions, depending on the brain region of action. For example, numerous neuropeptides arise from the hypothalamus and help maintain physiological homeostasis and energy balance by modulating motivated behaviors aimed at acquiring natural reinforces, such as food and water (Zeltser, Seeley, & Tschop, 2012). These hypothalamic neuropeptides can act locally within the hypothalamus, or project to other regions to interact with other systems, such as the mesolimbic dopamine reward system (Roseberry, Stuhrman, & Dunigan, 2015) which modulates the reinforcing properties of drugs and natural reward. Other neuropeptides are produced within the extended amygdala, an interconnected and functionally similar set of brain regions critical for integrating emotional behaviors (Koob & Le Moal, 2001). Here again, neuropeptides can function locally within the extended amygdala or project to, and interact with, other system.

A growing body of literature has emerged indicating important roles for neuropeptides in modulating neurobiological responses to alcohol and other drugs of abuse, and several of these neuropeptides are described in the following chapters. Repeated drug and alcohol abuse are thought to produce plastic alterations in neuropeptide systems, creating allostatic shifts in activity and function, above or below, homeostatic levels maintained under baseline conditions (Koob & Le Moal, 2001). Alterations of neuropeptide activity, in systems such as those that modulate energy balance, reward, or emotional responses, are thought to contribute to escalating drug and alcohol use as one transitions to dependence. Years ago we proposed the idea that overlapping neuropeptide systems may modulate excessive alcohol intake and eating disorders, based on converging evidence that numerous neuropeptide systems were similarly influence by ethanol drinking and energy balance, and that manipulating these systems similarly influence ethanol intake and feeding (Thiele et al., 2003). These observations raise the interesting possibility that neuropeptides involved with feeding and alcohol intake influence the consumption of these stimuli through modulation of their reinforcing properties. Interestingly, every neuropeptide reviewed in this chapter has been implicated in modulating both feeding behavior and consumption or seeking of drugs/alcohol.

Drug- and alcohol-induced alterations of neuropeptide systems within the extended amygdala promote dysregulation of emotions which contributes to a negative affective state (associated with symptoms such as anxiety and depression) that is characteristic to those dependent on drugs or alcohol, particularly during withdrawal. Negative affect, in turn, is thought to motivate drug and alcohol intake as a way to reduce these adverse symptoms (an example of negative reinforcement), reflecting excessive drug and alcohol intake stemming from dependence (Koob & Le Moal, 2001). Finally, neuropeptides have been implicated in other neurobiological responses to drugs and alcohol that have been proposed to influence their intake, including sensitivity to the drug's intoxicating effects (Schuckit, 2009) and the modulation of pain (Egli, Koob, & Edwards, 2012).

While it would not be practical to review all neuropeptides involved, the neuropeptides that are reviewed in this chapter represent some of the most well-studied targets in the current drug and alcohol abuse literature. We hope that this chapter will convey the significant roles that neuropeptides play in neurobiological responses to alcohol and drugs of abuse, and reinforce the idea that targeting neuropeptides and their receptors continue to be attractive avenues for treating drug and alcohol use disorders, as well as eating disorders.

REFERENCES

Egli, M., Koob, G. F., & Edwards, S. (2012). Alcohol dependence as a chronic pain disorder. *Neuroscience and Biobehavioral Reviews*, *36*(10), 2179–2192. http://dx.doi.org/10.1016/j.neubiorev.2012.07.010.

Koob, G. F., & Le Moal, M. (2001). Drug addiction, dysregulation of reward, and allostasis. *Neuropsychopharmacology*, *24*(2), 97–129.

Roseberry, A. G., Stuhrman, K., & Dunigan, A. I. (2015). Regulation of the mesocorticolimbic and mesostriatal dopamine systems by alpha-melanocyte stimulating hormone and agouti-related protein. *Neuroscience and Biobehavioral Reviews*, *56*, 15–25. http://dx.doi.org/10.1016/j.neubiorev.2015.06.020.

Schuckit, M. A. (2009). An overview of genetic influences in alcoholism. *Journal of Substance Abuse Treatment*, *36*(1), S5–14.

Thiele, T. E., Navarro, M., Sparta, D. R., Fee, J. R., Knapp, D. J., & Cubero, I. (2003). Alcoholism and obesity: Overlapping neuropeptide pathways? *Neuropeptides*, *37*, 321–337.

van den Pol, A. N. (2012). Neuropeptide transmission in brain circuits. *Neuron*, *76*(1), 98–115. http://dx.doi.org/10.1016/j.neuron.2012.09.014.

Zeltser, L. M., Seeley, R. J., & Tschop, M. H. (2012). Synaptic plasticity in neuronal circuits regulating energy balance. *Nature Neuroscience*, *15*(10), 1336–1342. http://dx.doi.org/10.1038/nn.3219.

CHAPTER TWO

Corticotropin-Releasing Factor (CRF) and Addictive Behaviors

Marisa Roberto[1], Samantha R. Spierling, Dean Kirson, Eric P. Zorrilla
The Scripps Research Institute, La Jolla, CA, United States
[1]Corresponding author: e-mail address: mroberto@scripps.edu

Contents

Abstract

Drug addiction is a complex disorder that is characterized by compulsivity to seek and take the drug, loss of control in limiting intake of the drug, and emergence of a withdrawal syndrome in the absence of the drug. The transition from casual drug use to dependence is mediated by changes in reward and brain stress functions and has been linked to a shift from positive reinforcement to negative reinforcement. The recruitment of brain stress systems mediates the negative emotional state produced by dependence that drives drug seeking through negative reinforcement mechanisms, defined as the "dark side" of addiction. In this chapter we focus on behavioral and cellular neuropharmacological studies that have implicated brain stress systems (i.e., corticotropin-releasing factor [CRF]) in the transition to addiction and the predominant brain regions involved. We also discuss the implication of CRF recruitment in compulsive eating disorders.

International Review of Neurobiology, Volume 136
ISSN 0074-7742
http://dx.doi.org/10.1016/bs.irn.2017.06.004

1. INTRODUCTION

Addiction is a complex, chronic relapsing disorder characterized by compulsive seeking and taking of a substance of abuse, loss of control in limiting intake, and the emergence of a withdrawal syndrome with negative emotional components upon cessation of use (American Psychiatric Association, 1994; World Health Organization, 1992). Clinically, the occasional and limited use of an abused substance is distinct from the escalated, uncontrolled use, and compulsive seeking that defines addiction (Koob & Volkow, 2010). Each year, about half of all American adults suffer from an addictive disorder (Sussman, Lisha, & Griffiths, 2011). Addiction imposes a major public health burden with devastating medical and socioeconomic consequences (Kessler et al., 2005). Alcohol misuse alone has an annual prevalence of 10% in the United States (Sussman et al., 2011) and accounts for 10% of total disability in developed countries (Rehm et al., 2009). In 2015, 20.5 million Americans age 12 or older had a substance use disorder; of these, 2 million involved prescription pain relievers and 591,000 involved heroin (Center for Behavioral Health Statistics and Quality, 2016). Furthermore, drug overdose is the leading cause of accidental death in the United States, with 52,404 lethal drug overdoses in 2015, more than 60% involving opioids (Rudd, Seth, David, & Scholl, 2016).

Addiction has been conceptualized as a cycle with three stages—*binge/intoxication*, *withdrawal/negative affect*, and *preoccupation/anticipation*—that worsen over time and use and culminate in a severe neurobiological disorder. Much ongoing research seeks to identify the molecular and cellular neuroadaptive changes that result from exposure to substances of abuse and which promote the transition from casual use to addiction. Chronic substance use, even if begun for its rewarding effects, progressively leads to anxiety, irritability, and depressed mood during abstinence, resulting in negatively reinforced use in order to "self-medicate" the negative emotional state. An extension of the "opponent process theory of affective regulation" (Solomon & Corbit, 1974), this hypothesis of addiction proposes that substances with abuse potential initially activate brain structures that subserve positive emotional states (e.g., pleasure, contentment, well-being), such as dopaminergic inputs to the nucleus accumbens (NAc) from the ventral tegmental area (VTA) and opioid inputs to the extended amygdala from the arcuate nucleus of the hypothalamus. To restore emotional homeostasis, however, a counter-regulatory opponent process then decreases mood and increases vigilance/tension via downregulation of brain reward systems (e.g., NAc) and upregulation of brain stress systems (Breese, Sinha, & Heilig, 2011; George et al., 2012; Heilig, Goldman, Berrettini, & O'Brien, 2011; Heilig & Koob, 2007; Heilig, Thorsell, et al., 2010; Koob & Volkow, 2010; Logrip, Koob, & Zorrilla, 2011). With continued cycles of intoxication/withdrawal, the neuroadaptive opponent process predominates over the primary rewarding process. Then, more of the substance of abuse is needed just to reattain euthymia. If drug use stops, negative emotional symptoms emerge during acute withdrawal. As a result, associated environmental cues and stress can trigger the development of preoccupation with obtaining the substance in anticipation of its alleviating effects. With enough drug use history, symptoms of dysphoria may episodically reappear even weeks or months after detoxification as components of the protracted withdrawal syndrome. Exaggerated reactivity to otherwise mild stressors also can occur. Accordingly, fMRI activation responses to negative affective pictures are sensitized in detoxified alcoholics (Gilman & Hommer, 2008). Under this conceptual framework, substance use escalates and becomes compulsive because it mitigates the counter-regulatory, long-term emotional disturbances that persist despite abstinence (Heilig & Koob, 2007; Koob & Volkow, 2010; Zorrilla, Heilig, de Wit, & Shaham, 2013). As will be reviewed, corticotropin-releasing factor (CRF) stress systems are hypothesized to play a key role in all three stages of the addiction cycle but particularly in the *withdrawal/negative affect* stage.

2. CRF STRESS SYSTEMS

Since the successive discovery by Wylie Vale and his colleagues of the 41-residue stress-related peptide CRF (also known as corticotropin-releasing hormone or CRH) (Vale, Spiess, Rivier, & Rivier, 1981), the structurally related urocortins (Ucn 1, Ucn 2, Ucn 3), and their cognate receptors (CRF_1, CRF_2) (Bale & Vale, 2004; Fekete & Zorrilla, 2007), CRF systems have received attention as therapeutic targets for substance abuse. CRF_1 and CRF_2 receptors are class B1 (secretin-like) G-protein coupled receptors with ~70% sequence identity for one another. Though most functional significance is attributed to the $CRF_{1(a)}$ subtype, there are multiple CRF_1 receptor isoforms (e.g., CRF_{1a}–CRF_{1h}). In humans, the CRF_2 receptor has three known membrane-associated functional subtypes—$CRF_{2(a)}$, $CRF_{2(b)}$, and $CRF_{2(c)}$. CRF is a preferential agonist for CRF_1 over CRF_2 receptors. Ucn 1 is a high-affinity agonist at both receptor subtypes, and the type 2 urocortins (Ucn 2 and Ucn 3) are selective CRF_2 receptor agonists. A CRF-binding protein (CRF-BP) binds CRF and Ucn 1 with equal or greater affinity than do CRF receptors, and Ucn 2 with somewhat lesser affinity, and has been suggested to act to sequester peptide (inhibiting activity) or alternatively to chaperone/complex with peptide to provide a different signaling method. Most CRF receptor antagonists do not bind to the CRF-BP, confirming the different structural requirements for binding to CRF receptors vs the CRF-BP (Fekete & Zorrilla, 2007; Zorrilla & Koob, 2004). Pharmacological and transgenic studies indicate that brain and pituitary CRF_1 receptors mediate endocrine, behavioral, and autonomic responses to stress (Heinrichs & Koob, 2004; Koob & Heinrichs, 1999; Zorrilla et al., 2013; Zorrilla & Koob, 2010), and CRF_2 receptor activation has been associated with decreased feeding (Fekete & Zorrilla, 2007; Pelleymounter et al., 2000; Spina et al., 1996).

3. CRF, THE HPA AXIS, AND ADDICTION

CRF was first identified as the central initiator of the classic hypothalamic–pituitary–adrenal (HPA) axis stress neuroendocrine response. Stressors elicit CRF synthesis in the paraventricular nucleus of the hypothalamus and its release, via the median eminence, into the portal blood. Then, CRF activates CRF_1 receptors on anterior pituitary corticotrophs, stimulating ACTH synthesis and release into circulation, which culminates in the production and secretion of the glucocorticoid cortisol (structurally similar to corticosterone in rodents) by the adrenal cortex.

Most drugs with abuse potential, including opioids (Buckingham, 1982), amphetamine (Swerdlow, Koob, Cador, Lorang, & Hauger, 1993), cocaine (Calogero, Gallucci, Kling, Chrousos, & Gold, 1989), nicotine (Buckingham & Hodges, 1979), marijuana (Δ^9-tetrahydrocannabinol; Weidenfeld, Feldman, & Mechoulam, 1994), and alcohol (Rivier, Rivier, & Vale, 1984), acutely activate the HPA axis via hypothalamic CRF (Rivier, Grigoriadis, & Rivier, 2003; Sarnyai, Biro, Penke, & Telegdy, 1992). This HPA axis activation, via glucocorticoids, has been hypothesized to facilitate activity in brain motivational circuits, drug reward, and the acquisition of drug-seeking behavior (Fahlke, Hard, & Hansen, 1996; Goeders, 1997; Piazza et al., 1993; Piazza & Le Moal, 1997). For example, acute HPA activation is involved in the development of drug-induced locomotor sensitization (Cole et al., 1990), and inhibition of CRF_1 receptors blocks drug-induced behavioral sensitization (Fee, Sparta, Picker, & Thiele, 2007; Przegalinski, Filip, Frankowska,

Zaniewska, & Papla, 2005). Genetic deletion of CRF_1, but not CRF_2 receptors, similarly reduces the development of ethanol-induced behavioral sensitization (Pastor et al., 2008). Supporting a role for glucocorticoids in this effect, adrenalectomy or corticosteroid synthesis inhibitors reduce not only cocaine-induced locomotor activation but also the acquisition of cocaine self-administration behavior (Goeders & Guerin, 1996). Importantly, withdrawal from all drugs of abuse studied to date also activates the HPA axis and glucocorticoid release. Thereby, both exposure- and withdrawal-induced CRF-induced glucocorticoid release may promote the development of self-administration of substances of abuse.

Perhaps more relevant to neuroadaptive aspects of addiction, progressive changes in the HPA axis occur during the transition from acute to chronic administration of drugs of abuse. Whereas acute administration of most drugs of abuse activates the HPA axis, repeated administration of cocaine, opioids, nicotine, and alcohol leads to blunted or dysregulated responses (Goeders, 2002; Koob & Kreek, 2007; Kreek & Koob, 1998; Rasmussen et al., 2000; Semba, Wakuta, Maeda, & Suhara, 2004; Sharp & Matta, 1993). Atypical HPA reactivity to stressors putatively contributes to the persistence and relapse to cycles of opioid dependence, a hypothesis that has been extended to other substances of abuse (Kreek & Koob, 1998).

High circulating levels of glucocorticoids exert negative feedback on the hypothalamus to shut-off the HPA axis neuroendocrine stress response. However, relevant to drug-induced neuroadaptation, high levels of glucocorticoids appear to "sensitize" CRF systems in the CeA and BLA, which subserve behavioral stress responses (Imaki, Nahan, Rivier, Sawchenko, & Vale, 1991; Makino, Gold, & Schulkin, 1994; Schulkin, McEwen, & Gold, 1994; Shepard, Barron, & Myers, 2000; Swanson & Simmons, 1989). Perhaps accordingly, glucocorticoid receptor antagonists have been shown to reduce the development and expression of excessive alcohol self-administration that results from repeated, intermittent intoxication (Vendruscolo et al., 2012, 2015). A double-blind, placebo-controlled human clinical study found that mifepristone (a glucocorticoid receptor and progesterone receptor antagonist) reduced alcohol-cued craving in the laboratory and naturalistic measures of alcohol consumption (Vendruscolo et al., 2015).

In sum, CRF-driven HPA axis activation may result from initial drug use and contribute to the *binge/intoxication* stage of addiction; continued activation may recruit extrahypothalamic brain stress systems that are involved in the *withdrawal/negative affect* stage (Koob & Kreek, 2007; Koob & Le Moal, 2005; Kreek & Koob, 1998).

4. EXTRAHYPOTHALAMIC CRF SYSTEMS

As noted, outside the hypothalamus, CRF-immunoreactive cell bodies also are found in many other brain regions, with particularly strong expression in the cortex, lateral septum, and a macrostructure known as the extended amygdala (which includes the central nucleus of the amygdala [CeA] and medial amygdala [MeA], bed nucleus of the stria terminalis [BNST], and a transition area in the medial [shell] part of the NAc) (Pilcher & Joseph, 1984), all of which are activated by stressors (Day, Nebel, Sasse, & Campeau, 2005; Hammack, Richey, Watkins, & Maier, 2004; Heinrichs, Menzaghi, Merlo Pich, Britton, & Koob, 1995; Melia, Sananes, & Davis, 1992). Accordingly, extrahypothalamic CRF is recognized to subserve behavioral stress responses (Heinrichs et al., 1995) independent of HPA axis activation (reviewed in Koob, 1999). Thus, central administration of CRF mimics the behavioral response to stress in rodents, and competitive CRF receptor antagonists generally have anti-stress-like effects in rodent models (Heinrichs et al., 1994; Spina et al., 2000; for reviews, see Dunn & Berridge, 1990; Koob & Le Moal, 2001; Koob, Rassnick, Heinrichs, & Weiss, 1994; Logrip et al., 2011; Sarnyai, Shaham, & Heinrichs, 2001). The distribution of the CRF-binding protein (CRF-BP) overlaps partly with that of CRF, notably in the cortex and amygdala (Potter et al., 1992), where terminals containing CRF-BP colocalize with CRF-positive cell bodies (Potter et al., 1992) consistent with the regulatory interaction posited for these molecules.

CRF receptors are distributed even more widely than the CRF-BP (De Souza & Battaglia, 1988), affirming a role for CRF family peptides in modulating the development (Swinny et al., 2003) and excitability (Fu & Neugebauer, 2008; Gallagher, Orozco-Cabal, Liu, & Shinnick-Gallagher, 2008; Miyata, Okada, Hashimoto, Kano, & Ito, 1999; Rainnie et al., 2004; Schierloh, Deussing, Wurst, Zieglgansberger, & Rammes, 2007; Schmolesky, De Ruiter, De Zeeuw, & Hansel, 2007) of many neuronal populations. CRF_1 receptors are highly expressed in stress-responsive forebrain regions, including the neocortex, BLA, BNST, medial septum, hippocampus, and thalamus, with moderate expression in the NAc and VTA as well as marked expression in autonomic midbrain and hindbrain nuclei (Grigoriadis et al., 1996; Primus, Yevich, Baltazar, & Gallager, 1997; Sanchez, Young, Plotsky, & Insel, 1999; Van Pett et al., 2000). The CRF_1 receptor distribution resembles that of its natural ligands CRF and Ucn 1 and accounts for the dissociable role of extrahypothalamic CRF_1 systems (i.e., outside the HPA axis) to mediate behavioral and autonomic stress responses (Fekete & Zorrilla, 2007; Kozicz, Yanaihara, & Arimura, 1998; Swanson, Sawchenko, Rivier, & Vale, 1983; Zorrilla & Koob, 2004). In contrast, extrahypothalamic, forebrain CRF_2 receptors are primarily confined to the medial extended amygdala (MeA and medial BNST), lateral septum, and ventromedial hypothalamus (Chalmers, Lovenberg, & De Souza, 1995).

5. ROLE FOR CRF–CRF_1 SYSTEMS IN ANIMAL MODELS OF ADDICTION

Several preclinical lines of evidence support the hypothesis that central CRF systems contribute to the addiction cycle. For example, the anxiogenic-like response that results during withdrawal from repeated administration of cocaine, alcohol, nicotine, cannabinoids, and benzodiazepines can be reversed by CRF receptor antagonists, including after intracerebroventricular administration in rodent models of addiction (Baldwin, Rassnick, Rivier, Koob, & Britton, 1991; Basso, Spina, Rivier, Vale, & Koob, 1999; Knapp, Overstreet, Moy, & Breese, 2004; Overstreet, Knapp, & Breese, 2004; Rodriguez de Fonseca, Carrera, Navarro, Koob, & Weiss, 1997; Sarnyai et al., 1995; Tucci, Cheeta, Seth, & File, 2003). The increased ethanol (EtOH) self-administration that results from exposure to repeated cycles of chronic EtOH vapor (O'Dell, Roberts, Smith, & Koob, 2004; Rimondini, Arlinde, Sommer, & Heilig, 2002) likewise can be blocked by intracerebroventricular administration of a CRF_1/CRF_2 antagonist at doses that do not influence EtOH self-administration in nondependent animals (Valdez, Sabino, & Koob, 2004).

Many findings in animal models implicate CRF_1 receptors specifically in these "dark side" actions of CRF (or other endogenous CRF_1 agonists) to produce a negative emotional state. For example, the decreased brain reward function seen in animal models of withdrawal from nicotine, alcohol, or opioids, evident as increased current thresholds for intracranial self-stimulation behavior, can be blocked by CRF_1 antagonists (Bruijnzeel et al., 2012; Bruijnzeel, Prado, & Isaac, 2009; Bruijnzeel, Small, Pasek, & Yamada, 2010; Bruijnzeel, Zislis, Wilson, & Gold, 2007). A CRF_1 antagonist also prevented the development of a conditioned place aversion following precipitated withdrawal in opioid-dependent rats (Stinus, Cador, Zorrilla, & Koob, 2005). Similarly, *Crhr1* knockout mice did not show opioid withdrawal-induced conditioned place aversion (Contarino & Papaleo, 2005; Garcia-Carmona, Almela, Baroja-Mazo, Milanes, & Laorden, 2012). Finally, systemic injections of brain-penetrant CRF_1 antagonists reduced the heightened anxiety-like behavior of dependent rodents acutely withdrawn from alcohol or other substances of abuse (Breese, Overstreet, & Knapp, 2005; Breese, Overstreet, Knapp, & Navarro, 2005; Gehlert et al., 2007; Knapp et al., 2004; Overstreet et al., 2004; Sommer et al., 2008).

Perhaps because they reduce the negative emotional state, CRF_1 receptor antagonists also reduce the seeking and taking of substances of abuse in dependence models. For example, systemic administration of CRF_1 antagonists reduced the escalated EtOH self-administration of dependent rats and mice acutely withdrawn from alcohol at doses that did not alter the intake of nondependent animals (Chu, Koob, Cole, Zorrilla, & Roberts, 2007; Funk, Zorrilla, Lee, Rice, & Koob, 2007; Gehlert et al., 2007; Gilpin, Richardson, & Koob, 2008; Richardson et al., 2008; Sabino et al., 2006). Similarly, small-molecule CRF_1 antagonists differentially reduced the increase in drug self-administration associated with extended access to cocaine (Specio et al., 2008), nicotine (George et al., 2007), alcohol (Roberto et al., 2010), and heroin (Greenwell et al., 2009). Furthermore, both global (Chu et al., 2007) and conditional brain-specific *Crhr1* knockout mice (Molander et al., 2012) show reduced EtOH intake during withdrawal compared to their wild-type littermates. Consistent with a role for CRF systems in stress-induced substance use, the CRF_1 antagonist CP-154,526 reduced stress-induced potentiation of EtOH intake in nondependent mice (Lowery, Sparrow, Breese, Knapp, & Thiele, 2008), and *Crhr1* knockout mice were resistant to the ability of repeated forced swim stress to increase deprivation-induced EtOH intake (Pastor et al., 2011).

These salutary actions of CRF_1 antagonists in animal models extend to the postdependent phase, suggesting that CRF–CRF_1 activation can persist past detoxification into protracted withdrawal. For example, CRF_1 antagonists reduced potentiated stress-induced anxiogenic-like responses and EtOH intake during protracted withdrawal (Rimondini et al., 2002; Sommer et al., 2008; Valdez et al., 2002; Valdez, Zorrilla, Roberts, & Koob, 2003). CRF_1 antagonists also reduced the spontaneously greater levels of anxiety-like behavior (Breese, Overstreet, & Knapp, 2005; Breese, Overstreet, Knapp, & Navarro, 2005; Sommer et al., 2008; Valdez et al., 2002; Zhao, Valdez, et al., 2007; Zhao, Weiss, & Zorrilla, 2007) and EtOH drinking that occur in postdependent rats even in the absence of external stressors (Rimondini et al., 2002; Sommer et al.,

2008; Valdez et al., 2002). Both actions may have implications for relapse prevention since negative emotional states predict relapse in alcohol use disorders (Mossberg, Liljeberg, & Borg, 1985; Pickens, Hatsukami, Spicer, & Svikis, 1985).

Though the CRF-BP had initially been regarded as exclusively serving an inhibitory role in the CRF system, it has increasingly been recognized to have other modulatory roles in the brain, some of which may be relevant to the neurobiology of excessive consumption (Westphal & Seasholtz, 2006). For example, CRF_2 can interact with the CRF-BP to produce actions independent from CRF_1 (Milan-Lobo et al., 2009; Slater, Cerda, Pereira, Andres, & Gysling, 2016; Ungless et al., 2003; Wang, You, Rice, & Wise, 2007). The CRF-BP has recently received attention as a potential target for its role in alcohol use disorder (Haass-Koffler et al., 2016; Ketchesin, Stinnett, & Seasholtz, 2016), and its role in the escalation of alcohol drinking may involve its interaction with CRF_2 (Albrechet-Souza et al., 2015; Quadros, Macedo, Domingues, & Favoretto, 2016). Intriguingly, recent gene variant studies in humans have shown that the *CRHBP* rs1875999 locus was associated with risk for both cocaine and heroin addiction in African Americans in a study of heroin addicts ($n=314$), cocaine addicts ($n=281$), and healthy controls ($n=208$) (Levran, Peles, et al., 2014; Levran, Randesi, et al., 2014). SNPs in the *CRHBP* (10 kDa) fragment, *rs10055255*, *rs10062367*, and *rs7728378*, were each associated with increased risk of alcohol drinking and/or anxiety in patients with alcohol use disorder (Haass-Koffler et al., 2016).

Additional data that implicate CRF–CRF_1 activation in relapse behavior come from preclinical models of stress-induced reinstatement of substance seeking (Shaham, Shalev, Lu, De Wit, & Stewart, 2003; Stewart & de Wit, 1987). For example, systemic injections of the CRF_1 antagonist CP-154,526-attenuated footshock-induced reinstatement of EtOH seeking in nondependent rats (Le et al., 2000), and the CRF_1 antagonist antalarmin-attenuated reinstatement of EtOH seeking induced by yohimbine, an alpha-2 adrenergic antagonist with anxiogenic properties (Marinelli et al., 2007). Intracranial infusion of nonselective CRF antagonists or systemic administration of small-molecule CRF_1 antagonists also blocked stressor-induced reinstatement of cocaine-, opiate-, nicotine-, and methamphetamine-seeking behavior (Bruijnzeel et al., 2009; Le & Shaham, 2002; Lu, Shepard, Hall, & Shaham, 2003; Nawata, Kitaichi, & Yamamoto, 2012; Plaza-Zabala, Martin-Garcia, de Lecea, Maldonado, & Berrendero, 2010; Shaham, Erb, Leung, Buczek, & Stewart, 1998;

Shaham et al., 2003). CP-154,526 also inhibited stress-induced reinstatement of conditioned place preference to morphine and cocaine (Lu, Ceng, & Huang, 2000; Lu, Liu, & Ceng, 2001; Lu, Liu, Ceng, & Ma, 2000). A CRF_1 antagonist also reduced social defeat stress-induced locomotor sensitization to cocaine and escalated "binge" operant self-administration of cocaine (Boyson, Miguel, Quadros, Debold, & Miczek, 2011) as well as stress-induced reinstatement of cocaine-seeking behavior in rats with a history of extended access to cocaine (Blacktop et al., 2011). Similarly, the CRF_1 antagonists antalarmin and MTIP were effective in attenuating footshock-induced reinstatement of EtOH seeking in alcohol-dependent rats or in genetically selected, Marchigian Sardinian alcohol-preferring rats, which show innately upregulated CRF_1 mRNA expression in several brain areas (Gehlert et al., 2007; Hansson et al., 2006). Importantly, CRF_1 antagonists do not prevent cue-, substance-, or context-induced reinstatement. This specificity is consistent with the unique neuroanatomical and neuropharmacological bases for stress-induced reinstatement (see Bossert, Marchant, Calu, & Shaham, 2013; Koob & Le Moal, 2008; Steketee & Kalivas, 2011 for reviews).

Similar to the reviewed antianxiety and antidrug taking actions of CRF receptor antagonists, the ability of CRF_1 antagonists to reduce stress-induced reinstatement appears to involve nonneuroendocrine (i.e., extrahypothalamic) CRF. For example, intermittent fooshock-induced reinstatement is not affected by adrenalectomy (Le et al., 2000), and effective doses of antalarmin did not alter yohimbine-induced corticosterone levels (Marinelli et al., 2007). Accordingly, many findings affirm that stress-induced reinstatement of EtOH, cocaine, and heroin seeking is are mediated by stress-induced activation of extrahypothalamic CRF sites (Blacktop et al., 2011; Erb, Shaham, & Stewart, 1998; Le, Funk, Coen, Li, & Shaham, 2013; Le, Harding, Juzytsch, Fletcher, & Shaham, 2002; Le & Shaham, 2002; Shaham et al., 1997; Wang, Fang, Liu, & Lu, 2006; Wang et al., 2005)

6. CENTRAL EXTENDED AMYGDALA

Many of the long-term emotional disturbances associated with the withdrawal/negative affect stage of the addiction cycle have been attributed to activity within the central subdivision of the extended amygdala (Koob, 2008), which consists of the CeA, lateral BNST, and shell of the NAc (Heimer & Alheid, 1991). These regions share afferents from limbic cortices, the hippocampus, and basolateral amygdala (BLA) and have common effector targets, including the lateral hypothalamus and brainstem regions that produce behavioral and autonomic responses that have been associated with stress, fear, and anxiety (Davis, Rainnie, & Cassell, 1994).

In particular, the CeA is a brain region uniquely situated to function as an interface between stress- and addiction-related processes. The CeA is not only critical in aversive (e.g., fear) conditioning but also in the negative emotional states that define drug dependence and withdrawal. The CeA is composed mostly of inhibitory γ-aminobutyric acid (GABA) projection neurons and interneurons (McDonald & Augustine, 1993; Pare & Smith, 1993; Sun & Cassell, 1993; Veinante & Freund-Mercier, 1998), and modulation of local intra-CeA inhibitory gating is involved in the control of fear and anxiety-like behavior (Davis, Walker, Miles, & Grillon, 2010; Ehrlich et al., 2009; Pape & Pare, 2010).

The CeA is the major output region of the amygdaloid complex; as such, it receives complex inputs from other amygdaloid nuclei, as well as from other brain regions that integrate sensory information from the external environment (e.g., cortex, thalamus). The CeA is functionally and anatomically divided into lateral (CeL) and medial (CeM) subdivisions that themselves are interconnected and mainly populated by both inhibitory interneurons and GABAergic projection neurons. Additionally, both the CeL and CeM subdivisions project to the lateral division of the BNST (Davis et al., 2010; Dong, Petrovich, & Swanson, 2001; Petrovich, Risold, & Swanson, 1996; Sun, Roberts, & Cassell, 1991), another extended amygdala structure dominated by inhibitory neurotransmission (Sun & Cassell, 1993; Veinante & Freund-Mercier, 1998). Importantly, there are also reciprocal connections between the CeA and BNST, and the CeA is a major source of CRF in the BNST (Sakanaka, Shibasaki, & Lederis, 1986). As the CeM subdivision projects to regions that produce behavioral and physiological responses to emotionally relevant events (Le Gal LaSalle, Paxinos, & Ben-Ari, 1978; Pitkänen, 2000), it likely plays a major role in the behavioral changes associated with extrahypothalamic CRF signaling. However, more recent data suggest that the CeL also sends GABAergic projections to behavioral and physiological effector regions (Penzo, Robert, & Li, 2014).

Substantial preclinical data implicate extended amygdala CRF–CRF_1 systems in excessive or addiction-like consumption. While acute drug exposure only yields a transient elevation of CRF expression (Maj, Turchan, Smialowska, & Przewlocka, 2003; Zhou et al., 1996), chronic exposure leads to more sustained overactivation following cessation of drug use (Caberlotto, Rimondini, Hansson, Eriksson, & Heilig, 2004; George et al., 2007; Sommer et al., 2008; Zhou, Spangler, Ho, & Kreek, 2003; Zorrilla, Valdez, & Weiss, 2001). For example, in vivo microdialysis showed that acute withdrawal from chronic administration or self-administration of drugs of abuse increases extracellular CRF in the extended amygdala, a stress-like response (Merlo Pich et al., 1995; Richter, Zorrilla, Basso, Koob, & Weiss, 2000). Extracellular levels of CRF increase in the CeA and BNST of dependent rats withdrawn from EtOH (Merlo Pich et al., 1995; Olive, Koenig, Nannini, & Hodge, 2002). Similarly, extracellular CRF levels increased in the CeA during precipitated withdrawal from chronic nicotine (George et al., 2007), opioids (Weiss et al., 2001), or cannabinoids (Rodriguez de Fonseca et al., 1997) as well as during withdrawal from binge cocaine self-administration (Richter & Weiss, 1999). Increased CeA CRF mRNA also has been seen at various withdrawal time points (Caberlotto et al., 2004; Maj et al., 2003; Roberto et al., 2010; Sommer et al., 2008). The high levels of CRF release at early withdrawal time points can paradoxically yield low tissue content of CRF peptide (Funk, O'Dell, Crawford, & Koob, 2006; Maj et al., 2003; Sarnyai et al., 1995; Zorrilla et al., 2001) because CRF synthesis may lag behind release, resulting in transient depletion of CRF levels.

Consistent with a functional role for central extended amygdala CRF receptor activation in the negative affect/withdrawal stage, site-specific injections of CRF receptor antagonists into the CeA reduce anxiety-like behavior, motivational deficits for other reinforcers, and excessive self-administration of addictive substances during acute withdrawal in animal models (Funk et al., 2006, 2007; Heilig, Egli, Crabbe, & Becker, 2010; Heilig & Koob, 2007; Koob & Zorrilla, 2010; Logrip et al., 2011; Parylak, Koob, & Zorrilla, 2011; Rassnick, Heinrichs, Britton, & Koob, 1993). Furthermore, intra-CeA infusion of D-Phe CRF_{12-41}, a CRF receptor antagonist, reduced the elevations in brain reward stimulation thresholds that result from withdrawal from nicotine (Marcinkiewcz et al., 2009; Bruijnzeel et al., 2007 #96) or EtOH (Bruijnzeel et al., 2010). Intra-CeA administration of α-helical CRF_{9-41} also blocked the formation of a conditioned place aversion to precipitated morphine withdrawal (Heinrichs et al., 1995). Finally, administration of CRF receptor antagonists into the BNST, where CRF mRNA levels increase following stress and which also receives CRFergic projections from the CeA, reduces stress-induced reinstatement of drug seeking as well as stress-induced reinstatement of conditioned place preference for substances of abuse (Erb, Salmaso, Rodaros, & Stewart, 2001; Erb, Shaham, & Stewart, 2001; Erb & Stewart, 1999; McReynolds et al., 2014; Shalev, Morales, Hope, Yap, & Shaham, 2001; Wang et al., 2006).

Altered expression in amygdala CRF–CRF_1 systems has been seen at both the peptide and receptor level several weeks after detoxification from repeated EtOH intoxication/withdrawal in animal models (Sommer et al., 2008; Zorrilla et al., 2001). With respect to receptors, contrasting changes in CRF_1 vs CRF_2 were reported 3-week postcessation of alcohol exposure (Sommer et al., 2008). Within the BLA and MeA, CRF_1 levels were reportedly elevated, whereas BLA CRF_2 levels decreased. The generality of these changes across substances of abuse and withdrawal time points is unclear, however. For example, precipitated morphine withdrawal acutely reduced CRF_1 mRNA expression in the BLA and NAc (Iredale et al., 2000). With respect to CRF peptide levels, both EtOH- and cocaine-withdrawn animals initially exhibited a reduction of CRF-like tissue content in the amygdala, followed by a progressive increase that culminated in elevated levels by 6-week postwithdrawal (Funk et al., 2006; Maj et al., 2003; Sarnyai et al., 1995; Zorrilla et al., 2001). Thus, dysregulation of extended amygdala CRF systems continues into protracted abstinence and is thereby also hypothesized to contribute to the *preoccupation/anticipation* stage of the addiction cycle.

6.1 Cellular Mechanisms of CRF

Much of the research into the cellular actions of CRF in addictive disorders has focused on interactions with alcohol in extrahypothalamic stress systems. Therefore, although we will briefly discuss cellular CRF interactions with other drugs of abuse later, we will focus mainly on CRF produced in the CeA and the effects of acute and chronic alcohol on neurotransmission and plasticity in CeA and the role of these effects in the mediation of alcohol-related behaviors.

The CeA is particularly enriched with neuropeptides that regulate anxiety- and alcohol-related behaviors (Koob, 2008). These neuropeptides can be conceptually divided into prostress and antistress peptides, which, respectively, promote and rescue negative affective disturbances during drug abstinence following heavy drug use. Prostress peptides include CRF, dynorphin, and orexin; antistress peptides include neuropeptide Y, oxytocin, and nociception/orphanin FQ (N/OFQ). Notably these peptides interact in a complex way in the extended amygdala to modulate inhibitory

GABAergic and excitatory glutamatergic transmission. Acute and chronic alcohol dysregulate these peptidergic systems and their modulatory role on neurotransmitter systems.

At the cellular level, acute alcohol and CRF increase presynaptic GABA release in both rat (Roberto et al., 2010; Roberto, Madamba, Moore, Tallent, & Siggins, 2003) and mouse (Nie et al., 2004) CeA neurons via activation of CRF_1 (Nie et al., 2009; Roberto et al., 2010). Many studies on chronic alcohol effects on CeA neurotransmission now utilize a chronic intermittent ethanol (CIE) vapor inhalation model to induce alcohol dependence in rodents (Gilpin, Richardson, Cole, & Koob, 2008). Of note for electrophysiological studies, alcohol-dependent rats are maintained in the EtOH vapor chamber until the preparation of CeA slices under EtOH-free conditions. Thus, recordings of CeA neurons are performed at 2–8 h of EtOH withdrawal for these neurons. In the CeA of alcohol-dependent rats, baseline frequency of miniature inhibitory postsynaptic currents (mIPSCs) and spontaneous inhibitory postsynaptic currents (sIPSCs), as well as baseline-evoked inhibitory postsynaptic currents (eIPSCs), are significantly higher. Additionally, baseline paired pulse facilitation ratios of eIPSCs are significantly lower, suggesting that basal CeA GABA release is augmented by chronic EtOH exposure (Roberto, Madamba, Stouffer, Parsons, & Siggins, 2004; Roberto et al., 2010, 2003; Varodayan, de Guglielmo, Logrip, George, & Roberto, 2017). Interestingly, acute EtOH application significantly increases eIPSCs and s/mIPSC frequencies in slices from dependent rats to an extent equivalent to that in slices from nondependent rats, suggesting a lack of tolerance for these acute EtOH effects (Roberto et al., 2004; Varodayan et al., 2017). Consistent with in vitro electrophysiological results at the same time points (2–8 h of withdrawal), in vivo EtOH administration directly into the CeA via a microdialysis probe reveals a significant elevation in basal dialysate GABA levels in the CeA of alcohol-dependent rats compared with alcohol-naive rats, further suggesting increased GABAergic tone during alcohol dependence (Roberto et al., 2004). A lack of tolerance to the alcohol-induced increases in dialysate GABA content in alcohol-dependent rats is also observed in vivo (Roberto et al., 2010).

CRF increases GABA release in the CeA of alcohol-dependent rats beyond levels already significantly enhanced during withdrawal, along with concomitant increases in CRF and CRF_1 mRNA levels, and increases in CRF release (Merlo Pich et al., 1995; Roberto et al., 2010). This supports the electrophysiological data indicating upregulation of the CRF–CRF_1

system with alcohol dependence. Importantly, the application of CRF_1 antagonists alone decreases GABA release, suggesting tonic facilitation of GABA release by endogenous CRF (or other CRF-like molecules) via these receptors in the CeA (Nie et al., 2004; Roberto et al., 2010). CRF_1 antagonists also block the acute alcohol-induced increase in GABAergic transmission in the CeA. Additionally, in vivo intra-CeA administration of a CRF_1 antagonist via retro-microdialysis reverses dependence-related elevations in extracellular GABA dialysates and blocks alcohol-induced increases in GABA in CeA of both dependent and nondependent rats (Roberto et al., 2010). Ongoing electrophysiological studies reveal that the increased sensitivity of CeA GABAergic synapses to exogenous application of CRF persist after 2 weeks of withdrawal from alcohol vapor exposure (unpublished observations from the Roberto lab), suggesting that the neuroadaptive changes associated with chronic alcohol are long-lasting. Given the prominent action of acute CRF_1 receptor antagonist administration in reducing alcohol self-administration in dependent rats, dependent rats chronically treated with a CRF_1 receptor antagonist prevent the development of dependence-induced increases in alcohol drinking (Roberto et al., 2010). These results strongly point to the critical role of CeA CRF system not only in the expression, but also the development, of alcohol dependence in animal models.

Similarly, binge-like alcohol drinking increases CRF_1 signaling in the CeA of C57BL/6J mice (Chu et al., 2007), and CRF_1 antagonists reduce binge-like alcohol drinking in nondependent rodents without affecting nonbinge-like alcohol intake (Funk et al., 2006, 2007; Lowery-Gionta et al., 2012). However, the ability of CRF to increase GABAergic transmission in CeA is blunted in mice with a history of binge-like alcohol drinking (Chu et al., 2007), in contrast to the upregulated CRF/CRF_1 signaling observed in CeA of alcohol-dependent rats (Roberto et al., 2010). Lowery-Gionta et al. (2012) suggest that binge alcohol drinking may abolish CRF effects on GABAergic transmission in CeA via functional downregulation or desensitization of CRF_1 (Lowery-Gionta et al., 2012), as seen in the dorsal raphe following stress (Cippitelli et al., 2012; Waselus, Nazzaro, Valentino, & Van Bockstaele, 2009).

Although the precise mechanisms by which alcohol and CRF enhance GABA release have yet to be completely identified, previous studies have examined the role of intracellular signaling pathways, such as adenylyl cyclase and protein kinase C (PKC), in the facilitatory effect of acute alcohol on GABAergic transmission. The ability of acute alcohol and CRF to augment GABAergic transmission in the CeA is contingent on the integrity of PKCε intracellular signaling pathways (Bajo, Cruz, Siggins, Messing, & Roberto, 2008). The alcohol- and CRF-induced increase in GABA release is abolished in the CeA of mice that lacked PKCε (Bajo et al., 2008), suggesting that PKCε facilitates vesicular GABA release. Protein kinase A (PKA) activity is initiated following CRF_1 activation, via G_s and G_q protein cascades. Interestingly, type 7 adenylate cyclase plays a role in the modulation of presynaptic GABA release in the mouse CeA by alcohol and CRF (Cruz et al., 2011), and a PKA antagonist blocked CRF from regulating spontaneous GABA release in rat CeA (Cruz, Herman, Kallupi, & Roberto, 2012; Varodayan et al., 2017), suggesting that the presynaptic PKA pathway plays an essential role in CRF-induced GABA release.

More recently, L-type voltage-gated calcium channels (LTCCs) have been implicated in the EtOH-induced increases in CeA action-potential-dependent activity (neuronal firing rates and GABA release) in naïve rats, and alcohol dependence reduces CeA LTCC membrane abundance (Varodayan et al., 2017). Nifedipine, an LTCC antagonist, prevents alcohol's potentiation of sIPSC frequencies and firing in naïve rats, but not in alcohol-dependent rats where a CRF_1 antagonist (R121919) did. This switch from an LTCC- to a CRF_1-based mechanism with alcohol dependence is accompanied by a shift from a role for inositol triphosphate receptor (IP_3R)-mediated calcium-induced calcium release to the involvement of ryanodine receptors (Varodayan et al., 2017). Notably, intra-CeA LTCC blockade reduces alcohol intake in nondependent rats, and R121919 blocks the escalated alcohol intake of alcohol-dependent rats. Collectively, these data indicate that alcohol dependence functionally alters the molecular mechanisms underlying the CeA's response to alcohol (from LTCC- to CRF_1-driven). This mechanistic switch contributes to and reflects the prominent role of the CeA in the negative emotional state that drives excessive drinking. Critically, P/Q-type voltage-gated calcium channels mediate alcohol-induced CeA vescicular GABA release in a PKA and PKC-dependent manner in both naïve and EtOH-dependent rats (unpublished data from the Roberto lab).

Given the importance of CRF_1 signaling, transgenic mice expressing green fluorescent protein (GFP) in CRF_1 expressing neurons (CRF_1+) have been used to examine the effects of chronic EtOH exposure on inhibitory signaling in CRF_1+ CeA neurons, and, more recently on the excitability and acute EtOH sensitivity of CeA projection neurons that output to downstream targets in the extended amygdala (Herman, Contet, Justice,

Vale, & Roberto, 2013; Herman, Contet, & Roberto, 2016). CeA neurons display two types of inhibition: phasic, which involves IPSCs that reflect "point to point" transmission; and tonic, which involves persistent inhibitory currents resulting from ambient GABA acting at highly sensitized $GABA_A$ receptors (Belelli et al., 2009; Glykys & Mody, 2007). Tonic inhibition regulates neural network activity (Semyanov, Walker, Kullmann, & Silver, 2004) and is modulated by both acute and chronic alcohol (Jia, Chandra, Homanics, & Harrison, 2008; Liang et al., 2008; Mody, Glykys, & Wei, 2007). As described earlier, acute alcohol reversibly increases phasic GABA release in a dose-dependent fashion in the CeA (Roberto et al., 2003, 2004), independent of $GABA_BR$ blockade (Roberto et al., 2008). Acute alcohol also increases phasic and tonic inhibition in a population of CeA neurons that synapse onto CeA output neurons, resulting in disinhibition of CeA output to BNST (Herman et al., 2013).

CRF_1+ neurons make up a distinct component of CeA circuitry that are not directly activated by acute EtOH but instead are engaged via a local microcircuit involving increased inhibition at local CRF_1- interneurons that synapse onto CRF_1+ neurons (Herman et al., 2013, 2016), thus CRF_1+ neurons are normally under tonic inhibitory control. This inhibitory control is lost with chronic EtOH exposure. In particular, the loss of tonic inhibition in CRF_1+ neurons that project into the dorsal lateral BNST occurs in parallel with an increase in phasic and tonic signaling in a subpopulation of late-spiking (LS) CRF_1- neurons. This loss of tonic inhibition is also accompanied by an increase in baseline firing and loss of sensitivity to acute EtOH. Together, these data suggest that chronic EtOH exposure produces a functional switch in tonic signaling in the CeA, such that the balance of inhibitory control shifts from CRF_1+ projection neurons to LS CRF_1- interneurons. This switch results in increased output of the CeA CRF_1 system via disinhibition of projection neurons and includes a loss of sensitivity to acute EtOH (Herman et al., 2013, 2016). Interestingly, many of these changes persist into EtOH withdrawal (5–7 days). Thus, they may represent long-term neuroadaptations that play a role in alcohol-dependent behaviors, like escalated alcohol intake and the susceptibility to relapse.

As previously mentioned, CRF has been shown to augment GABA transmission in the CeA via actions at CRF_1 (Nie et al., 2004; Roberto et al., 2010) and CeA CRF neurons employ GABA as a cotransmitter as well as displaying increased activity after chronic stress (Partridge et al., 2016). It is important to note that the CRF_1 mouse model just discussed only reflects neurons that express CRF_1 (Justice, Yuan, Sawchenko, & Vale, 2008) and provides no information on CRF peptide actions or circuitry. It is not currently known if the CRF_1 + CeA neurons in the CRF_1:GFP mouse display any overlap with CRF neurons in the CeA. Although progress has been made with other reporter mice (Silberman, Matthews, & Winder, 2013), more sophisticated animal models and/or technological approaches are required to more directly pinpoint the intersection of the CRF and CRF_1 systems in the CeA. Recently, the development of a transgenic *Crh*-Cre rat has permitted genetic access to CRF neurons, thereby allowing direct investigation of their anatomy and roles in physiology and behavior (Pomrenze et al., 2015). The CRF-expressing CeA neurons identified in this transgenic rat project to several brain regions including the basal ganglia, extended amygdala, brainstem, and hypothalamus. In addition, these CRF-expressing neurons are mainly localized in the CeL subdivision of the CeA, where they act as local interneurons to provide both inhibitory (GABA) and excitatory (CRF) signals within the CeL and CeM (Pomrenze et al., 2015).

Although the CeA is a nucleus containing a heterogeneous mix of neuronal types, recent studies have begun to characterize distinct cell populations and their role in circuits governing specific behaviors and CRF effects. For example, circuitry-specific changes in CeA processing have been implicated in conditioned fear (Haubensak et al., 2010) and in anxiety (Botta et al., 2015). A common theme of these studies is the importance of local inhibitory signaling in the flow of information through distinct circuits. Specific inhibitory microcircuits in the CeA govern different aspects of fear learning (Ciocchi et al., 2010) and extrasynaptic inhibition regulates the changes in excitability underlying the encoding of generalized fear or anxiety (Botta et al., 2015). Disinhibition of CeA projection neurons is implicated in critical circuitry changes related to learning and memory in various brain regions and neurotransmitter/neuromodulator systems (Letzkus, Wolff, & Luthi, 2015). If the development of alcohol dependence can be conceptualized as a pathological expression of learning, then it is compelling that both processes would employ a similar mechanism of disinhibition for the engagement of specific components of brain circuitry. This view agrees with the current focus on the role of disease-specific alterations in distinct amygdala circuits as major contributing factors in the development of addiction (Koob & Volkow, 2016; Koob & Zorrilla, 2010) and the commonalities between addiction and pathological anxiety (Luthi & Luscher, 2014).

In other extended amygdala regions, such as the vBNST, chronic intermittent, but not continuous, ethanol vapor exposure increases the temporal summation of NMDAR-mediated synaptic transmission. This effect is not dependent on changes in glutamate release, but rather on an increase in the levels of NR2B-containing NMDARs. Furthermore, acute ethanol modulation of NMDARs in the vBNST is altered after intermittent alcohol exposure (Kash, Baucum, Conrad, Colbran, & Winder, 2009), supporting the concept of homeostatic regulation of the NR2B subunit in both acute ethanol inhibition of NMDARs and chronic alcohol withdrawal-induced effects on glutamatergic synaptic function in the BNST (Kash et al., 2009; Wills, Kash, & Winder, 2013; Wills et al., 2012). Specifically, CIE enhances long-term potentiation (LTP) within the BNST, an effect that is dependent upon GluN2B (Wills et al., 2013, 2012), and GluN2B is upregulated within the BNST following CIE (Kash et al., 2009). Interestingly, electrophysiological studies report that CRF exerts multiple actions in the BNST, including enhancement of glutamatergic signaling by increasing both the frequency of sEPSC (Kash, Nobis, Matthews, & Winder, 2008) and cell firing. Glutamate release is similarly enhanced by CRF_1-mediated signaling in mice exposed to CIE vapor (Silberman et al., 2013; Silberman & Winder, 2013).

Several studies, both functional (Meloni, Gerety, Knoll, Cohen, & Carlezon, 2006) and anatomical (Day, Curran, Watson, & Akil, 1999; Phelix, Liposits, & Paull, 1994; Rodaros, Caruana, Amir, & Stewart, 2007), have reported an interaction between DA and CRF in the BNST. DA increases excitatory glutamatergic transmission through activation of endogenous CRF/CRF_1 signaling in the BNST (Kash et al., 2008). Subsequent studies have reported that β-AR activation enhances excitatory transmission in the BNST, an effect that is dependent on intact CRF_1 signaling. Notably, this enhancement of excitatory transmission is disrupted by repeated cocaine administration, but not during withdrawal from this repeated cocaine administration. Furthermore, prior cocaine experience influences how CRF receptor signaling responds to subsequent cocaine challenges (Nobis, Kash, Silberman, & Winder, 2011).

Pharmacological experiments suggest that CRF and urocortin enhance GABAergic transmission through mainly postsynaptic responses to GABA via activation of the CRF_1 (Kash & Winder, 2006). Recent data suggest that CRF BNST neurons are exclusively GABAergic, and these GABAergic CRF neurons are highly interconnected with non CRF-expressing neurons within the BNST (Marcinkiewcz, Lowery-Gionta, & Kash, 2016; Mazzone et al., 2016; McElligott & Winder, 2009; Partridge et al., 2016; Vranjkovic, Pina, Kash, & Winder, 2017).

Francesconi et al. (2009) reported a form of LTP of the intrinsic excitability (LTP-IE) of jcBNST neurons in response to high-frequency stimulation in the BNST. This LTP-IE is impaired during protracted withdrawal from self-administration of alcohol, cocaine, and heroin and was altered in a CRF-dependent manner.

Considering other drugs of abuse, Pollandt et al. (2006) reported a long-lasting potentiation of glutamatergic transmission induced at lateral amygdala (LA)-to-CeA synapses by CRF in withdrawal from repeated intermittent cocaine administration. Bath application of CRF (12.5 nm) persistently increases amplitudes of fEPSPs in the LA–CeA pathway and this CRF-induced facilitation is greater after 2 weeks of withdrawal from chronic cocaine. CRF significantly decreases paired-pulse ratios at a 50-ms interstimulus interval, suggesting that presynaptic mechanisms play a role in CRF-induced LTP. In agreement with previous studies on the BLA-to-CeA synapse from the same lab (Liu et al., 2004), CRF-induced LTP in the LA–CeA pathway is mediated by pre- and postsynaptic mechanisms, primarily through activation of CRF_2. After 2 weeks of withdrawal from chronic cocaine, CRF-induced LTP depends on both CRF_1 and CRF_2 and the enhanced LTP in response to CRF is mediated by an upregulation of CRF_1 receptor function. This switch was correlated to CRF_2 linked to PKC that became predominantly CRF_1 receptors linked to PKA during cocaine withdrawal. In addition, the increased response to CRF during withdrawal from chronic cocaine was accompanied by an increase in NMDA-mediated synaptic responses and Cav2.3 channels.

7. VENTRAL TEGMENTAL AREA

Unlike the reviewed, typically negative, findings with CRF_1 antagonists in nondependent rats, some other behavioral data show that CRF_1 may contribute to excessive consumption in a manner that also suggests a role in the binge/intoxication phase of the addiction cycle. For example, *Crhr1* knockout mice drank less 20% *v/v* EtOH under basal conditions (Pastor et al., 2011), and *Crh* knockout mice failed to develop a conditioned place preference to a moderate dose of alcohol (2 g/kg) (Olive et al., 2003), Moreover, both *Crh* and *Crhr1* knockout mice showed reduced EtOH intake and blood EtOH concentrations in a mouse model of scheduled, limited access to EtOH (drinking in the dark (DID)) that can lead to binge-like intake (Kaur, Li, Stenzel-Poore, & Ryabinin, 2012). As mentioned earlier, systemic administration of small-molecule CRF_1 antagonists reduced binge-like, but not nonbinge-like, EtOH intake in C57BL/6J mice and outbred rats (Cippitelli et al., 2012; Lowery et al., 2010; Simms, Nielsen, Li, & Bartlett, 2014; Sparta et al., 2008; but see Giardino & Ryabinin, 2013, data suggesting that the results may not be specific to substances of abuse). The neuroanatomical basis of these effects remains under investigation, but site-specific infusion of CRF_1 antagonists into the CeA or VTA likewise could reduce heightened EtOH intake in rodent models under intermittent access schedules (Hwa, Debold, & Miczek, 2013; Lowery-Gionta et al., 2012; Sparta et al., 2013).

Indeed, the involvement of CRF signaling in the VTA in the neurobiology of substance use disorders has received more attention within the past decade. Dopaminergic VTA neurons receive CRF inputs from fibers that originate in the BNST, the CeA and, to a lesser extent, the hypothalamic PVN (Rodaros et al., 2007), and VTA dopaminergic neurons also synthesize CRF (Grieder et al., 2014). In the VTA, CRF increases dopamine neuron firing (Wanat, Hopf, Stuber, Phillips, & Bonci, 2008), and both stress (Wang et al., 2005) and drug withdrawal (Grieder et al., 2014; Zhao-Shea et al., 2015a, 2015b) have been associated with increased VTA CRF signaling. Given the known role of VTA dopaminergic efferents in stress-induced seeking and taking of substances of abuse (Holly & Miczek, 2016), it is unsurprising that stress-modulated VTA CRF systems have been implicated in reinstatement of drug seeking and excessive consumption. For example, social defeat stress phasically increased CRF levels in anterior vs posterior compartments of the VTA, with repeated social defeat increasing levels in both subregions; furthermore, intra-VTA administration of CRF_1 or CRF_2 antagonists reduced cocaine-seeking behavior in previously defeated rats (Holly et al., 2016; Holly & Miczek, 2016). The recruitment of VTA CRF signaling, through a CRF_1 mechanism, also appears to play a key role in the aversive and anxiogenic-like effects of nicotine withdrawal and in the escalation of nicotine self-administration that occurs with intermittent, extended drug exposure (Grieder et al., 2014).

Whereas the bulk of reviewed data implicate CRF_1 signaling in the extended amygdala as contributing to addiction-related phenotypes, CRF_2 have been suggested also to play a role in the VTA (Wang et al., 2007), particularly in reinstatement of drug-seeking behavior, as observed in cocaine-experienced rats. For example, intra-VTA administration of the preferential CRF_2 antagonist antisauvagine-30 blocked reinstatement of both cocaine-conditioned place preference (Lu et al., 2001) and cocaine-seeking (Wang et al., 2007) following footshock stress. On the other hand, the CRF_2 mediation of antisauvagine-30s actions is now uncertain because anti-sauvagine-30 still has behavioral actions in *Crhr2* knockout mice (Zorrilla et al., 2013). Furthermore, others who used the more selective CRF_2 antagonist astressin$_2$-B did not observe a reversal of stress-induced reinstatement of substance seeking (Bruijnzeel et al., 2009), and very little CRF_2 mRNA is detected in the rodent VTA by in situ hybridization under basal conditions (Chalmers et al., 1995; Van Pett et al., 2000). Still, a role for VTA CRF_2 in addiction-related phenotypes, if confirmed, may also implicate possible roles for the urocortins, CRF peptide family members with higher affinity than CRF for CRF_2 (Gysling, 2012).

In a more general way, the role of CRF_2 in addiction-related phenotypes and excessive consumption remains unclear. Early studies indicated that i.c.v. infusion of Ucn 3, a selective CRF_2 agonist, reduced heightened anxiety-like behavior and increased EtOH self-administration in dependent rats during acute withdrawal (Valdez et al., 2004) as well as the alcohol intake of mice that receive limited, binge-like access to alcohol (Lowery et al., 2010; Sharpe & Phillips, 2009). Intra-CeA infusion of Ucn 3 likewise reduced EtOH self-administration in withdrawn, dependent rats (Funk & Koob, 2007), and infusion of Ucn 1 into the lateral septum, which abundantly expresses CRF_2, potently reduced the acquisition and expression of alcohol intake in rats (Ryabinin, Yoneyama, Tanchuck, Mark, & Finn, 2008). On the other hand, CRF_2 antagonists did not influence withdrawal-associated behavioral changes in other studies (Lu, Liu, et al., 2000; Overstreet et al., 2004). Furthermore, several studies have implicated CRF_2 activation in opiate withdrawal. Genetic deletion of the CRF_2 receptor blocked somatic signs of opiate withdrawal (Papaleo et al., 2008) as well as dysphoria- or anhedonic-like behaviors (Ingallinesi, Rouibi, Le Moine, Papaleo, & Contarino, 2012). Perhaps the discrepant results reflect different roles of CRF_2 with respect to different substances of abuse or, alternatively, that the actions of CRF_2 activation can depend on the particular CRF_2 ligand, concentration, or brain site (Fekete & Zorrilla, 2007; Ho et al., 2001; Takahashi, Ho, Livanov, Graciani, & Arneric, 2001; Zhao, Valdez, et al., 2007).

With regards to cellular physiology, CRF potentiates NMDA receptor currents on VTA dopamine (DA) neurons (Sparta et al., 2013; Ungless et al., 2003). Application of (100 nM) CRF induces a facilitation of NMDA currents in putative VTA DA neurons in brain slices of juvenile mice that consumed EtOH in the DID procedure. This CRF-induced potentiation of NMDAR currents in EtOH-drinking mice is blocked by administration of the selective CRF_1 receptor antagonist CP-154,526. Furthermore, intra-VTA infusion of CP-154,526 significantly reduces binge EtOH consumption in adult mice. Notably, no alteration in VTA NMDAR number or function was observed, but CRF_1-mediated signaling was increased suggesting that binge drinking may enhance VTA CRF_1 receptor-mediated effects onto NMDARs (Sparta et al., 2013).

Another electrophysiological study reported a dose-dependent CRF-induced increase in VTA DA neuron firing, an effect that was prevented by antagonism of CRF_1 and was mimicked by CRF_1 agonists. This increase in DA firing was prevented by inhibition of the phospholipase C (PLC)–PKC signaling cascade, but not the cAMP–PKA pathway. In addition, PKC enhanced the *I*(h) current and increased the firing rate in VTA dopamine neurons (Wanat et al., 2008). Interestingly, the Brodie lab has reported that CRF and the CRF receptor agonist urocortin reversed inhibition produced by the D2 agonist quinpirole via increased glutamatergic signaling in the VTA (Nimitvilai et al., 2014).

In rat brain slices, CRF, in a concentration-dependent manner, facilitates (at low concentrations: 3–100 nm) and decreases (at higher concentrations: 300 nm) EPSCs in VTA neurons via presynaptic CRF_1 and CRF_2 receptors, respectively. A $GABA_B$ receptor antagonist (CGP55843) blocks the CRF_2 attenuation (Williams, Buchta, & Riegel, 2014). Notably in the same study, CRF_2 activation facilitates presynaptic release of GABA, suggesting that CRF_2 may regulate glutamate release via heterosynaptic facilitation of GABA synapses. Although, chronic cocaine self-administration does not alter the sensitivity of glutamate and GABA receptors, the ability of CRF_2 agonists to depress glutamate and potentiate GABA responses of VTA neurons is diminished. However, yohimbine plus cue reinstatement reverses the actions of CRF_2 on GABA and glutamate release. CRF_2 activation increases EPSCs as a result of a reduction in tonic GABA-dependent inhibition (Williams et al., 2014). These studies demonstrate that presynaptic CRF-R_1/R_2 tightly regulate glutamate transmission in the VTA in a concerted, heterosynaptic manner that is altered by stress-related pathologies, such as addiction.

A population of VTA DA neurons that synthesize CRF has been identified, and chronic nicotine exposure upregulates CRF mRNA in DA neurons of the posterior VTA, activates local CRF_1, and blocks nicotine-induced activation of transient GABAergic input to DA neurons (Grieder et al., 2014). Local downregulation of CRF mRNA and specific pharmacological blockade of CRF_1 in the VTA reverses the effect of nicotine on GABAergic input to DA neurons, prevents the aversive effects of nicotine withdrawal, and limits the escalation of nicotine intake. These results link the brain reward and stress systems in the same brain region to signaling of the negative motivational effects of nicotine withdrawal (Grieder et al., 2014).

8. PREFRONTAL CORTEX

Emerging data also suggest that CRF systems in the medial prefrontal cortex (mPFC) may contribute to excessive substance use. Extended access to cocaine elicits escalated, compulsive-like self-administration as well as impaired working memory (George, Mandyam, Wee, & Koob, 2008). The degree of escalated cocaine intake correlates directly with impaired delayed nonmatching to sample working memory performance under long delay conditions. Perhaps relatedly, rats receiving intermittent, extended access to two-bottle choice EtOH intake, which leads to escalated drinking (Simms et al., 2008; Wise, 1973), show increased activation of CRF and GABAergic interneurons in the mPFC during abstinence. As with cocaine, working memory impairments in these rats correlated directly with greater alcohol intake following abstinence (George et al., 2012). Abstinence from escalated drinking was associated with functional disconnection of the mPFC and CeA, perhaps reflecting impaired inhibitory control over motivated behavior. CRF tissue content in the frontal cortex also is altered during acute or protracted withdrawal from EtOH or binge-like cocaine use, respectively (Zorrilla et al., 2001). The results have been proposed to indicate that CRF-mediated dysregulation of mPFC interneurons may contribute to disturbed executive control over self-administration of substances.

9. COMPULSIVE EATING

A final domain of study suggests that addiction-like changes in CRF systems may be similarly implicated in the neurobiology of compulsive eating. So-called food addiction has been hypothesized to underlie many cases of obesity and eating disorders, which may collectively comprise the most prevalent and deadliest form of addictive behavior (Agh et al., 2016; Fichter & Quadflieg, 2016; Flegal, Kit, Orpana, & Graubard, 2013; Flegal, Kruszon-Moran, Carroll, Fryar, & Ogden, 2016; Hoang, Goldacre, & James, 2014; Kroes, Osei-Assibey, Baker-Searle, & Huang, 2016; Micali et al., 2017; Mitchell, 2016; Olguin et al., 2017; Perez,

Ohrt, & Hoek, 2016; Smink, van Hoeken, & Hoek, 2013; Westmoreland, Krantz, & Mehler, 2016). In the United States, ~three of four men and two of three women were overweight ($25 <$ body mass index [BMI] < 30) or obese (BMI > 30) in 2013–14, and ~33% of overweight/obese people (and ~54% of patients undergoing bariatric surgery for obesity) met diagnostic criteria of the Yale Food Addiction Scale (Flegal et al., 2016; Long, Blundell, & Finlayson, 2015; Pursey, Stanwell, Gearhardt, Collins, & Burrows, 2014). As with substances of abuse, CRF_1 antagonists reduce stress-induced reinstatement of palatable food seeking (Ghitza, Gray, Epstein, Rice, & Shaham, 2006). Furthermore, rats acutely withdrawn from intermittent access to a high-sucrose, chocolate-flavored diet (diet cycling) showed increased anxiety-like behavior (Cottone, Sabino, Steardo, & Zorrilla, 2009) in association with increased mRNA and peptide expression of CRF in the CeA; similar molecular changes were seen by Bale and colleagues in mice withdrawn from high-fat diet (Teegarden, Scott, & Bale, 2009). Systemic pretreatment with R121919, a selective CRF_1 antagonist blocked food withdrawal-associated anxiety at doses that did not alter the behavior of chow-fed controls. Furthermore, the CRF_1 antagonist reduced the motivational deficits that emerged for otherwise acceptable reinforcers. Finally, CRF_1 antagonist pretreatment decreased the magnitude of overeating of the palatable sucrose-rich diet by diet-cycled animals at doses that did not alter the food intake of chow-fed controls or of animals fed the sucrose-rich diet, but without a history of diet cycling (Cottone, Sabino, Roberto, et al., 2009). When diet-cycled animals had access to the preferred, sucrose-rich diet, both anxiety-like behavior and CeA CRF levels normalized, supporting the hypothesis that activation of the amygdala CRF–CRF_1 system contributed to the withdrawal-like state. Subsequent studies identified an increased number of CRF-positive neurons in the CeA of diet-cycled rats irrespective of their current diet, and intra-CeA administration of CRF_1 antagonists normalized anxiety-like behavior and overeating of the palatable diet. Furthermore, intra-BLA administration of a CRF_1 antagonist reduced the rejection of otherwise acceptable, but less preferred, chow (Iemolo et al., 2013). The unique involvement of CRF systems may be greater with high-sucrose diets, because CRF_1 antagonists did not reduce binge-like intake of a comparably preferred diet that also contained fat (Parylak, Cottone, Sabino, Rice, & Zorrilla, 2012).

Interestingly, whereas initial clinical trials with CRF_1 antagonists for alcohol use disorders have yielded disappointingly negative results so far (see Spierling & Zorrilla, 2017 for review), a recent double-blind, placebo-controlled trial in a small sample of healthy individuals with restrained eating yielded promising results for stress-related eating that appear to warrant further study. Because the study was stopped prematurely by the NIH IRB for reasons unrelated to adverse drug effects or efficacy (reinterpretation of the Common Rule for human subject protection under HHS, 45 CFR 46A), it was underpowered (30% power) to detect a priori effect sizes of interest. Nonetheless, the CRF_1 antagonist pexacerfont produced effect sizes more consistent with reduction of food craving and laboratory stress-induced eating than with a null result. The effect size for pexacerfont's reduction of laboratory stress-induced eating was $r=0.30$ (counternull $r=0.55$; Rosenthal & Rubin, 1994), and its reduction of craving for sweet foods (brownies and Swedish fish) ranged from $r=0.28$–0.49 (counternull $r=0.52$–0.79). Furthermore, in bogus taste tests that masked the true-dependent measure of interest (intake), pexacerfont reduced palatable food intake independent of which imagery script was presented before food access (neutral, food cue, stress) with an effect size of $r=0.34$ (counternull $r=0.61$). Finally, nightly Yale Food Addiction Scores were lower in subjects that received pexacerfont (vs placebo) beginning the evening after the first loading dose of pexacerfont with an effect size of $r=0.39$ (counternull $r=0.68$). Bayes factor analysis, a ratio that relates to the relative probability of an effect actually being present vs the null effect (Goodman, 1999), and counternull analysis, which describes the effect size as likely to be true as the null, collectively indicated a strong positive potential of this CRF_1 antagonist to reduce palatable food craving and eating in restrained eaters (Epstein et al., 2016) and appear to justify a well-powered clinical trial in this domain. A caveat with these results is that the YFAS scores changed as soon as 24 h posttreatment, when the degree of CNS exposure obtained is unclear, casting uncertainty on the CRF_1 antagonist mechanism of action.

At the cellular level, our labs have examined the effect of the CRF_1 antagonist (R121919) on GABAergic transmission in the CeA and found that withdrawal from palatable food, similar to alcohol withdrawal, upregulates CRF/CRF_1 signaling in CeA (Cottone, Sabino, Roberto, et al., 2009). Although diet history did not alter baseline CeA GABAergic transmission, application of R121919 induced a greater reduction in evoked GABAergic responses in CeA neurons of diet-cycled rats vs chow-fed controls, supporting the hypothesis that a similar overactivation of the amygdala CRF–CRF_1 system during drug withdrawal is also seen during food withdrawal (Cottone, Sabino, Roberto, et al., 2009).

10. CONCLUSION

In this chapter we reviewed the importance of both hypothalamic and extrahypothalamic CRF systems and their roles in the development of dependence for all substances of abuse, as well as for compulsive eating. Notably, activation of brain stress systems contributes to the negative motivational state associated with acute withdrawal and could also contribute to vulnerability to stressors observed during protracted abstinence in humans. Thus, CRFergic circuits that normally subserve appropriate responses to acute stressors appear to pathologically contribute to the aversive emotional state that drives the negative reinforcement of addiction. In the development of drug dependence and withdrawal, the CRF system is recruited to produce stress-like, aversive states (Aston-Jones & Druhan, 1999; Koob & Le Moal, 2001). This state of compulsive substance seeking and use involves numerous neurobiological changes, including increased function of the CRF brain stress system in the extended amygdala (Koob, 2008; Roberto et al., 2010), as well as roles for CRF circuitry in the VTA and prefrontal cortex.

ACKNOWLEDGMENTS

This publication was supported by National Institutes of Health Grants AA015566, AA006420, AA017447, U01 AA013498, F32 AA025262, and R01 AA021491 from the National Institute on Alcohol Abuse and Alcoholism, and the Pearson Center for Alcoholism and Addiction Research. The content is solely the responsibility of the authors and does not necessarily represent the official views of the National Institutes of Health. This is Scripps manuscript number 29509.

REFERENCES

Agh, T., Kovacs, G., Supina, D., Pawaskar, M., Herman, B. K., Voko, Z., et al. (2016). A systematic review of the health-related quality of life and economic burdens of anorexia nervosa, bulimia nervosa, and binge eating disorder. *Eating and Weight Disorders*, *21*(3), 353–364. http://dx.doi.org/10.1007/s40519-016-0264-x.

Albrechet-Souza, L., Hwa, L. S., Han, X., Zhang, E. Y., DeBold, J. F., & Miczek, K. A. (2015). Corticotropin releasing factor binding protein and CRF2 receptors in the ventral tegmental area: Modulation of ethanol binge drinking in C57BL/6J mice. *Alcoholism, Clinical and Experimental Research*, *39*(9), 1609–1618. http://dx.doi.org/10.1111/acer.12825.

American Psychiatric Association. (1994). *Diagnostic and statistical manual of mental disorders* (4th ed.). . Washington, DC: American Psychiatric Press.

Aston-Jones, G., & Druhan, J. (1999). Behavioural pharmacology. Breaking the chain of addiction. *Nature*, *400*(6742), 317–319. http://dx.doi.org/10.1038/22432.

Bajo, M., Cruz, M. T., Siggins, G. R., Messing, R., & Roberto, M. (2008). Protein kinase C epsilon mediation of CRF- and ethanol-induced GABA release in central amygdala. *Proceedings of the National Academy of Sciences of the United States of America*, *105*(24), 8410–8415. http://dx.doi.org/10.1073/pnas.0802302105.

Baldwin, H. A., Rassnick, S., Rivier, J., Koob, G. F., & Britton, K. T. (1991). CRF antagonist reverses the "anxiogenic" response to ethanol withdrawal in the rat. *Psychopharmacology, 103*(2), 227–232.

Bale, T. L., & Vale, W. W. (2004). CRF and CRF receptors: Role in stress responsivity and other behaviors. *Annual Review of Pharmacology and Toxicology, 44*, 525–557.

Basso, A. M., Spina, M., Rivier, J., Vale, W., & Koob, G. F. (1999). Corticotropin-releasing factor antagonist attenuates the "anxiogenic-like" effect in the defensive burying paradigm but not in the elevated plus-maze following chronic cocaine in rats. *Psychopharmacology, 145*(1), 21–30.

Belelli, D., Harrison, N. L., Maguire, J., Macdonald, R. L., Walker, M. C., & Cope, D. W. (2009). Extrasynaptic GABAA receptors: Form, pharmacology, and function. *The Journal of Neuroscience, 29*(41), 12757–12763. http://dx.doi.org/10.1523/jneurosci.3340-09.2009.

Blacktop, J. M., Seubert, C., Baker, D. A., Ferda, N., Lee, G., Graf, E. N., et al. (2011). Augmented cocaine seeking in response to stress or CRF delivered into the ventral tegmental area following long-access self-administration is mediated by CRF receptor type 1 but not CRF receptor type 2. *The Journal of Neuroscience, 31*(31), 11396–11403. http://dx.doi.org/10.1523/jneurosci.1393-11.2011.

Bossert, J. M., Marchant, N. J., Calu, D. J., & Shaham, Y. (2013). The reinstatement model of drug relapse: Recent neurobiological findings, emerging research topics, and translational research. *Psychopharmacology, 229*(3), 453–476. http://dx.doi.org/10.1007/s00213-013-3120-y.

Botta, P., Demmou, L., Kasugai, Y., Markovic, M., Xu, C., Fadok, J. P., et al. (2015). Regulating anxiety with extrasynaptic inhibition. *Nature Neuroscience, 18*(10), 1493–1500. http://dx.doi.org/10.1038/nn.4102.

Boyson, C. O., Miguel, T. T., Quadros, I. M., Debold, J. F., & Miczek, K. A. (2011). Prevention of social stress-escalated cocaine self-administration by CRF-R1 antagonist in the rat VTA. *Psychopharmacology, 218*(1), 257–269. http://dx.doi.org/10.1007/s00213-011-2266-8.

Breese, G. R., Overstreet, D. H., & Knapp, D. J. (2005). Conceptual framework for the etiology of alcoholism: A "kindling"/stress hypothesis. *Psychopharmacology, 178*(4), 367–380. http://dx.doi.org/10.1007/s00213-004-2016-2.

Breese, G. R., Overstreet, D. H., Knapp, D. J., & Navarro, M. (2005). Prior multiple ethanol withdrawals enhance stress-induced anxiety-like behavior: Inhibition by CRF1- and benzodiazepine-receptor antagonists and a 5-HT1a-receptor agonist. *Neuropsychopharmacology, 30*(9), 1662–1669. http://dx.doi.org/10.1038/sj.npp.1300706.

Breese, G. R., Sinha, R., & Heilig, M. (2011). Chronic alcohol neuroadaptation and stress contribute to susceptibility for alcohol craving and relapse. *Pharmacology and Therapeutics, 129*(2), 149–171. http://dx.doi.org/10.1016/j.pharmthera.2010.09.007.

Bruijnzeel, A. W., Ford, J., Rogers, J. A., Scheick, S., Ji, Y., Bishnoi, M., et al. (2012). Blockade of CRF1 receptors in the central nucleus of the amygdala attenuates the dysphoria associated with nicotine withdrawal in rats. *Pharmacology, Biochemistry, and Behavior, 101*(1), 62–68. http://dx.doi.org/10.1016/j.pbb.2011.12.001.

Bruijnzeel, A. W., Prado, M., & Isaac, S. (2009). Corticotropin-releasing factor-1 receptor activation mediates nicotine withdrawal-induced deficit in brain reward function and stress-induced relapse. *Biological Psychiatry, 66*(2), 110–117. http://dx.doi.org/10.1016/j.biopsych.2009.01.010.

Bruijnzeel, A. W., Small, E., Pasek, T. M., & Yamada, H. (2010). Corticotropin-releasing factor mediates the dysphoria-like state associated with alcohol withdrawal in rats. *Behavioural Brain Research, 210*(2), 288–291. http://dx.doi.org/10.1016/j.bbr.2010.02.043.

Bruijnzeel, A. W., Zislis, G., Wilson, C., & Gold, M. S. (2007). Antagonism of CRF receptors prevents the deficit in brain reward function associated with precipitated nicotine withdrawal in rats. *Neuropsychopharmacology, 32*(4), 955–963. http://dx.doi.org/10.1038/sj.npp.1301192.

Buckingham, J. C. (1982). Secretion of corticotrophin and its hypothalamic releasing factor in response to morphine and opioid peptides. *Neuroendocrinology, 35*(2), 111–116.

Buckingham, J. C., & Hodges, J. R. (1979). Hypothalamic receptors influencing the secretion of corticotrophin releasing hormone in the rat. *The Journal of Physiology, 290*(2), 421–431.

Caberlotto, L., Rimondini, R., Hansson, A., Eriksson, S., & Heilig, M. (2004). Corticotropin-releasing hormone (CRH) mRNA expression in rat central amygdala in cannabinoid tolerance and withdrawal: Evidence for an allostatic shift? *Neuropsychopharmacology, 29*(1), 15–22. http://dx.doi.org/10.1038/sj.npp.1300296.

Calogero, A. E., Gallucci, W. T., Kling, M. A., Chrousos, G. P., & Gold, P. W. (1989). Cocaine stimulates rat hypothalamic corticotropin-releasing hormone secretion in vitro. *Brain Research, 505*(1), 7–11.

Center for Behavioral Health Statistics and Quality (2016). Key substance use and mental health indicators in the United States: Results from the 2015 National Survey on Drug Use and Health.

Chalmers, D. T., Lovenberg, T. W., & De Souza, E. B. (1995). Localization of novel corticotropin-releasing factor receptor (CRF2) mRNA expression to specific subcortical nuclei in rat brain: Comparison with CRF1 receptor mRNA expression. *The Journal of Neuroscience, 15*(10), 6340–6350.

Chu, K., Koob, G. F., Cole, M., Zorrilla, E. P., & Roberts, A. J. (2007). Dependence-induced increases in ethanol self-administration in mice are blocked by the CRF1 receptor antagonist antalarmin and by CRF1 receptor knockout. *Pharmacology, Biochemistry, and Behavior, 86*(4), 813–821. http://dx.doi.org/10.1016/j.pbb.2007.03.009.

Ciocchi, S., Herry, C., Grenier, F., Wolff, S. B., Letzkus, J. J., Vlachos, I., et al. (2010). Encoding of conditioned fear in central amygdala inhibitory circuits. *Nature, 468*(7321), 277–282. http://dx.doi.org/10.1038/nature09559.

Cippitelli, A., Damadzic, R., Singley, E., Thorsell, A., Ciccocioppo, R., Eskay, R. L., et al. (2012). Pharmacological blockade of corticotropin-releasing hormone receptor 1 (CRH1R) reduces voluntary consumption of high alcohol concentrations in non-dependent Wistar rats. *Pharmacology, Biochemistry, and Behavior, 100*(3), 522–529. http://dx.doi.org/10.1016/j.pbb.2011.10.016.

Cole, B. J., Cador, M., Stinus, L., Rivier, C., Rivier, J., Vale, W., et al. (1990). Critical role of the hypothalamic pituitary adrenal axis in amphetamine-induced sensitization of behavior. *Life Sciences, 47*(19), 1715–1720.

Contarino, A., & Papaleo, F. (2005). The corticotropin-releasing factor receptor-1 pathway mediates the negative affective states of opiate withdrawal. *Proceedings of the National Academy of Sciences of the United States of America, 102*(51), 18649–18654. http://dx.doi.org/10.1073/pnas.0506999102.

Cottone, P., Sabino, V., Roberto, M., Bajo, M., Pockros, L., Frihauf, J. B., et al. (2009). CRF system recruitment mediates dark side of compulsive eating. *Proceedings of the National Academy of Sciences of the United States of America, 106*(47), 20016–20020. http://dx.doi.org/10.1073/pnas.0908789106.

Cottone, P., Sabino, V., Steardo, L., & Zorrilla, E. P. (2009). Consummatory, anxiety-related and metabolic adaptations in female rats with alternating access to preferred food. *Psychoneuroendocrinology, 34*(1), 38–49. http://dx.doi.org/10.1016/j.psyneuen.2008.08.010.

Cruz, M. T., Bajo, M., Magnoli, E. M., Tabakoff, B., Siggins, G. R., & Roberto, M. (2011). Type 7 adenylyl cyclase is involved in the ethanol and CRF sensitivity of GABAergic synapses in mouse central amygdala. *Frontiers in Neuroscience*, *4*, 207. http://dx.doi.org/10.3389/fnins.2010.00207.

Cruz, M. T., Herman, M. A., Kallupi, M., & Roberto, M. (2012). Nociceptin/orphanin FQ blockade of corticotropin-releasing factor-induced gamma-aminobutyric acid release in central amygdala is enhanced after chronic ethanol exposure. *Biological Psychiatry*, *71*(8), 666–676. http://dx.doi.org/10.1016/j.biopsych.2011.10.032.

Davis, M., Rainnie, D., & Cassell, M. (1994). Neurotransmission in the rat amygdala related to fear and anxiety. *Trends in Neurosciences*, *17*(5), 208–214.

Davis, M., Walker, D. L., Miles, L., & Grillon, C. (2010). Phasic vs sustained fear in rats and humans: Role of the extended amygdala in fear vs anxiety. *Neuropsychopharmacology*, *35*(1), 105–135.

Day, H. E., Curran, E. J., Watson, S. J., Jr., & Akil, H. (1999). Distinct neurochemical populations in the rat central nucleus of the amygdala and bed nucleus of the stria terminalis: Evidence for their selective activation by interleukin-1beta. *The Journal of Comparative Neurology*, *413*(1), 113–128.

Day, H. E., Nebel, S., Sasse, S., & Campeau, S. (2005). Inhibition of the central extended amygdala by loud noise and restraint stress. *European Journal of Neurology*, *21*(2), 441–454. http://dx.doi.org/10.1111/j.1460-9568.2005.03865.x.

De Souza, E. B., & Battaglia, G. (1988). Corticotropin-releasing hormone (CRH) receptors in brain. *Advances in Experimental Medicine and Biology*, *245*, 123–136.

Dong, H. W., Petrovich, G. D., & Swanson, L. W. (2001). Topography of projections from amygdala to bed nuclei of the stria terminalis. *Brain Research. Brain Research Reviews*, *38*(1–2), 192–246.

Dunn, A. J., & Berridge, C. W. (1990). Physiological and behavioral responses to corticotropin-releasing factor administration: Is CRF a mediator of anxiety or stress responses? *Brain Research. Brain Research Reviews*, *15*(2), 71–100.

Ehrlich, I., Humeau, Y., Grenier, F., Ciocchi, S., Herry, C., & Luthi, A. (2009). Amygdala inhibitory circuits and the control of fear memory. *Neuron*, *62*(6), 757–771. http://dx.doi.org/10.1016/j.neuron.2009.05.026.

Epstein, D. H., Kennedy, A. P., Furnari, M., Heilig, M., Shaham, Y., Phillips, K. A., et al. (2016). Effect of the CRF1-receptor antagonist pexacerfont on stress-induced eating and food craving. *Psychopharmacology*, *233*(23–24), 3921–3932. http://dx.doi.org/10.1007/s00213-016-4424-5.

Erb, S., Salmaso, N., Rodaros, D., & Stewart, J. (2001). A role for the CRF-containing pathway from central nucleus of the amygdala to bed nucleus of the stria terminalis in the stress-induced reinstatement of cocaine seeking in rats. *Psychopharmacology*, *158*(4), 360–365. http://dx.doi.org/10.1007/s002130000642.

Erb, S., Shaham, Y., & Stewart, J. (1998). The role of corticotropin-releasing factor and corticosterone in stress- and cocaine-induced relapse to cocaine seeking in rats. *The Journal of Neuroscience*, *18*(14), 5529–5536.

Erb, S., Shaham, Y., & Stewart, J. (2001). Stress-induced relapse to drug seeking in the rat: Role of the bed nucleus of the stria terminalis and amygdala. *Stress*, *4*(4), 289–303.

Erb, S., & Stewart, J. (1999). A role for the bed nucleus of the stria terminalis, but not the amygdala, in the effects of corticotropin-releasing factor on stress-induced reinstatement of cocaine seeking. *The Journal of Neuroscience*, *19*(20), RC35.

Fahlke, C., Hard, E., & Hansen, S. (1996). Facilitation of ethanol consumption by intracerebroventricular infusions of corticosterone. *Psychopharmacology*, *127*(2), 133–139.

Fee, J. R., Sparta, D. R., Picker, M. J., & Thiele, T. E. (2007). Corticotropin releasing factor-1 receptor antagonist, CP-154,526, blocks the expression of ethanol-induced behavioral sensitization in DBA/2J mice. *Neuroscience*, *150*(1), 14–21. http://dx.doi.org/10.1016/j.neuroscience.2007.08.027.

Fekete, E. M., & Zorrilla, E. P. (2007). Physiology, pharmacology, and therapeutic relevance of urocortins in mammals: Ancient CRF paralogs. *Frontiers in Neuroendocrinology*, *28*(1), 1–27. http://dx.doi.org/10.1016/j.yfrne.2006.09.002.

Fichter, M. M., & Quadflieg, N. (2016). Mortality in eating disorders—Results of a large prospective clinical longitudinal study. *The International Journal of Eating Disorders*, *49*(4), 391–401. http://dx.doi.org/10.1002/eat.22501.

Flegal, K. M., Kit, B. K., Orpana, H., & Graubard, B. I. (2013). Association of all-cause mortality with overweight and obesity using standard body mass index categories: A systematic review and meta-analysis. *JAMA*, *309*(1), 71–82. http://dx.doi.org/10.1001/jama.2012.113905.

Flegal, K. M., Kruszon-Moran, D., Carroll, M. D., Fryar, C. D., & Ogden, C. L. (2016). Trends in obesity among adults in the United States, 2005 to 2014. *JAMA*, *315*(21), 2284–2291. http://dx.doi.org/10.1001/jama.2016.6458.

Francesconi, W., Berton, F., Repunte-Canonigo, V., Hagihara, K., Thurbon, D., Lekic, D., et al. (2009). Protracted withdrawal from alcohol and drugs of abuse impairs long-term potentiation of intrinsic excitability in the juxtacapsular bed nucleus of the stria terminalis. *The Journal of Neuroscience*, *29*(17), 5389–5401. http://dx.doi.org/10.1523/jneurosci.5129-08.2009.

Fu, Y., & Neugebauer, V. (2008). Differential mechanisms of CRF1 and CRF2 receptor functions in the amygdala in pain-related synaptic facilitation and behavior. *The Journal of Neuroscience*, *28*(15), 3861–3876. http://dx.doi.org/10.1523/JNEUROSCI.0227-08.2008.

Funk, C. K., & Koob, G. F. (2007). A CRF(2) agonist administered into the central nucleus of the amygdala decreases ethanol self-administration in ethanol-dependent rats. *Brain Research*, *1155*, 172–178. http://dx.doi.org/10.1016/j.brainres.2007.04.009.

Funk, C. K., O'Dell, L. E., Crawford, E. F., & Koob, G. F. (2006). Corticotropin-releasing factor within the central nucleus of the amygdala mediates enhanced ethanol self-administration in withdrawn, ethanol-dependent rats. *The Journal of Neuroscience*, *26*(44), 11324–11332. http://dx.doi.org/10.1523/jneurosci.3096-06.2006.

Funk, C. K., Zorrilla, E. P., Lee, M. J., Rice, K. C., & Koob, G. F. (2007). Corticotropin-releasing factor 1 antagonists selectively reduce ethanol self-administration in ethanol-dependent rats. *Biological Psychiatry*, *61*(1), 78–86. http://dx.doi.org/10.1016/j.biopsych.2006.03.063.

Gallagher, J. P., Orozco-Cabal, L. F., Liu, J., & Shinnick-Gallagher, P. (2008). Synaptic physiology of central CRH system. *European Journal of Pharmacology*, *583*(2–3), 215–225.

Garcia-Carmona, J. A., Almela, P., Baroja-Mazo, A., Milanes, M. V., & Laorden, M. L. (2012). Restricted role of CRF1 receptor for the activity of brainstem catecholaminergic neurons in the negative state of morphine withdrawal. *Psychopharmacology*, *220*(2), 379–393. http://dx.doi.org/10.1007/s00213-011-2478-y.

Gehlert, D. R., Cippitelli, A., Thorsell, A., Le, A. D., Hipskind, P. A., Hamdouchi, C., et al. (2007). 3-(4-Chloro-2-morpholin-4-yl-thiazol-5-yl)-8-(1-ethylpropyl)-2,6-dimethyl-imidazo [1,2-b]pyridazine: A novel brain-penetrant, orally available corticotropin-releasing factor receptor 1 antagonist with efficacy in animal models of alcoholism. *Journal of Neuroscience*, *27*(10), 2718–2726. http://dx.doi.org/10.1523/JNEUROSCI.4985-06.2007.

George, O., Ghozland, S., Azar, M. R., Cottone, P., Zorrilla, E. P., Parsons, L. H., et al. (2007). CRF-CRF1 system activation mediates withdrawal-induced increases in nicotine self-administration in nicotine-dependent rats. *Proceedings of the National Academy of Sciences of the United States of America*, *104*(43), 17198–17203. http://dx.doi.org/10.1073/pnas.0707585104.

George, O., Mandyam, C. D., Wee, S., & Koob, G. F. (2008). Extended access to cocaine self-administration produces long-lasting prefrontal cortex-dependent working memory impairments. *Neuropsychopharmacology*, *33*(10), 2474–2482. http://dx.doi.org/10.1038/sj.npp.1301626.

George, O., Sanders, C., Freiling, J., Grigoryan, E., Vu, S., Allen, C. D., et al. (2012). Recruitment of medial prefrontal cortex neurons during alcohol withdrawal predicts cognitive impairment and excessive alcohol drinking. *Proceedings of the National Academy of Sciences of the United States of America*, *109*(44), 18156–18161. http://dx.doi.org/10.1073/pnas.1116523109.

Ghitza, U. E., Gray, S. M., Epstein, D. H., Rice, K. C., & Shaham, Y. (2006). The anxiogenic drug yohimbine reinstates palatable food seeking in a rat relapse model: A role of CRF1 receptors. *Neuropsychopharmacology*, *31*(10), 2188–2196. http://dx.doi.org/10.1038/sj.npp.1300964.

Giardino, W. J., & Ryabinin, A. E. (2013). CRF1 receptor signaling regulates food and fluid intake in the drinking-in-the-dark model of binge alcohol consumption. *Alcoholism, Clinical and Experimental Research*, *37*(7), 1161–1170. http://dx.doi.org/10.1111/acer.12076.

Gilman, J. M., & Hommer, D. W. (2008). Modulation of brain response to emotional images by alcohol cues in alcohol-dependent patients. *Addiction Biology*, *13*(3–4), 423–434. http://dx.doi.org/10.1111/j.1369-1600.2008.00111.x.

Gilpin, N. W., Richardson, H. N., Cole, M., & Koob, G. F. (2008). Vapor inhalation of alcohol in rats. *Current Protocols in Neuroscience*. http://dx.doi.org/10.1002/0471142301.ns0929s44. chapter 9, Unit 9.29.

Gilpin, N. W., Richardson, H. N., & Koob, G. F. (2008). Effects of CRF1-receptor and opioid-receptor antagonists on dependence-induced increases in alcohol drinking by alcohol-preferring (P) rats. *Alcoholism, Clinical and Experimental Research*, *32*(9), 1535–1542. http://dx.doi.org/10.1111/j.1530-0277.2008.00745.x.

Glykys, J., & Mody, I. (2007). Activation of GABAA receptors: Views from outside the synaptic cleft. *Neuron*, *56*(5), 763–770. http://dx.doi.org/10.1016/j.neuron.2007.11.002.

Goeders, N. E. (1997). A neuroendocrine role in cocaine reinforcement. *Psychoneuroendocrinology*, *22*(4), 237–259.

Goeders, N. E. (2002). The HPA axis and cocaine reinforcement. *Psychoneuroendocrinology*, *27*(1–2), 13–33.

Goeders, N. E., & Guerin, G. F. (1996). Effects of surgical and pharmacological adrenalectomy on the initiation and maintenance of intravenous cocaine self-administration in rats. *Brain Research*, *722*(1–2), 145–152.

Goodman, S. N. (1999). Toward evidence-based medical statistics. 2: The Bayes factor. *Annals of Internal Medicine*, *130*(12), 1005–1013.

Greenwell, T. N., Funk, C. K., Cottone, P., Richardson, H. N., Chen, S. A., Rice, K. C., et al. (2009). Corticotropin-releasing factor-1 receptor antagonists decrease heroin self-administration in long- but not short-access rats. *Addiction Biology*, *14*(2), 130–143. http://dx.doi.org/10.1111/j.1369-1600.2008.00142.x.

Grieder, T. E., Herman, M. A., Contet, C., Tan, L. A., Vargas-Perez, H., Cohen, A., et al. (2014). VTA CRF neurons mediate the aversive effects of nicotine withdrawal and promote intake escalation. *Nature Neuroscience*, *17*(12), 1751–1758. http://dx.doi.org/10.1038/nn.3872.

Grigoriadis, D. E., Liu, X. J., Vaughn, J., Palmer, S. F., True, C. D., Vale, W. W., et al. (1996). 125I-Tyro-sauvagine: A novel high affinity radioligand for the pharmacological and biochemical study of human corticotropin-releasing factor 2 alpha receptors. *Molecular Pharmacology*, *50*(3), 679–686.

Gysling, K. (2012). Relevance of both type-1 and type-2 corticotropin releasing factor receptors in stress-induced relapse to cocaine seeking behaviour. *Biochemical Pharmacology*, *83*(1), 1–5. http://dx.doi.org/10.1016/j.bcp.2011.07.101.

Haass-Koffler, C. L., Henry, A. T., Melkus, G., Simms, J. A., Naemmuddin, M., Nielsen, C. K., et al. (2016). Defining the role of corticotropin releasing factor binding protein in alcohol consumption. *Translational Psychiatry*, *6*(11), e953. http://dx.doi.org/10.1038/tp.2016.208.

Hammack, S. E., Richey, K. J., Watkins, L. R., & Maier, S. F. (2004). Chemical lesion of the bed nucleus of the stria terminalis blocks the behavioral consequences of uncontrollable stress. *Behavioral Neuroscience, 118*(2), 443–448. http://dx.doi.org/10.1037/0735-7044.118.2.443,2004-12681-022.

Hansson, A. C., Cippitelli, A., Sommer, W. H., Fedeli, A., Bjork, K., Soverchia, L., et al. (2006). Variation at the rat Crhr1 locus and sensitivity to relapse into alcohol seeking induced by environmental stress. *Proceedings of the National Academy of Sciences of the United States of America, 103*(41), 15236–15241. http://dx.doi.org/10.1073/pnas.0604419103.

Haubensak, W., Kunwar, P. S., Cai, H., Ciocchi, S., Wall, N. R., Ponnusamy, R., et al. (2010). Genetic dissection of an amygdala microcircuit that gates conditioned fear. *Nature, 468*(7321), 270–276. http://dx.doi.org/10.1038/nature09553.

Heilig, M., Egli, M., Crabbe, J. C., & Becker, H. C. (2010). Acute withdrawal, protracted abstinence and negative affect in alcoholism: Are they linked? *Addiction Biology, 15*(2), 169–184. http://dx.doi.org/10.1111/j.1369-1600.2009.00194.x.

Heilig, M., Goldman, D., Berrettini, W., & O'Brien, C. P. (2011). Pharmacogenetic approaches to the treatment of alcohol addiction. *Nature Reviews Neuroscience, 12*(11), 670–684. http://dx.doi.org/10.1038/nrn3110.

Heilig, M., & Koob, G. F. (2007). A key role for corticotropin-releasing factor in alcohol dependence. *Trends in Neurosciences, 30*(8), 399–406. http://dx.doi.org/10.1016/j.tins.2007.06.006.

Heilig, M., Thorsell, A., Sommer, W. H., Hansson, A. C., Ramchandani, V. A., George, D. T., et al. (2010). Translating the neuroscience of alcoholism into clinical treatments: From blocking the buzz to curing the blues. *Neuroscience and Biobehavioral Reviews, 35*(2), 334–344. http://dx.doi.org/10.1016/j.neubiorev.2009.11.018.

Heimer, L., & Alheid, G. (1991). Piecing together the puzzle of basal forebrain anatomy. In T. C. Napier, P. W. Kalivas, & I. Hanin (Eds.), *Advances in experimental medicine and biology: Vol. 295. The basal forebrain: Anatomy to function* (pp. 1–42). New York: Plenum Press.

Heinrichs, S. C., & Koob, G. F. (2004). Corticotropin-releasing factor in brain: A role in activation, arousal, and affect regulation. *The Journal of Pharmacology and Experimental Therapeutics, 311*(2), 427–440. http://dx.doi.org/10.1124/jpet.103.052092.

Heinrichs, S. C., Menzaghi, F., Merlo Pich, E., Britton, K. T., & Koob, G. F. (1995). The role of CRF in behavioral aspects of stress. *Annals of the New York Academy of Sciences, 771*, 92–104.

Heinrichs, S. C., Menzaghi, F., Pich, E. M., Baldwin, H. A., Rassnick, S., Britton, K. T., et al. (1994). Anti-stress action of a corticotropin-releasing factor antagonist on behavioral reactivity to stressors of varying type and intensity. *Neuropsychopharmacology, 11*(3), 179–186. http://dx.doi.org/10.1038/sj.npp.1380104.

Herman, M. A., Contet, C., Justice, N. J., Vale, W., & Roberto, M. (2013). Novel subunit-specific tonic GABA currents and differential effects of ethanol in the central amygdala of CRF receptor-1 reporter mice. *The Journal of Neuroscience, 33*(8), 3284–3298. http://dx.doi.org/10.1523/jneurosci.2490-12.2013.

Herman, M. A., Contet, C., & Roberto, M. (2016). A functional switch in tonic GABA currents alters the output of central amygdala Corticotropin releasing factor receptor-1 neurons following chronic ethanol exposure. *The Journal of Neuroscience, 36*(42), 10729–10741. http://dx.doi.org/10.1523/jneurosci.1267-16.2016.

Ho, S. P., Takahashi, L. K., Livanov, V., Spencer, K., Lesher, T., Maciag, C., et al. (2001). Attenuation of fear conditioning by antisense inhibition of brain corticotropin releasing factor-2 receptor. *Brain Research. Molecular Brain Research, 89*(1–2), 29–40. doi: S0169328X0100050X [pii].

Hoang, U., Goldacre, M., & James, A. (2014). Mortality following hospital discharge with a diagnosis of eating disorder: National record linkage study, England, 2001-2009. *The International Journal of Eating Disorders, 47*(5), 507–515. http://dx.doi.org/10.1002/eat.22249.

Holly, E. N., Boyson, C. O., Montagud-Romero, S., Stein, D. J., Gobrogge, K. L., DeBold, J. F., et al. (2016). Episodic social stress-escalated cocaine self-administration: Role of phasic and tonic Corticotropin releasing factor in the anterior and posterior ventral tegmental area. *The Journal of Neuroscience*, *36*(14), 4093–4105. http://dx.doi.org/10.1523/jneurosci.2232-15.2016.

Holly, E. N., & Miczek, K. A. (2016). Ventral tegmental area dopamine revisited: Effects of acute and repeated stress. *Psychopharmacology*, *233*(2), 163–186. http://dx.doi.org/10.1007/s00213-015-4151-3.

Hwa, L. S., Debold, J. F., & Miczek, K. A. (2013). Alcohol in excess: CRF(1) receptors in the rat and mouse VTA and DRN. *Psychopharmacology*, *225*(2), 313–327. http://dx.doi.org/10.1007/s00213-012-2820-z.

Iemolo, A., Blasio, A., St Cyr, S. A., Jiang, F., Rice, K. C., Sabino, V., et al. (2013). CRF-CRF1 receptor system in the central and basolateral nuclei of the amygdala differentially mediates excessive eating of palatable food. *Neuropsychopharmacology*, *38*(12), 2456–2466. http://dx.doi.org/10.1038/npp.2013.147.

Imaki, T., Nahan, J. L., Rivier, C., Sawchenko, P. E., & Vale, W. (1991). Differential regulation of corticotropin-releasing factor mRNA in rat brain regions by glucocorticoids and stress. *The Journal of Neuroscience*, *11*(3), 585–599.

Ingallinesi, M., Rouibi, K., Le Moine, C., Papaleo, F.; & Contarino, A. (2012). CRF2 receptor-deficiency eliminates opiate withdrawal distress without impairing stress coping. *Molecular Psychiatry*, *17*(12), 1283–1294. http://dx.doi.org/10.1038/mp.2011.119.

Iredale, P. A., Alvaro, J. D., Lee, Y., Terwilliger, R., Chen, Y. L., & Duman, R. S. (2000). Role of corticotropin-releasing factor receptor-1 in opiate withdrawal. *Journal of Neurochemistry*, *74*(1), 199–208.

Jia, F., Chandra, D., Homanics, G. E., & Harrison, N. L. (2008). Ethanol modulates synaptic and extrasynaptic GABAA receptors in the thalamus. *The Journal of Pharmacology and Experimental Therapeutics*, *326*(2), 475–482. http://dx.doi.org/10.1124/jpet.108.139303.

Justice, N. J., Yuan, Z. F., Sawchenko, P. E., & Vale, W. (2008). Type 1 corticotropin-releasing factor receptor expression reported in BAC transgenic mice: Implications for reconciling ligand-receptor mismatch in the central corticotropin-releasing factor system. *The Journal of Comparative Neurology*, *511*(4), 479–496. http://dx.doi.org/10.1002/cne.21848.

Kash, T. L., Baucum, A. J., 2nd, Conrad, K. L., Colbran, R. J., & Winder, D. G. (2009). Alcohol exposure alters NMDAR function in the bed nucleus of the stria terminalis. *Neuropsychopharmacology*, *34*(11), 2420–2429. http://dx.doi.org/10.1038/npp.2009.69.

Kash, T. L., Nobis, W. P., Matthews, R. T., & Winder, D. G. (2008). Dopamine enhances fast excitatory synaptic transmission in the extended amygdala by a CRF-R1-dependent process. *The Journal of Neuroscience*, *28*(51), 13856–13865. http://dx.doi.org/10.1523/jneurosci.4715-08.2008.

Kash, T. L., & Winder, D. G. (2006). Neuropeptide Y and corticotropin-releasing factor bi-directionally modulate inhibitory synaptic transmission in the bed nucleus of the stria terminalis. *Neuropharmacology*, *51*(5), 1013–1022. http://dx.doi.org/10.1016/j.neuropharm.2006.06.011.

Kaur, S., Li, J., Stenzel-Poore, M. P., & Ryabinin, A. E. (2012). Corticotropin-releasing factor acting on corticotropin-releasing factor receptor type 1 is critical for binge alcohol drinking in mice. *Alcoholism, Clinical and Experimental Research*, *36*(2), 369–376. http://dx.doi.org/10.1111/j.1530-0277.2011.01610.x.

Kessler, R. C., Birnbaum, H., Demler, O., Falloon, I. R., Gagnon, E., Guyer, M., et al. (2005). The prevalence and correlates of nonaffective psychosis in the National Comorbidity Survey Replication (NCS-R). *Biological Psychiatry*, *58*(8), 668–676.

Ketchesin, K. D., Stinnett, G. S., & Seasholtz, A. F. (2016). Binge drinking decreases corticotropin-releasing factor-binding protein expression in the medial prefrontal cortex of mice. *Alcoholism, Clinical and Experimental Research*, *40*(8), 1641–1650. http://dx.doi.org/10.1111/acer.13119.

Knapp, D. J., Overstreet, D. H., Moy, S. S., & Breese, G. R. (2004). SB242084, flumazenil, and CRA1000 block ethanol withdrawal-induced anxiety in rats. *Alcohol*, *32*(2), 101–111. http://dx.doi.org/10.1016/j.alcohol.2003.08.007.

Koob, G. F. (1999). Corticotropin-releasing factor, norepinephrine, and stress. *Biological Psychiatry*, *46*(9), 1167–1180.

Koob, G. F. (2008). A role for brain stress systems in addiction. *Neuron*, *59*(1), 11–34.

Koob, G. F., & Heinrichs, S. C. (1999). A role for corticotropin releasing factor and urocortin in behavioral responses to stressors. *Brain Research*, *848*(1–2), 141–152.

Koob, G., & Kreek, M. J. (2007). Stress, dysregulation of drug reward pathways, and the transition to drug dependence. *The American Journal of Psychiatry*, *164*(8), 1149–1159. http://dx.doi.org/10.1176/appi.ajp.2007.05030503.

Koob, G. F., & Le Moal, M. (2001). Drug addiction, dysregulation of reward, and allostasis. *Neuropsychopharmacology*, *24*(2), 97–129. http://dx.doi.org/10.1016/s0893-133x(00)00195-0.

Koob, G. F., & Le Moal, M. (2005). Plasticity of reward neurocircuitry and the 'dark side' of drug addiction. *Nature Neuroscience*, *8*(11), 1442–1444. http://dx.doi.org/10.1038/nn1105-1442.

Koob, G. F., & Le Moal, M. (2008). Review. Neurobiological mechanisms for opponent motivational processes in addiction. *Philosophical Transactions of the Royal Society of London. Series B, Biological Sciences*, *363*(1507), 3113–3123. http://dx.doi.org/10.1098/rstb.2008.0094.

Koob, G. F., Rassnick, S., Heinrichs, S., & Weiss, F. (1994). Alcohol, the reward system and dependence. *EXS*, *71*, 103–114.

Koob, G. F., & Volkow, N. D. (2010). Neurocircuitry of addiction. *Neuropsychopharmacology*, *35*(1), 217–238. http://dx.doi.org/10.1038/npp.2009.110.

Koob, G. F., & Volkow, N. D. (2016). Neurobiology of addiction: A neurocircuitry analysis. *The Lancet. Psychiatry*, *3*(8), 760–773. http://dx.doi.org/10.1016/S2215-0366(16)00104-8.

Koob, G. F., & Zorrilla, E. P. (2010). Neurobiological mechanisms of addiction: Focus on corticotropin-releasing factor. *Current Opinion in Investigational Drugs*, *11*(1), 63–71.

Kozicz, T., Yanaihara, H., & Arimura, A. (1998). Distribution of urocortin-like immunoreactivity in the central nervous system of the rat. *The Journal of Comparative Neurology*, *391*(1), 1–10. http://dx.doi.org/10.1002/(SICI)1096-9861(19980202)391:1<1::AID-CNE1>3.0.CO;2-6.

Kreek, M. J., & Koob, G. F. (1998). Drug dependence: Stress and dysregulation of brain reward pathways. *Drug and Alcohol Dependence*, *51*(1–2), 23–47.

Kroes, M., Osei-Assibey, G., Baker-Searle, R., & Huang, J. (2016). Impact of weight change on quality of life in adults with overweight/obesity in the United States: A systematic review. *Current Medical Research and Opinion*, *32*(3), 485–508. http://dx.doi.org/10.1185/03007995.2015.1128403.

Le, A. D., Funk, D., Coen, K., Li, Z., & Shaham, Y. (2013). Role of corticotropin-releasing factor in the median raphe nucleus in yohimbine-induced reinstatement of alcohol seeking in rats. *Addiction Biology*, *18*(3), 448–451. http://dx.doi.org/10.1111/j.1369-1600.2011.00374.x.

Le Gal LaSalle, G., Paxinos, G., & Ben-Ari, Y. (1978). Neurochemical mapping of GABAergic systems in the amygdaloid complex and bed nucleus of the stria terminalis. *Brain Research*, *155*(2), 397–403.

Le, A. D., Harding, S., Juzytsch, W., Fletcher, P. J., & Shaham, Y. (2002). The role of corticotropin-releasing factor in the median raphe nucleus in relapse to alcohol. *The Journal of Neuroscience*, *22*(18), 7844–7849.

Le, A. D., Harding, S., Juzytsch, W., Watchus, J., Shalev, U., & Shaham, Y. (2000). The role of corticotrophin-releasing factor in stress-induced relapse to alcohol-seeking behavior in rats. *Psychopharmacology*, *150*(3), 317–324.

Le, A., & Shaham, Y. (2002). Neurobiology of relapse to alcohol in rats. *Pharmacology & Therapeutics, 94*(1–2), 137–156.

Letzkus, J. J., Wolff, S. B., & Luthi, A. (2015). Disinhibition, a circuit mechanism for associative learning and memory. *Neuron, 88*(2), 264–276. http://dx.doi.org/10.1016/j.neuron.2015.09.024.

Levran, O., Peles, E., Randesi, M., Li, Y., Rotrosen, J., Ott, J., et al. (2014). Stress-related genes and heroin addiction: A role for a functional FKBP5 haplotype. *Psychoneuroendocrinology, 45*, 67–76. http://dx.doi.org/10.1016/j.psyneuen.2014.03.017.

Levran, O., Randesi, M., Li, Y., Rotrosen, J., Ott, J., Adelson, M., et al. (2014). Drug addiction and stress-response genetic variability: Association study in African Americans. *Annals of Human Genetics, 78*(4), 290–298. http://dx.doi.org/10.1111/ahg.12064.

Liang, J., Suryanarayanan, A., Chandra, D., Homanics, G. E., Olsen, R. W., & Spigelman, I. (2008). Functional consequences of GABAA receptor alpha 4 subunit deletion on synaptic and extrasynaptic currents in mouse dentate granule cells. *Alcoholism, Clinical and Experimental Research, 32*(1), 19–26. http://dx.doi.org/10.1111/j.1530-0277.2007.00564.x.

Liu, J., Yu, B., Neugebauer, V., Grigoriadis, D. E., Rivier, J., Vale, W. W., et al. (2004). Corticotropin-releasing factor and Urocortin I modulate excitatory glutamatergic synaptic transmission. *Journal of Neuroscience, 24*(16), 4020–4029. http://dx.doi.org/10.1523/JNEUROSCI.5531-03.2004.

Logrip, M. L., Koob, G. F., & Zorrilla, E. P. (2011). Role of corticotropin-releasing factor in drug addiction: Potential for pharmacological intervention. *CNS Drugs, 25*(4), 271–287. http://dx.doi.org/10.2165/11587790-000000000-00000.

Long, C. G., Blundell, J. E., & Finlayson, G. (2015). A systematic review of the application and correlates of YFAS-diagnosed 'food addiction' in humans: Are eating-related 'addictions' a cause for concern or empty concepts? *Obesity Facts, 8*(6), 386–401. http://dx.doi.org/10.1159/000442403.

Lowery, E. G., Spanos, M., Navarro, M., Lyons, A. M., Hodge, C. W., & Thiele, T. E. (2010). CRF-1 antagonist and CRF-2 agonist decrease binge-like ethanol drinking in C57BL/6J mice independent of the HPA axis. *Neuropsychopharmacology, 35*(6), 1241–1252. http://dx.doi.org/10.1038/npp.2009.209.

Lowery, E. G., Sparrow, A. M., Breese, G. R., Knapp, D. J., & Thiele, T. E. (2008). The CRF-1 receptor antagonist, CP-154,526, attenuates stress-induced increases in ethanol consumption by BALB/cJ mice. *Alcoholism, Clinical and Experimental Research, 32*(2), 240–248. http://dx.doi.org/10.1111/j.1530-0277.2007.00573.x.

Lowery-Gionta, E. G., Navarro, M., Li, C., Pleil, K. E., Rinker, J. A., Cox, B. R., et al. (2012). Corticotropin releasing factor signaling in the central amygdala is recruited during binge-like ethanol consumption in C57BL/6J mice. *Journal of Neuroscience, 32*(10), 3405–3413. http://dx.doi.org/10.1523/JNEUROSCI.6256-11.2012.

Lu, L., Ceng, X., & Huang, M. (2000). Corticotropin-releasing factor receptor type I mediates stress-induced relapse to opiate dependence in rats. *Neuroreport, 11*(11), 2373–2378.

Lu, L., Liu, D., & Ceng, X. (2001). Corticotropin-releasing factor receptor type 1 mediates stress-induced relapse to cocaine-conditioned place preference in rats. *European Journal of Pharmacology, 415*(2–3), 203–208.

Lu, L., Liu, D., Ceng, X., & Ma, L. (2000). Differential roles of corticotropin-releasing factor receptor subtypes 1 and 2 in opiate withdrawal and in relapse to opiate dependence. *The European Journal of Neuroscience, 12*(12), 4398–4404.

Lu, L., Shepard, J. D., Hall, F. S., & Shaham, Y. (2003). Effect of environmental stressors on opiate and psychostimulant reinforcement, reinstatement and discrimination in rats: A review. *Neuroscience and Biobehavioral Reviews, 27*(5), 457–491.

Luthi, A., & Luscher, C. (2014). Pathological circuit function underlying addiction and anxiety disorders. *Nature Neuroscience, 17*(12), 1635–1643. http://dx.doi.org/10.1038/nn.3849.

Maj, M., Turchan, J., Smialowska, M., & Przewlocka, B. (2003). Morphine and cocaine influence on CRF biosynthesis in the rat central nucleus of amygdala. *Neuropeptides*, *37*(2), 105–110. doi: S0143417903000210 [pii].

Makino, S., Gold, P. W., & Schulkin, J. (1994). Corticosterone effects on corticotropin-releasing hormone mRNA in the central nucleus of the amygdala and the parvocellular region of the paraventricular nucleus of the hypothalamus. *Brain Research*, *640*(1–2), 105–112.

Marcinkiewcz, C. A., Lowery-Gionta, E. G., & Kash, T. L. (2016). Serotonin's complex role in alcoholism: Implications for treatment and future research. *Alcoholism, Clinical and Experimental Research*, *40*(6), 1192–1201. http://dx.doi.org/10.1111/acer.13076.

Marcinkiewcz, C. A., Prado, M. M., Isaac, S. K., Marshall, A., Rylkova, D., & Bruijnzeel, A. W. (2009). Corticotropin-releasing factor within the central nucleus of the amygdala and the nucleus accumbens shell mediates the negative affective state of nicotine withdrawal in rats. *Neuropsychopharmacology*, *34*(7), 1743–1752. http://dx.doi.org/10.1038/npp.2008.231.

Marinelli, P. W., Funk, D., Juzytsch, W., Harding, S., Rice, K. C., Shaham, Y., et al. (2007). The CRF1 receptor antagonist antalarmin attenuates yohimbine-induced increases in operant alcohol self-administration and reinstatement of alcohol seeking in rats. *Psychopharmacology*, *195*(3), 345–355. http://dx.doi.org/10.1007/s00213-007-0905-x.

Mazzone, C. M., Pati, D., Michaelides, M., DiBerto, J., Fox, J. H., Tipton, G., et al. (2016). Acute engagement of Gq-mediated signaling in the bed nucleus of the stria terminalis induces anxiety-like behavior. *Molecular Psychiatry*. http://dx.doi.org/10.1038/mp.2016.218.

McDonald, A. J., & Augustine, J. R. (1993). Localization of GABA-like immunoreactivity in the monkey amygdala. *Neuroscience*, *52*(2), 281–294. doi: 0306-4522(93)90156-A [pii].

McElligott, Z. A., & Winder, D. G. (2009). Modulation of glutamatergic synaptic transmission in the bed nucleus of the stria terminalis. *Progress in Neuro-Psychopharmacology & Biological Psychiatry*, *33*(8), 1329–1335. http://dx.doi.org/10.1016/j.pnpbp.2009.05.022.

McReynolds, J. R., Vranjkovic, O., Thao, M., Baker, D. A., Makky, K., Lim, Y., et al. (2014). Beta-2 adrenergic receptors mediate stress-evoked reinstatement of cocaine-induced conditioned place preference and increases in CRF mRNA in the bed nucleus of the stria terminalis in mice. *Psychopharmacology*, *231*(20), 3953–3963. http://dx.doi.org/10.1007/s00213-014-3535-0.

Melia, K. R., Sananes, C. B., & Davis, M. (1992). Lesions of the central nucleus of the amygdala block the excitatory effects of septal ablation on the acoustic startle reflex. *Physiology & Behavior*, *51*(1), 175–180.

Meloni, E. G., Gerety, L. P., Knoll, A. T., Cohen, B. M., & Carlezon, W. A., Jr. (2006). Behavioral and anatomical interactions between dopamine and corticotropin-releasing factor in the rat. *The Journal of Neuroscience*, *26*(14), 3855–3863. http://dx.doi.org/10.1523/JNEUROSCI.4957-05.2006.

Merlo Pich, E., Lorang, M., Yeganeh, M., Rodriguez de Fonseca, F., Raber, J., Koob, G. F., et al. (1995). Increase of extracellular corticotropin-releasing factor-like immunoreactivity levels in the amygdala of awake rats during restraint stress and ethanol withdrawal as measured by microdialysis. *The Journal of Neuroscience*, *15*(8), 5439–5447.

Micali, N., Martini, M. G., Thomas, J. J., Eddy, K. T., Kothari, R., Russell, E., et al. (2017). Lifetime and 12-month prevalence of eating disorders amongst women in mid-life: A population-based study of diagnoses and risk factors. *BMC Medicine*, *15*(1), 12. http://dx.doi.org/10.1186/s12916-016-0766-4.

Milan-Lobo, L., Gsandtner, I., Gaubitzer, E., Runzler, D., Buchmayer, F., Kohler, G., et al. (2009). Subtype-specific differences in corticotropin-releasing factor receptor complexes detected by fluorescence spectroscopy. *Molecular Pharmacology*, *76*(6), 1196–1210. http://dx.doi.org/10.1124/mol.109.059139.

Mitchell, J. E. (2016). Medical comorbidity and medical complications associated with binge-eating disorder. *The International Journal of Eating Disorders*, *49*(3), 319–323. http://dx.doi.org/10.1002/eat.22452.

Miyata, M., Okada, D., Hashimoto, K., Kano, M., & Ito, M. (1999). Corticotropin-releasing factor plays a permissive role in cerebellar long-term depression. *Neuron*, *22*(4), 763–775 S0896-6273(00)80735-7 [pii].

Mody, I., Glykys, J., & Wei, W. (2007). A new meaning for "Gin & Tonic": Tonic inhibition as the target for ethanol action in the brain. *Alcohol*, *41*(3), 145–153. http://dx.doi.org/10.1016/j.alcohol.2007.03.009.

Molander, A., Vengeliene, V., Heilig, M., Wurst, W., Deussing, J. M., & Spanagel, R. (2012). Brain-specific inactivation of the Crhr1 gene inhibits post-dependent and stress-induced alcohol intake, but does not affect relapse-like drinking. *Neuropsychopharmacology*, *37*(4), 1047–1056. http://dx.doi.org/10.1038/npp.2011.297.

Mossberg, D., Liljeberg, P., & Borg, S. (1985). Clinical conditions in alcoholics during long-term abstinence: A descriptive, longitudinal treatment study. *Alcohol*, *2*(3), 551–553.

Nawata, Y., Kitaichi, K., & Yamamoto, T. (2012). Increases of CRF in the amygdala are responsible for reinstatement of methamphetamine-seeking behavior induced by footshock. *Pharmacology, Biochemistry, and Behavior*, *101*(2), 297–302. http://dx.doi.org/10.1016/j.pbb.2012.01.003.

Nie, Z., Schweitzer, P., Roberts, A. J., Madamba, S. G., Moore, S. D., & Siggins, G. R. (2004). Ethanol augments GABAergic transmission in the central amygdala via CRF1 receptors. *Science*, *303*(5663), 1512–1514. http://dx.doi.org/10.1126/science.1092550, 303/5663/1512.

Nie, Z., Zorrilla, E. P., Madamba, S. G., Rice, K. C., Roberto, M., & Siggins, G. R. (2009). Presynaptic CRF1 receptors mediate the ethanol enhancement of GABAergic transmission in the mouse central amygdala. *The ScientificWorldJournal*, *9*, 68–85. http://dx.doi.org/10.1100/tsw.2009.1.

Nimitvilai, S., Herman, M., You, C., Arora, D. S., McElvain, M. A., Roberto, M., et al. (2014). Dopamine D2 receptor desensitization by dopamine or corticotropin releasing factor in ventral tegmental area neurons is associated with increased glutamate release. *Neuropharmacology*, *82*, 28–40. http://dx.doi.org/10.1016/j.neuropharm.2014.03.006.

Nobis, W. P., Kash, T. L., Silberman, Y., & Winder, D. G. (2011). Beta-adrenergic receptors enhance excitatory transmission in the bed nucleus of the stria terminalis through a corticotrophin-releasing factor receptor-dependent and cocaine-regulated mechanism. *Biological Psychiatry*, *69*(11), 1083–1090. http://dx.doi.org/10.1016/j.biopsych.2010.12.030.

O'Dell, L. E., Roberts, A. J., Smith, R. T., & Koob, G. F. (2004). Enhanced alcohol self-administration after intermittent versus continuous alcohol vapor exposure. *Alcoholism, Clinical and Experimental Research*, *28*(11), 1676–1682.

Olguin, P., Fuentes, M., Gabler, G., Guerdjikova, A. I., Keck, P. E., Jr., & McElroy, S. L. (2017). Medical comorbidity of binge eating disorder. *Eating and Weight Disorders*, *22*(1), 13–26. http://dx.doi.org/10.1007/s40519-016-0313-5.

Olive, M. F., Koenig, H. N., Nannini, M. A., & Hodge, C. W. (2002). Elevated extracellular CRF levels in the bed nucleus of the stria terminalis during ethanol withdrawal and reduction by subsequent ethanol intake. *Pharmacology, Biochemistry, and Behavior*, *72*(1–2), 213–220. doi: S0091305701007481 [pii].

Olive, M. F., Mehmert, K. K., Koenig, H. N., Camarini, R., Kim, J. A., Nannini, M. A., et al. (2003). A role for corticotropin releasing factor (CRF) in ethanol consumption, sensitivity, and reward as revealed by CRF-deficient mice. *Psychopharmacology*, *165*(2), 181–187. http://dx.doi.org/10.1007/s00213-002-1248-2.

Overstreet, D. H., Knapp, D. J., & Breese, G. R. (2004). Modulation of multiple ethanol withdrawal-induced anxiety-like behavior by CRF and CRF1 receptors. *Pharmacology, Biochemistry, and Behavior*, 77(2), 405–413. http://dx.doi.org/10.1016/j.pbb.2003.11.010.

Papaleo, F., Ghozland, S., Ingallinesi, M., Roberts, A. J., Koob, G. F., & Contarino, A. (2008). Disruption of the CRF(2) receptor pathway decreases the somatic expression of opiate withdrawal. *Neuropsychopharmacology*, *33*(12), 2878–2887. http://dx.doi.org/10.1038/npp.2008.8.

Pape, H. C., & Pare, D. (2010). Plastic synaptic networks of the amygdala for the acquisition, expression, and extinction of conditioned fear. *Physiological Reviews*, *90*(2), 419–463. http://dx.doi.org/10.1152/physrev.00037.2009.

Pare, D., & Smith, Y. (1993). Distribution of GABA immunoreactivity in the amygdaloid complex of the cat. *Neuroscience*, *57*(4), 1061–1076. doi: 0306-4522(93)90049-L [pii].

Partridge, J. G., Forcelli, P. A., Luo, R., Cashdan, J. M., Schulkin, J., Valentino, R. J., et al. (2016). Stress increases GABAergic neurotransmission in CRF neurons of the central amygdala and bed nucleus stria terminalis. *Neuropharmacology*, *107*, 239–250. http://dx.doi.org/10.1016/j.neuropharm.2016.03.029.

Parylak, S. L., Cottone, P., Sabino, V., Rice, K. C., & Zorrilla, E. P. (2012). Effects of CB1 and CRF1 receptor antagonists on binge-like eating in rats with limited access to a sweet fat diet: Lack of withdrawal-like responses. *Physiology & Behavior*, *107*(2), 231–242. http://dx.doi.org/10.1016/j.physbeh.2012.06.017.

Parylak, S. L., Koob, G. F., & Zorrilla, E. P. (2011). The dark side of food addiction. *Physiology & Behavior*, *104*(1), 149–156. http://dx.doi.org/10.1016/j.physbeh.2011.04.063.

Pastor, R., McKinnon, C. S., Scibelli, A. C., Burkhart-Kasch, S., Reed, C., Ryabinin, A. E., et al. (2008). Corticotropin-releasing factor-1 receptor involvement in behavioral neuroadaptation to ethanol: A urocortin1-independent mechanism. *Proceedings of the National Academy of Sciences of the United States of America*, *105*(26), 9070–9075. http://dx.doi.org/10.1073/pnas.0710181105,0710181105.

Pastor, R., Reed, C., Burkhart-Kasch, S., Li, N., Sharpe, A. L., Coste, S. C., et al. (2011). Ethanol concentration-dependent effects and the role of stress on ethanol drinking in corticotropin-releasing factor type 1 and double type 1 and 2 receptor knockout mice. *Psychopharmacology*, *218*(1), 169–177. http://dx.doi.org/10.1007/s00213-011-2284-6.

Pelleymounter, M. A., Joppa, M., Carmouche, M., Cullen, M. J., Brown, B., Murphy, B., et al. (2000). Role of corticotropin-releasing factor (CRF) receptors in the anorexic syndrome induced by CRF. *The Journal of Pharmacology and Experimental Therapeutics*, *293*(3), 799–806.

Penzo, M. A., Robert, V., & Li, B. (2014). Fear conditioning potentiates synaptic transmission onto long-range projection neurons in the lateral subdivision of central amygdala. *The Journal of Neuroscience*, *34*(7), 2432–2437. http://dx.doi.org/10.1523/JNEUROSCI.4166-13.2014.

Perez, M., Ohrt, T. K., & Hoek, H. W. (2016). Prevalence and treatment of eating disorders among Hispanics/Latino Americans in the United States. *Current Opinion in Psychiatry*, *29*(6), 378–382. http://dx.doi.org/10.1097/yco.0000000000000277.

Petrovich, G. D., Risold, P. Y., & Swanson, L. W. (1996). Organization of projections from the basomedial nucleus of the amygdala: A PHAL study in the rat. *The Journal of Comparative Neurology*, *374*(3), 387–420. http://dx.doi.org/10.1002/(SICI)1096-9861(19961021)374:3<387::AID-CNE6>3.0.CO;2-Y.

Phelix, C. F., Liposits, Z., & Paull, W. K. (1994). Catecholamine-CRF synaptic interaction in a septal bed nucleus: Afferents of neurons in the bed nucleus of the stria terminalis. *Brain Research Bulletin*, *33*(1), 109–119.

Piazza, P. V., Deroche, V., Deminiere, J. M., Maccari, S., Le Moal, M., & Simon, H. (1993). Corticosterone in the range of stress-induced levels possesses reinforcing properties: Implications for sensation-seeking behaviors. *Proceedings of the National Academy of Sciences of the United States of America*, *90*(24), 11738–11742.

Piazza, P. V., & Le Moal, M. (1997). Glucocorticoids as a biological substrate of reward: Physiological and pathophysiological implications. *Brain Research. Brain Research Reviews*, *25*(3), 359–372.

Pickens, R. W., Hatsukami, D. K., Spicer, J. W., & Svikis, D. S. (1985). Relapse by alcohol abusers. *Alcoholism, Clinical and Experimental Research, 9*(3), 244–247.

Pilcher, W. H., & Joseph, S. A. (1984). Co-localization of CRF-ir perikarya and ACTH-ir fibers in rat brain. *Brain Research, 299*(1), 91–102. doi: 0006-8993(84)90791-1 [pii].

Pitkänen, A. (2000). Connectivity of the rat amygdaloid complex. In J. P. Aggleton (Ed.), *The amygdala* (pp: 31–115). New York: Oxford University Press.

Plaza-Zabala, A., Martin-Garcia, E., de Lecea, L., Maldonado, R., & Berrendero, F. (2010). Hypocretins regulate the anxiogenic-like effects of nicotine and induce reinstatement of nicotine-seeking behavior. *The Journal of Neuroscience, 30*(6), 2300–2310. http://dx.doi.org/10.1523/jneurosci.5724-09.2010.

Pollandt, S., Liu, J., Orozco-Cabal, L., Grigoriadis, D. E., Vale, W. W., Gallagher, J. P., et al. (2006). Cocaine withdrawal enhances long-term potentiation induced by corticotropin-releasing factor at central amygdala glutamatergic synapses via CRF, NMDA receptors and PKA. *The European Journal of Neuroscience, 24*(6), 1733–1743.

Pomrenze, M. B., Millan, E. Z., Hopf, F. W., Keiflin, R., Maiya, R., Blasio, A., et al. (2015). A transgenic rat for investigating the anatomy and function of corticotrophin releasing factor circuits. *Frontiers in Neuroscience, 9*, 487. http://dx.doi.org/10.3389/fnins.2015.00487.

Potter, E., Behan, D. P., Linton, E. A., Lowry, P. J., Sawchenko, P. E., & Vale, W. W. (1992). The central distribution of a corticotropin-releasing factor (CRF)-binding protein predicts multiple sites and modes of interaction with CRF. *Proceedings of the National Academy of Sciences of the United States of America, 89*(9), 4192–4196.

Primus, R. J., Yevich, E., Baltazar, C., & Gallager, D. W. (1997). Autoradiographic localization of CRF1 and CRF2 binding sites in adult rat brain. *Neuropsychopharmacology, 17*(5), 308–316. http://dx.doi.org/10.1016/S0893-133X(97)00071-7.

Przegalinski, E., Filip, M., Frankowska, M., Zaniewska, M., & Papla, I. (2005). Effects of CP 154,526, a CRF1 receptor antagonist, on behavioral responses to cocaine in rats. *Neuropeptides, 39*(5), 525–533. http://dx.doi.org/10.1016/j.npep.2005.07.002.

Pursey, K. M., Stanwell, P., Gearhardt, A. N., Collins, C. E., & Burrows, T. L. (2014). The prevalence of food addiction as assessed by the Yale food addiction scale: A systematic review. *Nutrients, 6*(10), 4552–4590. http://dx.doi.org/10.3390/nu6104552.

Quadros, I. M., Macedo, G. C., Domingues, L. P., & Favoretto, C. A. (2016). An update on CRF mechanisms underlying alcohol use disorders and dependence. *Frontiers in Endocrinology, 7*, 134. http://dx.doi.org/10.3389/fendo.2016.00134.

Rainnie, D. G., Bergeron, R., Sajdyk, T. J., Patil, M., Gehlert, D. R., & Shekhar, A. (2004). Corticotrophin releasing factor-induced synaptic plasticity in the amygdala translates stress into emotional disorders. *The Journal of Neuroscience, 24*(14), 3471–3479. http://dx.doi.org/10.1523/JNEUROSCI.5740-03.2004,24/14/3471.

Rasmussen, D. D., Boldt, B. M., Bryant, C. A., Mitton, D. R., Larsen, S. A., & Wilkinson, C. W. (2000). Chronic daily ethanol and withdrawal: 1. Long-term changes in the hypothalamo-pituitary-adrenal axis. *Alcoholism, Clinical and Experimental Research, 24*(12), 1836–1849.

Rassnick, S., Heinrichs, S. C., Britton, K. T., & Koob, G. F. (1993). Microinjection of a corticotropin-releasing factor antagonist into the central nucleus of the amygdala reverses anxiogenic-like effects of ethanol withdrawal. *Brain Research, 605*(1), 25–32. doi: 0006-8993(93)91352-S [pii].

Rehm, J., Mathers, C., Popova, S., Thavorncharoensap, M., Teerawattananon, Y., & Patra, J. (2009). Global burden of disease and injury and economic cost attributable to alcohol use and alcohol-use disorders. *Lancet, 373*(9682), 2223–2233. http://dx.doi.org/10.1016/S0140-6736(09)60746-7.

Richardson, H. N., Zhao, Y., Fekete, E. M., Funk, C. K., Wirsching, P., Janda, K. D., et al. (2008). MPZP: A novel small molecule corticotropin-releasing factor type 1 receptor (CRF1) antagonist. *Pharmacology, Biochemistry, and Behavior, 88*(4), 497–510. http://dx.doi.org/10.1016/j.pbb.2007.10.008.
Richter, R. M., & Weiss, F. (1999). In vivo CRF release in rat amygdala is increased during cocaine withdrawal in self-administering rats. *Synapse, 32*(4), 254–261.
Richter, R. M., Zorrilla, E. P., Basso, A. M., Koob, G. F., & Weiss, F. (2000). Altered amygdalar CRF release and increased anxiety-like behavior in Sardinian alcohol-preferring rats: A microdialysis and behavioral study. *Alcoholism, Clinical and Experimental Research, 24*(12), 1765–1772.
Rimondini, R., Arlinde, C., Sommer, W., & Heilig, M. (2002). Long-lasting increase in voluntary ethanol consumption and transcriptional regulation in the rat brain after intermittent exposure to alcohol. *The FASEB Journal, 16*(1), 27–35. http://dx.doi.org/10.1096/fj.01-0593com,16/1/27.
Rivier, C. L., Grigoriadis, D. E., & Rivier, J. E. (2003). Role of corticotropin-releasing factor receptors type 1 and 2 in modulating the rat adrenocorticotropin response to stressors. *Endocrinology, 144*(6), 2396–2403.
Rivier, J., Rivier, C., & Vale, W. (1984). Synthetic competitive antagonists of corticotropin-releasing factor: Effect on ACTH secretion in the rat. *Science, 224*(4651), 889–891.
Roberto, M., Cruz, M. T., Gilpin, N. W., Sabino, V., Schweitzer, P., Bajo, M., et al. (2010). Corticotropin releasing factor-induced amygdala gamma-aminobutyric acid release plays a key role in alcohol dependence. *Biological Psychiatry, 67*(9), 831–839. http://dx.doi.org/10.1016/j.biopsych.2009.11.007.
Roberto, M., Gilpin, N. W., O'Dell, L. E., Cruz, M. T., Morse, A. C., Siggins, G. R., et al. (2008). Cellular and behavioral interactions of gabapentin with alcohol dependence. *Journal of Neuroscience, 28*(22), 5762–5771. http://dx.doi.org/10.1523/JNEUROSCI.0575-08.2008.
Roberto, M., Madamba, S. G., Moore, S. D., Tallent, M. K., & Siggins, G. R. (2003). Ethanol increases GABAergic transmission at both pre- and postsynaptic sites in rat central amygdala neurons. *Proceedings of the National Academy of Sciences of the United States of America, 100*(4), 2053–2058.
Roberto, M., Madamba, S. G., Stouffer, D. G., Parsons, L. H., & Siggins, G. R. (2004). Increased GABA release in the central amygdala of ethanol-dependent rats. *The Journal of Neuroscience, 24*(45), 10159–10166.
Rodaros, D., Caruana, D. A., Amir, S., & Stewart, J. (2007). Corticotropin-releasing factor projections from limbic forebrain and paraventricular nucleus of the hypothalamus to the region of the ventral tegmental area. *Neuroscience, 150*(1), 8–13. http://dx.doi.org/10.1016/j.neuroscience.2007.09.043.
Rodriguez de Fonseca, F., Carrera, M. R., Navarro, M., Koob, G. F., & Weiss, F. (1997). Activation of corticotropin-releasing factor in the limbic system during cannabinoid withdrawal. *Science, 276*(5321), 2050–2054.
Rosenthal, R., & Rubin, D. (1994). The counternull value of an effect size: A new statistic. *Psychological Science, 5*, 329–334.
Rudd, R. A., Seth, P., David, F., & Scholl, L. (2016). Increases in drug and opioid-involved overdose deaths—United States, 2010–2015. *MMWR. Morbidity and Mortality Weekly Report, 65*(5051), 1445–1452. http://dx.doi.org/10.15585/mmwr.mm655051e1.
Ryabinin, A. E., Yoneyama, N., Tanchuck, M. A., Mark, G. P., & Finn, D. A. (2008). Urocortin 1 microinjection into the mouse lateral septum regulates the acquisition and expression of alcohol consumption. *Neuroscience, 151*(3), 780–790. http://dx.doi.org/10.1016/j.neuroscience.2007.11.014.

Sabino, V., Cottone, P., Koob, G. F., Steardo, L., Lee, M. J., Rice, K. C., et al. (2006). Dissociation between opioid and CRF1 antagonist sensitive drinking in Sardinian alcohol-preferring rats. *Psychopharmacology, 189*(2), 175–186. http://dx.doi.org/10.1007/s00213-006-0546-5.
Sakanaka, M., Shibasaki, T., & Lederis, K. (1986). Distribution and efferent projections of corticotropin-releasing factor-like immunoreactivity in the rat amygdaloid complex. *Brain Research, 382*(2), 213–238. doi: 0006-8993(86)91332-6 [pii].
Sanchez, M. M., Young, L. J., Plotsky, P. M., & Insel, T. R. (1999). Autoradiographic and in situ hybridization localization of corticotropin-releasing factor 1 and 2 receptors in non-human primate brain. *The Journal of Comparative Neurology, 408*(3), 365–377.
Sarnyai, Z., Biro, E., Gardi, J., Vecsernyes, M., Julesz, J., & Telegdy, G. (1995). Brain corticotropin-releasing factor mediates 'anxiety-like' behavior induced by cocaine withdrawal in rats. *Brain Research, 675*(1–2), 89–97. doi: 0006-8993(95)00043-P [pii].
Sarnyai, Z., Biro, E., Penke, B., & Telegdy, G. (1992). The cocaine-induced elevation of plasma corticosterone is mediated by endogenous corticotropin-releasing factor (CRF) in rats. *Brain Research, 589*(1), 154–156. doi: 0006-8993(92)91176-F [pii].
Sarnyai, Z., Shaham, Y., & Heinrichs, S. C. (2001). The role of corticotropin-releasing factor in drug addiction. *Pharmacological Reviews, 53*(2), 209–243.
Schierloh, A., Deussing, J., Wurst, W., Zieglgansberger, W., & Rammes, G. (2007). Corticotropin-releasing factor (CRF) receptor type 1-dependent modulation of synaptic plasticity. *Neuroscience Letters, 416*(1), 82–86. http://dx.doi.org/10.1016/j.neulet.2007.01.047.
Schmolesky, M. T., De Ruiter, M. M., De Zeeuw, C. I., & Hansel, C. (2007). The neuropeptide corticotropin-releasing factor regulates excitatory transmission and plasticity at the climbing fibre-Purkinje cell synapse. *The European Journal of Neuroscience, 25*(5), 1460–1466.
Schulkin, J., McEwen, B. S., & Gold, P. W. (1994). Allostasis, amygdala, and anticipatory angst. *Neuroscience and Biobehavioral Reviews, 18*(3), 385–396.
Semba, J., Wakuta, M., Maeda, J., & Suhara, T. (2004). Nicotine withdrawal induces subsensitivity of hypothalamic-pituitary-adrenal axis to stress in rats: Implications for precipitation of depression during smoking cessation. *Psychoneuroendocrinology, 29*(2), 215–226.
Semyanov, A., Walker, M. C., Kullmann, D. M., & Silver, R. A. (2004). Tonically active GABA a receptors: Modulating gain and maintaining the tone. *Trends in Neurosciences, 27*(5), 262–269. http://dx.doi.org/10.1016/j.tins.2004.03.005.
Shaham, Y., Erb, S., Leung, S., Buczek, Y., & Stewart, J. (1998). CP-154,526, a selective, non-peptide antagonist of the corticotropin-releasing factor1 receptor attenuates stress-induced relapse to drug seeking in cocaine- and heroin-trained rats. *Psychopharmacology, 137*(2), 184–190.
Shaham, Y., Funk, D., Erb, S., Brown, T. J., Walker, C. D., & Stewart, J. (1997). Corticotropin-releasing factor, but not corticosterone, is involved in stress-induced relapse to heroin-seeking in rats. *The Journal of Neuroscience, 17*(7), 2605–2614.
Shaham, Y., Shalev, U., Lu, L., De Wit, H., & Stewart, J. (2003). The reinstatement model of drug relapse: History, methodology and major findings. *Psychopharmacology, 168*(1–2), 3–20.
Shalev, U., Morales, M., Hope, B., Yap, J., & Shaham, Y. (2001). Time-dependent changes in extinction behavior and stress-induced reinstatement of drug seeking following withdrawal from heroin in rats. *Psychopharmacology, 156*(1), 98–107.
Sharp, B. M., & Matta, S. G. (1993). Detection by in vivo microdialysis of nicotine-induced norepinephrine secretion from the hypothalamic paraventricular nucleus of freely moving rats: Dose-dependency and desensitization. *Endocrinology, 133*(1), 11–19. http://dx.doi.org/10.1210/endo.133.1.8391419.
Sharpe, A. L., & Phillips, T. J. (2009). Central urocortin 3 administration decreases limited-access ethanol intake in nondependent mice. *Behavioural Pharmacology, 20*(4), 346–351. http://dx.doi.org/10.1097/FBP.0b013e32832f01ba.

Shepard, J. D., Barron, K. W., & Myers, D. A. (2000). Corticosterone delivery to the amygdala increases corticotropin-releasing factor mRNA in the central amygdaloid nucleus and anxiety-like behavior. *Brain Research*, *861*(2), 288–295.

Silberman, Y., Matthews, R. T., & Winder, D. G. (2013). A corticotropin releasing factor pathway for ethanol regulation of the ventral tegmental area in the bed nucleus of the stria terminalis. *The Journal of Neuroscience*, *33*(3), 950–960. http://dx.doi.org/10.1523/JNEUROSCI.2949-12.2013.

Silberman, Y., & Winder, D. G. (2013). Corticotropin releasing factor and catecholamines enhance glutamatergic neurotransmission in the lateral subdivision of the central amygdala. *Neuropharmacology*, *70*, 316–323. http://dx.doi.org/10.1016/j.neuropharm.2013.02.014.

Simms, J. A., Nielsen, C. K., Li, R., & Bartlett, S. E. (2014). Intermittent access ethanol consumption dysregulates CRF function in the hypothalamus and is attenuated by the CRF-R1 antagonist, CP-376395. *Addiction Biology*, *19*(4), 606–611. http://dx.doi.org/10.1111/adb.12024.

Simms, J. A., Steensland, P., Medina, B., Abernathy, K. E., Chandler, L. J., Wise, R., et al. (2008). Intermittent access to 20% ethanol induces high ethanol consumption in long-Evans and Wistar rats. *Alcoholism, Clinical and Experimental Research*, *32*(10), 1816–1823. http://dx.doi.org/10.1111/j.1530-0277.2008.00753.x.

Slater, P. G., Cerda, C. A., Pereira, L. A., Andres, M. E., & Gysling, K. (2016). CRF binding protein facilitates the presence of CRF type 2alpha receptor on the cell surface. *Proceedings of the National Academy of Sciences of the United States of America*, *113*(15), 4075–4080. http://dx.doi.org/10.1073/pnas.1523745113.

Smink, F. R., van Hoeken, D., & Hoek, H. W. (2013). Epidemiology, course, and outcome of eating disorders. *Current Opinion in Psychiatry*, *26*(6), 543–548. http://dx.doi.org/10.1097/YCO.0b013e328365a24f.

Solomon, R. L., & Corbit, J. D. (1974). An opponent-process theory of motivation. I. Temporal dynamics of affect. *Psychological Review*, *81*(2), 119–145.

Sommer, W. H., Rimondini, R., Hansson, A. C., Hipskind, P. A., Gehlert, D. R., Barr, C. S., et al. (2008). Upregulation of voluntary alcohol intake, behavioral sensitivity to stress, and amygdala crhr1 expression following a history of dependence. *Biological Psychiatry*, *63*(2), 139–145. http://dx.doi.org/10.1016/j.biopsych.2007.01.010.

Sparta, D. R., Hopf, F. W., Gibb, S. L., Cho, S. L., Stuber, G. D., Messing, R. O., et al. (2013). Binge ethanol-drinking potentiates corticotropin releasing factor R1 receptor activity in the ventral tegmental area. *Alcoholism, Clinical and Experimental Research*, *37*(10), 1680–1687. http://dx.doi.org/10.1111/acer.12153.

Sparta, D. R., Sparrow, A. M., Lowery, E. G., Fee, J. R., Knapp, D. J., & Thiele, T. E. (2008). Blockade of the corticotropin releasing factor type 1 receptor attenuates elevated ethanol drinking associated with drinking in the dark procedures. *Alcoholism, Clinical and Experimental Research*, *32*(2), 259–265. http://dx.doi.org/10.1111/j.1530-0277.2007.00575.x.

Specio, S. E., Wee, S., O'Dell, L. E., Boutrel, B., Zorrilla, E. P., & Koob, G. F. (2008). CRF(1) receptor antagonists attenuate escalated cocaine self-administration in rats. *Psychopharmacology*, *196*(3), 473–482.

Spierling, S. R., & Zorrilla, E. P. (2017). Don't stress about CRF: Assessing the translational failures of CRF1 antagonists. *Psychopharmacology*. http://dx.doi.org/10.1007/s00213-017-4556-2.

Spina, M. G., Basso, A. M., Zorrilla, E. P., Heyser, C. J., Rivier, J., Vale, W., et al. (2000). Behavioral effects of central administration of the novel CRF antagonist astressin in rats. *Neuropsychopharmacology*, *22*(3), 230–239. http://dx.doi.org/10.1016/s0893-133x(99)00108-6.

Spina, M., Merlo-Pich, E., Chan, R. K., Basso, A. M., Rivier, J., Vale, W., et al. (1996). Appetite-suppressing effects of urocortin, a CRF-related neuropeptide. *Science*, *273*(5281), 1561–1564.

Steketee, J. D., & Kalivas, P. W. (2011). Drug wanting: Behavioral sensitization and relapse to drug-seeking behavior. *Pharmacological Reviews*, *63*(2), 348–365. http://dx.doi.org/10.1124/pr.109.001933.

Stewart, J., & de Wit, H. (1987). Reinstatement of drug-taking behavior as a method of assessing incentive motivational properties of drugs. In M. A. Bozarth (Ed.), *Methods of assessing the reinforcing properties of abused drugs* (pp. 211–227). New York: Springer-Verlag.

Stinus, L., Cador, M., Zorrilla, E. P., & Koob, G. F. (2005). Buprenorphine and a CRF1 antagonist block the acquisition of opiate withdrawal-induced conditioned place aversion in rats. *Neuropsychopharmacology*, *30*(1), 90–98. http://dx.doi.org/10.1038/sj.npp.1300487,1300487.

Sun, N., & Cassell, M. D. (1993). Intrinsic GABAergic neurons in the rat central extended amygdala. *The Journal of Comparative Neurology*, *330*(3), 381–404. http://dx.doi.org/10.1002/cne.903300308.

Sun, N., Roberts, L., & Cassell, M. D. (1991). Rat central amygdaloid nucleus projections to the bed nucleus of the stria terminalis. *Brain Research Bulletin*, *27*(5), 651–662. doi: 0361-9230(91)90041-H [pii].

Sussman, S., Lisha, N., & Griffiths, M. (2011). Prevalence of the addictions: A problem of the majority or the minority? *Evaluation & the Health Professions*, *34*(1), 3–56. http://dx.doi.org/10.1177/0163278710380124.

Swanson, L. W., Sawchenko, P. E., Rivier, J., & Vale, W. W. (1983). Organization of ovine corticotropin-releasing factor immunoreactive cells and fibers in the rat brain: An immunohistochemical study. *Neuroendocrinology*, *36*(3), 165–186.

Swanson, L. W., & Simmons, D. M. (1989). Differential steroid hormone and neural influences on peptide mRNA levels in CRH cells of the paraventricular nucleus: A hybridization histochemical study in the rat. *The Journal of Comparative Neurology*, *285*(4), 413–435. http://dx.doi.org/10.1002/cne.902850402.

Swerdlow, N. R., Koob, G. F., Cador, M., Lorang, M., & Hauger, R. L. (1993). Pituitary-adrenal axis responses to acute amphetamine in the rat. *Pharmacology, Biochemistry, and Behavior*, *45*(3), 629–637. doi: 0091-3057(93)90518-X [pii].

Swinny, J. D., Kalicharan, D., Blaauw, E. H., Ijkema-Paassen, J., Shi, F., Gramsbergen, A., et al. (2003). Corticotropin-releasing factor receptor types 1 and 2 are differentially expressed in pre- and post-synaptic elements in the post-natal developing rat cerebellum. *The European Journal of Neuroscience*, *18*(3), 549–562. doi:10.1046/j.1460-9568.2003.02776.x.

Takahashi, L. K., Ho, S. P., Livanov, V., Graciani, N., & Arneric, S. P. (2001). Antagonism of CRF(2) receptors produces anxiolytic behavior in animal models of anxiety. *Brain Research*, *902*(2), 135–142. doi: S0006-8993(01)02405-2 [pii].

Teegarden, S. L., Scott, A. N., & Bale, T. L. (2009). Early life exposure to a high fat diet promotes long-term changes in dietary preferences and central reward signaling. *Neuroscience*, *162*(4), 924–932. http://dx.doi.org/10.1016/j.neuroscience.2009.05.029.

Tucci, S., Cheeta, S., Seth, P., & File, S. E. (2003). Corticotropin releasing factor antagonist, alpha-helical CRF(9-41), reverses nicotine-induced conditioned, but not unconditioned, anxiety. *Psychopharmacology*, *167*(3), 251–256. http://dx.doi.org/10.1007/s00213-003-1403-4.

Ungless, M. A., Singh, V., Crowder, T. L., Yaka, R., Ron, D., & Bonci, A. (2003). Corticotropin-releasing factor requires CRF binding protein to potentiate NMDA receptors via CRF receptor 2 in dopamine neurons. *Neuron*, *39*(3), 401–407.

Valdez, G. R., Roberts, A. J., Chan, K., Davis, H., Brennan, M., Zorrilla, E. P., et al. (2002). Increased ethanol self-administration and anxiety-like behavior during acute ethanol withdrawal and protracted abstinence: Regulation by corticotropin-releasing factor. *Alcoholism, Clinical and Experimental Research*, *26*(10), 1494–1501. http://dx.doi.org/10.1097/01.alc.0000033120.51856.f0.

Valdez, G. R., Sabino, V., & Koob, G. F. (2004). Increased anxiety-like behavior and ethanol self-administration in dependent rats: Reversal via corticotropin-releasing factor-2 receptor activation. *Alcoholism, Clinical and Experimental Research, 28*(6), 865–872.

Valdez, G. R., Zorrilla, E. P., Roberts, A. J., & Koob, G. F. (2003). Antagonism of corticotropin-releasing factor attenuates the enhanced responsiveness to stress observed during protracted ethanol abstinence. *Alcohol, 29*(2), 55–60. doi: S074183290300020X [pii].

Vale, W., Spiess, J., Rivier, C., & Rivier, J. (1981). Characterization of a 41-residue ovine hypothalamic peptide that stimulates secretion of corticotropin and beta-endorphin. *Science, 213*(4514), 1394–1397.

Van Pett, K., Viau, V., Bittencourt, J. C., Chan, R. K., Li, H. Y., Arias, C., et al. (2000). Distribution of mRNAs encoding CRF receptors in brain and pituitary of rat and mouse. *The Journal of Comparative Neurology, 428*(2), 191–212. http://dx.doi.org/10.1002/1096-9861(20001211)428:2<191::AID-CNE1>3.0.CO;2-U.

Varodayan, F. P., de Guglielmo, G., Logrip, M. L., George, O., & Roberto, M. (2017). Alcohol dependence disrupts amygdalar L-type voltage-gated calcium channel mechanisms. *The Journal of Neuroscience*. http://dx.doi.org/10.1523/jneurosci.3721-16.2017.

Veinante, P., & Freund-Mercier, M. J. (1998). Intrinsic and extrinsic connections of the rat central extended amygdala: An in vivo electrophysiological study of the central amygdaloid nucleus. *Brain Research, 794*(2), 188–198.

Vendruscolo, L. F., Barbier, E., Schlosburg, J. E., Misra, K. K., Whitfield, T. W., Jr., Logrip, M. L., et al. (2012). Corticosteroid-dependent plasticity mediates compulsive alcohol drinking in rats. *The Journal of Neuroscience, 32*(22), 7563–7571. http://dx.doi.org/10.1523/jneurosci.0069-12.2012.

Vendruscolo, L. F., Estey, D., Goodell, V., Macshane, L. G., Logrip, M. L., Schlosburg, J. E., et al. (2015). Glucocorticoid receptor antagonism decreases alcohol seeking in alcohol-dependent individuals. *The Journal of Clinical Investigation, 125*(8), 3193–3197. http://dx.doi.org/10.1172/jci79828.

Vranjkovic, O., Pina, M., Kash, T. L., & Winder, D. G. (2017). The bed nucleus of the stria terminalis in drug-associated behavior and affect: A circuit-based perspective. *Neuropharmacology*. http://dx.doi.org/10.1016/j.neuropharm.2017.03.028.

Wanat, M. J., Hopf, F. W., Stuber, G. D., Phillips, P. E., & Bonci, A. (2008). Corticotropin-releasing factor increases mouse ventral tegmental area dopamine neuron firing through a protein kinase C-dependent enhancement of Ih. *The Journal of Physiology, 586*(8), 2157–2170. http://dx.doi.org/10.1113/jphysiol.2007.150078.

Wang, J., Fang, Q., Liu, Z., & Lu, L. (2006). Region-specific effects of brain corticotropin-releasing factor receptor type 1 blockade on footshock-stress- or drug-priming-induced reinstatement of morphine conditioned place preference in rats. *Psychopharmacology, 185*(1), 19–28. http://dx.doi.org/10.1007/s00213-005-0262-6.

Wang, B., Shaham, Y., Zitzman, D., Azari, S., Wise, R. A., & You, Z. B. (2005). Cocaine experience establishes control of midbrain glutamate and dopamine by corticotropin-releasing factor: A role in stress-induced relapse to drug seeking. *Journal of Neuroscience, 25*(22), 5389–5396. http://dx.doi.org/10.1523/JNEUROSCI.0955-05.2005.

Wang, B., You, Z. B., Rice, K. C., & Wise, R. A. (2007). Stress-induced relapse to cocaine seeking: Roles for the CRF(2) receptor and CRF-binding protein in the ventral tegmental area of the rat. *Psychopharmacology, 193*(2), 283–294. http://dx.doi.org/10.1007/s00213-007-0782-3.

Waselus, M., Nazzaro, C., Valentino, R. J., & Van Bockstaele, E. J. (2009). Stress-induced redistribution of corticotropin-releasing factor receptor subtypes in the dorsal raphe nucleus. *Biological Psychiatry, 66*(1), 76–83. http://dx.doi.org/10.1016/j.biopsych. 2009.02.014.

Weidenfeld, J., Feldman, S., & Mechoulam, R. (1994). Effect of the brain constituent anandamide, a cannabinoid receptor agonist, on the hypothalamo-pituitary-adrenal axis in the rat. *Neuroendocrinology, 59*(2), 110–112.

Weiss, F., Ciccocioppo, R., Parsons, L. H., Katner, S., Liu, X., Zorrilla, E. P., et al. (2001). Compulsive drug-seeking behavior and relapse. Neuroadaptation, stress, and conditioning factors. *Annals of the New York Academy of Sciences*, *937*, 1–26.

Westmoreland, P., Krantz, M. J., & Mehler, P. S. (2016). Medical complications of anorexia nervosa and bulimia. *The American Journal of Medicine*, *129*(1), 30–37. http://dx.doi.org/10.1016/j.amjmed.2015.06.031.

Westphal, N. J., & Seasholtz, A. F. (2006). CRH-BP: The regulation and function of a phylogenetically conserved binding protein. *Frontiers in Bioscience*, *11*, 1878–1891.

Williams, C. L., Buchta, W. C., & Riegel, A. C. (2014). CRF-R2 and the heterosynaptic regulation of VTA glutamate during reinstatement of cocaine seeking. *The Journal of Neuroscience*, *34*(31), 10402–10414. http://dx.doi.org/10.1523/JNEUROSCI. 0911-13.2014.

Wills, T. A., Kash, T. L., & Winder, D. G. (2013). Developmental changes in the acute ethanol sensitivity of glutamatergic and GABAergic transmission in the BNST. *Alcohol*, *47*(7), 531–537. http://dx.doi.org/10.1016/j.alcohol.2013.08.003.

Wills, T. A., Klug, J. R., Silberman, Y., Baucum, A. J., Weitlauf, C., Colbran, R. J., et al. (2012). GluN2B subunit deletion reveals key role in acute and chronic ethanol sensitivity of glutamate synapses in bed nucleus of the stria terminalis. *Proceedings of the National Academy of Sciences of the United States of America*, *109*(5), E278–E287. http://dx.doi.org/10.1073/pnas.1113820109.

Wise, R. A. (1973). Voluntary ethanol intake in rats following exposure to ethanol on various schedules. *Psychopharmacologia*, *29*(3), 203–210.

World Health Organization. (1992). *International statistical classification of diseases and related health problems, 10th revision*. Geneva: World Health Organization.

Zhao, Y., Valdez, G. R., Fekete, E. M., Rivier, J. E., Vale, W. W., Rice, K. C., et al. (2007). Subtype-selective corticotropin-releasing factor receptor agonists exert contrasting, but not opposite, effects on anxiety-related behavior in rats. *The Journal of Pharmacology and Experimental Therapeutics*, *323*(3), 846–854. http://dx.doi.org/10.1124/jpet.107.123208.

Zhao, Y., Weiss, F., & Zorrilla, E. P. (2007). Remission and resurgence of anxiety-like behavior across protracted withdrawal stages in ethanol-dependent rats. *Alcoholism, Clinical and Experimental Research*, *31*(9), 1505–1515. http://dx.doi.org/10.1111/j.1530-0277.2007.00456.x.

Zhao-Shea, R., DeGroot, S. R., Liu, L., Vallaster, M., Pang, X., Su, Q., et al. (2015a). Corrigendum: Increased CRF signalling in a ventral tegmental area-interpeduncular nucleus-medial habenula circuit induces anxiety during nicotine withdrawal. *Nature Communications*, *6*, 7625. http://dx.doi.org/10.1038/ncomms8625.

Zhao-Shea, R., DeGroot, S. R., Liu, L., Vallaster, M., Pang, X., Su, Q., et al. (2015b). Increased CRF signalling in a ventral tegmental area-interpeduncular nucleus-medial habenula circuit induces anxiety during nicotine withdrawal. *Nature Communications*, *6*, 6770. http://dx.doi.org/10.1038/ncomms7770.

Zhou, Y., Spangler, R., Ho, A., & Kreek, M. J. (2003). Increased CRH mRNA levels in the rat amygdala during short-term withdrawal from chronic 'binge' cocaine. *Brain Research Molecular Brain Research*, *114*(1), 73–79. doi: S0169328X03001396 [pii].

Zhou, Y., Spangler, R., LaForge, K. S., Maggos, C. E., Ho, A., & Kreek, M. J. (1996). Corticotropin-releasing factor and type 1 corticotropin-releasing factor receptor messenger RNAs in rat brain and pituitary during "binge"-pattern cocaine administration and chronic withdrawal. *The Journal of Pharmacology and Experimental Therapeutics*, *279*(1), 351–358.

Zorrilla, E. P., Heilig, M., de Wit, H., & Shaham, Y. (2013). Behavioral, biological, and chemical perspectives on targeting CRF(1) receptor antagonists to treat alcoholism. *Drug and Alcohol Dependence*, *128*(3), 175–186. http://dx.doi.org/10.1016/j.drugalcdep.2012.12.017.

Zorrilla, E. P., & Koob, G. F. (2004). The therapeutic potential of CRF1 antagonists for anxiety. *Expert Opinion on Investigational Drugs*, *13*(7), 799–828. http://dx.doi.org/10.1517/13543784.13.7.799.

Zorrilla, E. P., & Koob, G. F. (2010). Progress in corticotropin-releasing factor-1 antagonist development. *Drug Discovery Today*, *15*(9–10), 371–383. http://dx.doi.org/10.1016/j.drudis.2010.02.011.

Zorrilla, E. P., Valdez, G. R., & Weiss, F. (2001). Changes in levels of regional CRF-like-immunoreactivity and plasma corticosterone during protracted drug withdrawal in dependent rats. *Psychopharmacology*, *158*(4), 374–381. http://dx.doi.org/10.1007/s002130100773.

CHAPTER THREE

Dynorphin/Kappa Opioid Receptor Signaling in Preclinical Models of Alcohol, Drug, and Food Addiction

Anushree Karkhanis, Katherine M. Holleran, Sara R. Jones[1]
Wake Forest School of Medicine, Winston-Salem, NC, United States
[1]Corresponding author: e-mail address: srjones@wakehealth.edu

Contents

International Review of Neurobiology, Volume 136
ISSN 0074-7742
http://dx.doi.org/10.1016/bs.irn.2017.08.001

Abstract

The dynorphin/kappa opioid receptor (KOR) system is implicated in the "dark side" of addiction, in which stress exacerbates maladaptive responses to drug and alcohol exposure. For example, acute stress and acute ethanol exposure result in an elevation in dynorphin, the KOR endogenous ligand. Activation of KORs results in modulation of several neurotransmitters; however, this chapter will focus on its regulatory effects on dopamine in mesolimbic areas. Specifically, KOR activation has an inhibitory effect on dopamine release, thereby influencing reward processing. Repeated stimulation of KORs, for example, via chronic drug and/or stress exposure, results in increased function of the dynorphin/KOR system. This augmentation in KOR function shifts the homeostatic balance in favor of an overall reduction in dopamine signaling via either by reducing dopamine release or by increasing dopamine transporter function. This chapter examines the effects of chronic ethanol exposure on KOR function and the downstream effects on dopamine transmission. Additionally, the impact of chronic cocaine exposure and its effects on KOR function will be explored. Further, KORs may also be involved in driving excessive consumption of food, contributing to the risk of developing obesity. While some studies have shown that KOR agonists reduce drug intake, other studies have shown that antagonists reduce addiction-like behaviors, demonstrating therapeutic potential. For example, KOR inhibition reduces ethanol intake in dependent animals, motivation to self-administer cocaine in chronic stress-exposed animals, and food consumption in obese animals. This chapter will delve into the mechanisms by which modulation of the dynorphin/KOR system may be therapeutic.

1. INTRODUCTION

The endogenous opioid systems of the brain are involved in regulating nociception and emotional behaviors, and have been implicated in addiction to drugs, alcohol, and other disorders of excessive consumption such as obesity. There are three canonical classes of opioid receptors with different endogenous ligands: the mu receptor, which binds β-endorphin, the delta receptor, which binds enkephalin, and the kappa receptor, which binds dynorphin (Chavkin, James, & Goldstein, 1982; Fowler & Fraser, 1994). While activation of mu and delta opioid receptors with agonists such as morphine and heroin produces euphoria and is highly reinforcing, stimulation of kappa opioid receptors (KORs) is generally aversive. In fact, dynorphin is considered an integral part of the brain's stress response system, and release of dynorphin occurs during exposure to painful, noxious, or stressful stimuli (Chavkin, 2013). However, this is a somewhat simplistic view; the situation is more complex than "good" opioids vs "bad" opioids. In addition to producing aversion, dynorphin is also antinociceptive (Chavkin, 2013) similar to the other endogenous opioids, and in some cases, KOR agonists have been postulated to have antiaddictive properties because they have been shown to reduce drug seeking and consumption (Freeman, Naylor, Prisinzano, & Woolverton, 2014; Graziane, Polter, Briand, Pierce, & Kauer, 2013; Kivell, Ewald, & Prisinzano, 2014; Prisinzano, Tidgewell, & Harding, 2005; Shippenberg, Chefer, Zapata, & Heidbreder, 2001; Shippenberg, LeFevour, & Heidbreder, 1996; Simonson et al., 2015). On the other hand, there is growing support for the use of KOR antagonists in the prevention of relapse to drug and alcohol seeking (Chavkin, 2011; Crowley & Kash, 2015; Wee & Koob, 2010). Dynorphin, perhaps paradoxically, has been shown to be elevated both during acute exposure to abused drugs and during withdrawal from them (for review, see Wee & Koob, 2010). This suggests that the dynorphin/KOR system may be responsive to both positive and negative "stressors," defined as stimuli that lead to a deviation from homeostasis (Koob & Le Moal, 2001; Selye, 1998; Tejeda et al., 2017). Taken together, this suggests that the dynorphin/KOR system exerts complicated and sometimes opposing effects on reward-related behaviors, but understanding its multifaceted role in addiction may help highlight targets for interventions.

1.1 Stress, Addiction, and Kappa Opioid Receptors

Stress is known to promote maladaptive behaviors that often increase vulnerability to the development of affective disorders—such as anxiety and depression—and addiction to drugs and alcohol. Exposure to stressful adverse events often leads to negative affective states, which can contribute to the development of alcohol and substance use disorders, and these addictive disorders can in turn promote the development or worsening of affective disorders. While many neurobiological systems are involved in mediating stress effects, the dynorphin/KOR system plays critical role in behavioral stress responses. Human and rodent literature has shown that exposure to adverse events, particularly during adolescence, increases the vulnerability to alcohol and substance use disorders in adulthood (Burke & Miczek, 2014; Butler, Karkhanis, Jones, & Weiner, 2016; Dube et al., 2001, 2003; Karkhanis, Rose, Weiner, & Jones, 2016). Further, rodent studies have shown that stress results in increased dynorphin release (Chartoff et al., 2009; Shirayama et al., 2004) and anxiety-like behaviors (Karkhanis, Locke, McCool, Weiner, & Jones, 2014; Valdez & Harshberger, 2012; Van't Veer & Carlezon, 2013). Finally, repeated or prolonged activation of KORs often results in addiction-related behaviors including excessive drug seeking (Groblewski, Zietz, Willuhn, Phillips, & Chavkin, 2015). Clinically, stress is closely associated with craving and relapse risk (Sinha et al., 2009; Sinha, Fuse, Aubin, & O'Malley, 2000). For example, cocaine-dependent abstinent patients with a variant in the KOR gene *oprk1* exhibited heightened stress response and limbic brain activation, as well as increased cocaine craving, and relapse risk (Xu et al., 2013).

1.2 Modulation of Neurotransmitter Signaling by Kappa Opioid Receptors

The KOR is encoded by the *oprk1* gene and is a member of the seven transmembrane-spanning G protein-coupled receptor superfamily. When activated by its endogenous ligand, dynorphin, or an exogenous agonist (including, but not limited to, U50,488, U69,593, salvinorin A, mesyl salvinorin B, and nalfurafine), KORs can act via multiple downstream effectors, including activation of the G-$\alpha_{i/o}$ protein-coupled pathway (cAMP), G-$\beta\gamma$ modulation of various ion channels, and intracellular kinase pathways including G protein-coupled receptor kinases, beta-arrestin, and members of the mitogen-activated protein kinase family (p38, JNK, ERK1/2) (Bruchas & Chavkin, 2010). Recent results have shown that KORs can be guided by specific ligands, termed "biased agonists," to activate specific signaling pathways, and a recent trend in the KOR field has been to examine the nociceptive and antiaddiction therapeutic potential of synthetic biased agonists and antagonists that specifically enhance or suppress one of these (Bohn & Aube, 2017; Bruchas & Chavkin, 2010; Brust et al., 2016; Dogra & Yadav, 2015; Zhou et al., 2015).

KORs are ubiquitously present throughout the brain and modulate various neurotransmitter systems. Specifically, KORs are expressed in brain regions involved in affect, cognition, and motivated behaviors, including the ventral tegmental area (VTA), nucleus accumbens (NAc), dorsal striatum, prefrontal cortex, basolateral amygdala (BLA), pedunculopontine nucleus, locus coeruleus, substantia nigra, bed nucleus of the stria terminalis (BNST), hypothalamus, and the central amygdala, to name a few (Sirohi, Bakalkin, & Walker, 2012; Wee & Koob, 2010). The net effect of KOR activation on neurotransmission is variable and depends on the phenotype of the neuron that the receptor is on and whether the receptor is pre- or postsynaptic. As alluded to above, the primary focus of this chapter will be to examine the role of KORs in modulating mesolimbic dopaminergic signaling.

1.3 Kappa Opioid Receptors in the NAc

In the NAc, KORs are located on GABAergic medium spiny neurons, which are the primary output cells of the NAc, on cholinergic and GABAergic interneurons, and on afferent inputs from other brain areas, including glutamatergic, GABAergic, and dopaminergic nerve terminals. The dopamine projection from the VTA to the NAc has been linked to associative learning as well as motivational and reward processes (Saddoris, Sugam, Cacciapaglia, & Carelli, 2013; Saddoris, Sugam, & Carelli, 2017), and dopamine in the NAc is also involved in regulating responses to stress and anxiety (Abercrombie, Keefe, DiFrischia, & Zigmond, 1989; Radke & Gewirtz, 2012). KORs strongly inhibit dopamine release from dopamine terminals, and this likely underlies some aspects of KOR-mediated anhedonia, dysphoria, and aversion. For example, KOR activation, which results in a reduction of dopamine levels (Karkhanis, Huggins, Rose, & Jones, 2016), increases anxiety-like behaviors like avoiding the open arms in the elevated plus maze task (Valdez & Harshberger, 2012). The dopamine-decreasing effects of KOR activation in the NAc are due primarily due to a negative feedback loop in the NAc. Dopamine is released from presynaptic terminals and activates D1 receptors on medium spiny neurons, which depolarizes them and leads to increased intracellular production of dynorphin, which is then released by feedback collaterals synapsing onto dopamine terminals and inhibits dopamine release (Svingos, Chavkin, Colago, & Pickel, 2001; Thompson et al., 2000; Werling, Frattali, Portoghese, Takemori, & Cox, 1988). In addition, dopamine terminals receive strong inputs from local acetylcholine (ACh) interneurons which serve to boost dopamine release. Because the ACh interneurons also contain inhibitory KORs, when dynorphin is released from D1R medium spiny neurons, it inhibits ACh neuronal firing and therefore decreases dopamine release (Threlfell & Cragg, 2011). This "double whammy" of KOR inhibition on NAc dopamine release makes dynorphin one of the most powerful negative modulators of dopamine signaling and dopamine-regulated behaviors.

Direct modulation of the dynorphin/KOR system by microinjection of KOR agonists or optogenetic activation of KOR-containing neurons in the NAc has been shown, in general, to predictably alter place conditioning: KOR agonism favors aversion and KOR antagonism favors preference for the paired side, and similar results have been found with thresholds for intracranial self-stimulation (ICSS), which serves as a proxy for anhedonia (Bals-Kubik, Ableitner, Herz, & Shippenberg, 1993). This, however, is a simplified view of KOR activity in the NAc. There are regional differences in dynorphin responses in the NAc (Al-Hasani et al., 2015) as well as contributions from other systems, including but not limited to serotonin inputs from the dorsal raphe (Schindler et al., 2012). For this chapter, however, KOR regulation of the mesolimbic dopamine system in response to drugs of abuse, ethanol, and food reward will be the focus.

1.4 Kappa Opioid Receptors in the VTA

KOR activation directly inhibits both GABAergic and dopaminergic neurons in the VTA (Ford, Mark, & Williams, 2006; Margolis, Hjelmstad, Bonci, & Fields, 2003, 2005). A recent study showed that exposure to chronic stress abolished long-term potentiation of GABAergic synapses in the VTA in a KOR-dependent manner (Polter et al., 2017), suggesting that KORs mediate stress-induced synaptic plasticity. VTA dopamine neurons that project to various regions such as the NAc core, NAc shell, and BLA are largely nonoverlapping, and show distinct regional distribution within the VTA and intrinsic excitability properties based on projection target (Baimel, Lau, Qiao, & Borgland, 2017; Ford et al., 2006; Lammel et al., 2008). Some studies indicate that VTA dopamine neurons that project to the prefrontal cortex and the amygdala are inhibited by KOR activation, whereas NAc-projecting VTA dopamine neurons are unaffected (Margolis et al., 2006; Margolis, Mitchell, Ishikawa, Hjelmstad, & Fields, 2008). However, others have found that, in a different mouse strain (DBA-2J), NAc-projecting VTA neurons show greater inhibition in response to KOR agonists than amygdala-projecting VTA neurons (Ford et al., 2006). More recent work using optogenetics has demonstrated that while dynorphin is able to inhibit firing of VTA neurons projecting to either the lateral NAc shell, medial NAc shell, or BLA, each of these regions contains a different proportion of neurons responsive to dynorphin, with the lateral NAc shell containing the smallest proportion of responsive neurons of these regions (Baimel et al., 2017). While the focus of this chapter is on KOR modulation of mesolimbic dopamine signaling (Fig. 1), there are many other neurotransmitter systems and brain regions which participate in the regulation of affective behaviors by dynorphin.

1.5 Measuring Dynorphin Levels

While functional outcomes of dynorphin/KOR activation following chronic drug and ethanol exposure—such as augmented inhibition of

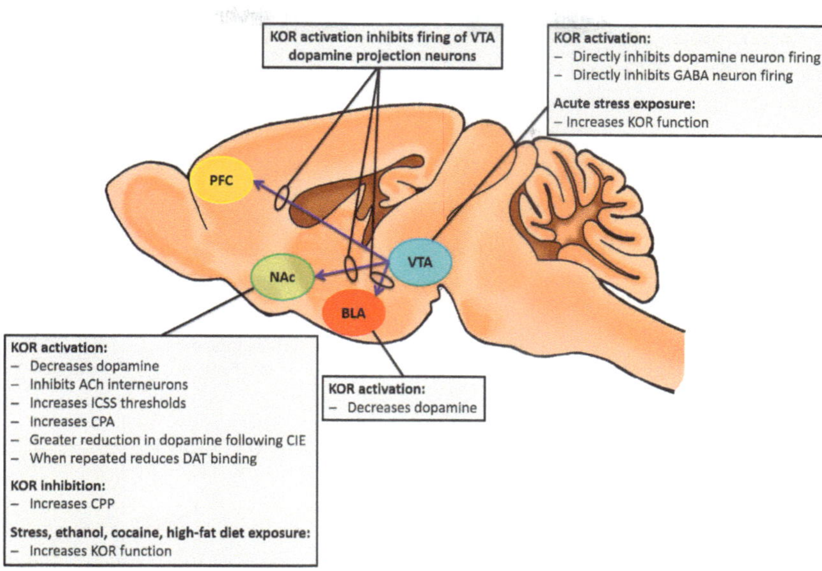

Fig. 1 Region-specific changes in response to kappa opioid receptor modulation. Representative sagittal cross section of a rodent brain showing regions impacted by alterations in dynorphin/KOR function within the mesolimbic dopamine system. Overall, stress, high-fat diet, and drug or alcohol exposure results in augmentations in KOR signaling. Activation of KOR is aversive and results in decreased dopamine release, while KOR inhibition may reverse some deficits observed in disorders of excess consumption or following chronic stress. *ACh*, acetylcholine; *BLA*, basolateral amygdala; *CIE*, chronic intermittent ethanol; *CPA*, conditioned place aversion; *CPP*, conditioned place preference; *DAT*, dopamine transporter; *ICSS*, intracranial self-stimulation; *KOR*, kappa opioid receptor; *NAc*, nucleus accumbens; *PFC*, prefrontal cortex; *VTA*, ventral tegmental area.

stimulated dopamine release and negative affect-like behaviors—are quite well documented and relatively clear, biochemically assessing alterations in dynorphin peptides in a temporally succinct manner has been more challenging. While measuring static dynorphin tissue content levels in dissected samples from specific brain region can be useful and is relatively simple using either ELISA or radioimmunoassay techniques, there are only a few reports measuring dynorphin levels in the extracellular space, using microdialysis on a timescale of 20–30 min (Lam, Marinelli, Bai, & Gianoulakis, 2008; Marinelli, Lam, Bai, Quirion, & Gianoulakis, 2006). However, faster dynorphin measurements are on the horizon, as Dr. Robert Kennedy's group at the University of Michigan has dramatically improved sensitivity for dynorphin detection (10 pM limit of detection) and temporal resolution (less than 5 min samples) using capillary liquid chromatography–multistage mass spectrometry (Zhou, Mabrouk, & Kennedy, 2013). Incorporating this new technology into the field of addiction neuroscience will aid in our understanding of the function and dynamics of the dynorphin/KOR system.

1.6 Opponent Process Theory of Addiction

Mechanisms underlying the transition from social drug use to misuse to addiction have been classically defined by opponent process theory (Fig. 2). Briefly, process "a"—brief and temporally locked with alcohol or drug

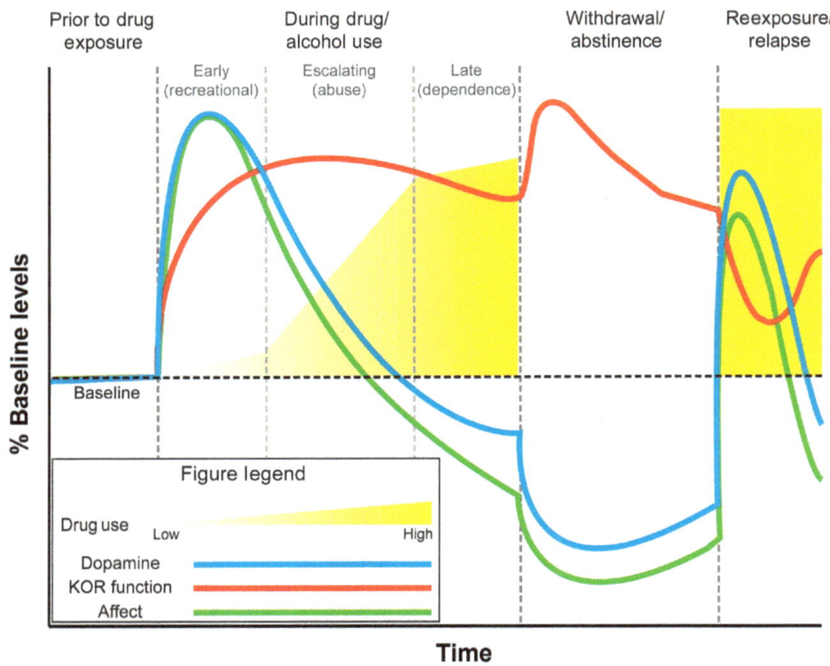

Fig. 2 Dynamics in dynorphin/kappa opioid receptor function, dopamine levels, and affective state as a function of the progression of drug and alcohol use to dependence. Early recreational use is categorized by drug- or alcohol-elicited positive affect and augmented mesolimbic dopamine ("a" process dominant). However, as drug use escalates into dependence, hedonic responding is gradually lost, and increased function of the dynorphin/KOR system contributes to a sustained negative affective state ("b" process dominant). Upon cessation of drug or alcohol use, this negative affective state is further augmented and withdrawal is characterized by low dopamine and sensitized KORs in the nucleus accumbens. Reexposure to the drug of abuse or alcohol is able to transiently reverse this negative affective state and is hypothesized to contribute to vulnerability to relapse in abstinent individuals with previous alcohol or substance use disorders. *KOR*, kappa opioid receptor; *horizontal black dotted line* indicates baseline levels prior to drug exposure; *vertical gray dotted lines* indicate stages of drug or alcohol use, dependence, withdrawal, and reexposure.

administration—is successively opposed by process "b," a slower-onset, longer-lasting compensatory mechanism initially engaged to bring the organism back to homeostasis. The "a" process in the case of alcohol and other drugs of abuse is pleasurable, anxiolytic, and fleeting and is believed to be directly associated with increases in accumbal dopamine and altered excitatory:inhibitory balance throughout the brain. The "b" process aims to normalize these neurochemical alterations, by blunting dopamine elevations and shifting the excitatory:inhibitory balance in the opposite direction of that induced by drug or alcohol use. During early nondependent drug or alcohol use, this "b" process is engaged in order to return the system to homeostasis. However, as an individual transitions to dependence, the "a" process becomes less and less pronounced, while the "b" process is augmented, and is believed to underlie the anhedonia and negative affect commonly associated with dependence. Thus, the nature of drug seeking shifts motivationally from positive reinforcement (attaining the "a" process, consuming the drug for pleasure) to negative reinforcement (drug use in order to avoid the "b" process, consuming the drug to prevent negative affect). The dynorphin/KOR receptor system is hypothesized to be involved in the shift to negative reinforcement (Walker & Koob, 2008). For example, Koob and colleagues have proposed that, initially, hedonic mu opioid receptors and dysphoric KORs function in balance. However, following chronic exposure to drugs or alcohol, overactive KORs overshadow hedonic opioid responses and prevent the system from returning to homeostasis (Walker & Koob, 2008).

1.7 Sex Differences in Kappa Opioid Receptor Function

In general, far more studies have been conducted in male than female animals in the KOR field. However, a few studies have reported important differences between males and females (for review, see Chartoff & Mavrikaki, 2015), although the results obtained from these studies have yielded mixed findings. Some studies have shown augmented KOR function in females. For example, expression of c-Fos induced by the KOR agonist U50,488 was increased in female rats compared to males in corticotropin-releasing factor (CRF)-containing neurons of the paraventricular nucleus of the hypothalamus, and in non-CRF-containing neurons in the BNST (Russell et al., 2014). Other groups have found divergent responses to KOR stimulation in males vs females. Social defeat stress exposure and administration of a KOR agonist resulted in increased neuronal activity in the NAc of female California mice—a pair-bonding species—whereas neural activity was reduced in NAc of male California mice (Laman-Maharg, Copeland, Sanchez, Campi, & Trainor, 2017). This same group showed greater KOR agonist-induced conditioned place aversion in female vs male California mice (Robles et al., 2014). One study showed an increased in ethanol intake in adult female rats following KOR blockade, whereas a reduction in ethanol intake in males was observed at the same dose (Morales, Anderson, Spear, & Varlinskaya, 2014). Finally, other studies have indicated that in some cases females show decreased KOR sensitivity compared to males. In opposition to the previously mentioned study by Morales et al., another group showed a reduction in ethanol intake following norbinaltorphimine (nBNI; a KOR antagonist) in male mice but no change in the intake of female mice (Zhou, Rubinstein, Low, & Kreek, 2017). Additionally, female rats were shown to be less sensitive than males to the depression-like effects of U50,488 (Russell et al., 2014), and this effect was independent of gonadal hormones. Moreover, female rats took significantly longer to discriminate a KOR agonist from vehicle using a fixed-ratio 10 schedule of food reinforcement compared to males (Craft, Kruzich, Boyer, Harding, & Hanesworth, 1998). Finally, administration of the KOR agonist spiradoline potentiated cocaine-induced hyperactivity in males, but not in females (Sershen, Hashim, & Lajtha, 1998). Sex differences have not only been observed in rodents but also in humans. An extensive PET imaging study showed higher KOR availability in males compared to females in the anterior cingulate cortex, insula, and the ventral pallidum among other areas (Vijay et al., 2016). Although, when examined as a whole, these studies are inconclusive, this is a burgeoning area of research, and more investigators are examining possible sex differences in KOR function.

2. ETHANOL AND KAPPA OPIOID RECEPTORS

Ethanol exposure transiently increases the firing rate of dopamine neurons within the VTA both directly (Brodie, Pesold, & Appel, 1999) and through alterations in neurotransmitter inputs (for review, see Morikawa & Morrisett, 2010), resulting in elevation of dopamine in the NAc (Yim & Gonzales, 2000; Yim, Schallert, Randall, & Gonzales, 1998). This elevation in dopamine results in activation of GABAergic D1 receptor-containing medium spiny neurons, which also release dynorphin. Dynorphin release from collaterals of medium spiny neurons activates KORs in the NAc and inhibits dopamine release. VTA dopamine neuron

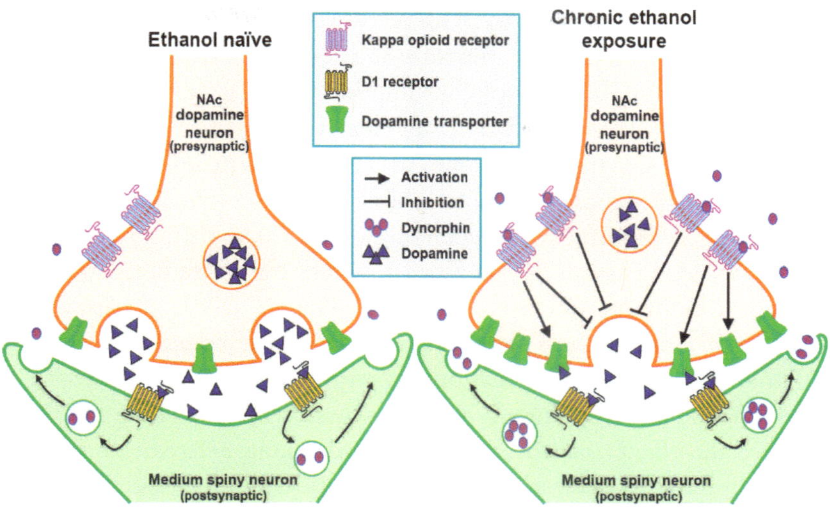

Fig. 3 Effects of chronic ethanol on kappa opioid receptors. In ethanol naïve animals, after dopamine release activates D1-containing medium spiny neurons, dynorphin is released resulting in activation of KORs which further leads to inhibition of dopamine release. Such a feedback maintains homeostatic balance. In chronic ethanol-exposed animals, similar to naïve animals, once dopamine is released and it activates D1-containing medium spiny neurons dynorphin is released, but it is hypothesized that this dynorphin release is greater. Furthermore, studies have shown that KOR function is upregulated following chronic ethanol exposure. Thus, when KOR is activated by dynorphin, there is an enhanced effect of KORs on the inhibition of dopamine release resulting in low dopamine in the extracellular space.

excitation is blunted to some degree by subsequent ethanol-mediated activation of GABAergic signaling (Theile, Morikawa, Gonzales, & Morrisett, 2011). These changes in VTA dopamine neuron firing then affect downstream signaling in the NAc. This section will explore the relationship between dynorphin/KOR function and chronic ethanol exposure in the mesolimbic dopamine system (Fig. 3).

2.1 Kappa Opioid Receptors and Ethanol Consumption

The dynorphin/KOR system plays an important role in ethanol drinking in both dependent and nondependent animals; however, whether KORs contribute to increased or decreased drinking behavior is dependent upon the state of the animal. The body of literature examining the role of dynorphin/KOR on ethanol consummatory behaviors can appear conflicting at first; however, when one takes into account the timing of pharmacological targeting of KORs with regard to ethanol administration and dependence, these studies show greater clarity. This section will explore the effects of activation and inhibition of KOR function at distinct time points and durations of ethanol exposure on ethanol consumption.

KOR activation, when occurring near the onset of ethanol intake, appears to be generally effective at reducing ethanol consumption. U50,488, a KOR agonist, is able to inhibit ethanol-conditioned place preference when administered simultaneously with ethanol (Logrip, Janak, & Ron, 2009), and chronic KOR agonist administration reduces free access drinking (Lindholm, Werme, Brene, & Franck, 2001). Indeed, even alcohol-preferring (P) rats decrease responding for ethanol when given U50,488, although this attenuation for responding was also observed for sucrose (Henderson-Redmond & Czachowski, 2014). These studies all examined the effect of KOR activation on drinking behavior during early acquisition. As indicated earlier, KOR stimulation with U50,488 is aversive, causing negative affect-like behavior similar to ethanol withdrawal. The pairing of an anxiogenic stimulus with ethanol or drug administration may reduce acquisition of drug taking or liking, such as development of preference for a drug-paired context or free access drinking in nondependent animals. However, one of the defining facets of alcohol (and drug) dependence is usage despite negative consequences. Accordingly, rats that have been made dependent on ethanol are resistant to decreased drinking as a result of "punishment," commonly the addition of quinine, an aversive tastant, to the ethanol solution (Vendruscolo et al., 2012), and show escalated or reinstated drinking behavior when exposed to a stressor (Mantsch, Baker, Funk, Le, & Shaham, 2016). In fact, U50,488 itself can be used as a stressor to drive ethanol reinstatement in rats after ethanol extinction training. In this case, the KOR agonist had not previously been paired with ethanol, instead serving as an aversive experience similar to the pharmacological stressor yohimbine (Funk, Coen, & Le, 2014).

Immediately after ethanol cessation and during ethanol withdrawal, the dynorphin/KOR system appears to be hyperactive (Karkhanis, Huggins, et al., 2016; Rose et al., 2016). This putatively contributes to negative affect-like states and may drive enhanced ethanol seeking during late dependence and withdrawal through negative reinforcement. At this point in the progression of initial ethanol intake to dependence, KOR system function is elevated, and further stimulation with an agonist would be expected to increase drinking behavior. In fact, a recent study has shown that KOR activation via U50,488 administration increases ethanol intake in dependent animals (Rose et al., 2016). Similarly, rats with prolonged (16 months) history of ethanol exposure display increased ethanol lever responding following administration of a KOR agonist (Holter, Henniger, Lipkowski, & Spanagel, 2000). Thus, in animals that have been made dependent on ethanol or show withdrawal-induced behavioral abnormalities, it is likely that KOR inhibition would be better suited to reduce ethanol seeking and drinking. Indeed, animals that have been chronically exposed to ethanol are responsive to treatment with KOR antagonists. nBNI, a KOR antagonist, reduces responding for ethanol in dependent P rats to a much greater extent than nondependent P rats (Cashman & Azar, 2014) and blocks voluntary drinking in dependent—but not nondependent—animals (Rose et al., 2016; Walker & Koob, 2008). In rats with a history of prolonged (16 months) exposure to ethanol, nBNI reduced withdrawal-induced augmentation of ethanol consumption (Holter et al., 2000). Stress-induced reinstatement of ethanol seeking through either U50,488 or yohimbine is blocked by prior administration of nBNI (Funk et al., 2014). On the other hand, nondependent animals showed no effect of nBNI on operant responding rates for ethanol (Doyon, Howard, Shippenberg, & Gonzales, 2006).

The collective findings described here with relation to KOR modulation of ethanol seeking and consumption can perhaps best be succinctly visualized by yet another study. The dynorphin knockout mouse (Dyn −/−) displays increased basal drinking behavior (Racz, Markert, Mauer, Stoffel-Wagner, & Zimmer, 2013). When examining this animal through the lens of opponent process theory, it is possible to imagine that these mice have a hampered aversive "b" process, as their KORs are unable to be endogenously activated. This fits nicely with the data from drinking acquisition above, showing that KOR agonists are able to reduce early ethanol drinking and seeking. On the other hand, Dyn −/− mice do not show stress-induced reinstatement of ethanol seeking. This mirrors the effects of nBNI described earlier and reiterates that KOR activation during withdrawal likely contributes to enhanced ethanol seeking during that time.

There is one important caveat to consider when targeting the KOR system to query drinking behavior. KOR activation alone is able to stimulate both feeding and water drinking behavior (Wee & Koob, 2010), and thus caution should be taken in any sort of model where no consummatory control is available. For instance, the classic drinking in the dark binge model for rodent drinking typically involves a single drinking tube containing ethanol.

In the case of KOR modulation, control water drinking measurements should always be provided in order to remove the confound of potential basal alterations in fluid intake.

2.2 Kappa Opioid Receptors and Chronic Ethanol Exposure

The dynorphin/KOR system plays a key role in the development of behavioral alterations consistent with a negative affect-like state. Recent studies have demonstrated that C57Bl/6J mice exhibit aberrant affective-like behaviors following cessation of ethanol administration. For example, mice exposed to chronic ethanol through liquid diet or chronic intermittent ethanol vapor show increased marble burying for up to 4 days following termination of ethanol exposure (Perez & De Biasi, 2015; Rose et al., 2016; Umathe, Bhutada, Dixit, & Shende, 2008), suggesting that chronic exposure to ethanol results in increased anxiety-like behavior. Additionally, mice exposed to relatively low levels of ethanol through a simple prolonged two bottle choice model of administration display long-lasting (up to 35 days after ethanol cessation) increases in negative affect-like behavior in the novelty-suppressed feeding test, a test that measures depression-like behavior in rodents (Holleran et al., 2016). Furthermore, 22 kHz ultrasonic vocalizations, believed to be an anxiety-related response, are increased both during ethanol withdrawal as well as during cue-induced escalation of intake in rats (Berger, Williams, McGinnis, & Walker, 2013). These negative affect-like behaviors driven by cessation of prolonged ethanol exposure mimic the negative affective state of individuals suffering from alcohol use disorder.

Modulation of the dynorphin/KOR system influences behaviors that are observed during withdrawal, further supporting the notion that altered KORs may, in part, underlie the negative affect experienced by individuals with alcohol use disorder during withdrawal. nBNI (antagonist) is able to reverse marble-burying behavior and reduce the 22 kHz ultrasonic vocalizations of ethanol withdrawn rodents back to baseline levels (Berger et al., 2013; Rose et al., 2016). Conversely, systemic administration of a KOR agonist phenocopies behaviors observed in ethanol withdrawn mice. Thus, it is suggested that the ethanol withdrawal-induced augmentation of negative affect-like behaviors observed in rodents is due, at least in part, to an increase in KOR function. The remainder of this chapter will examine effects of KOR modulation on ethanol intake and on neural substrates that may be involved in alcohol use disorder.

2.3 Ethanol Effects on the Dynorphin/Kappa Opioid Receptor System

Acute and chronic ethanol exposure has been shown to have positively regulate dynorphin levels across multiple brain regions. For example, systemic administration of ethanol results in a transient increase in extracellular dynorphin levels in the NAc (Lindholm, Ploj, Franck, & Nylander, 2000; Marinelli et al., 2006). Dynorphin mRNA levels have been observed to be greater in the central nucleus of the amygdala and the hypothalamus in P rats that are voluntarily drinking compared to NP rats (Zhou, Colombo, Gessa, & Kreek, 2013). Further, during acute withdrawal following 1 month of ethanol drinking, mRNA encoding prodynorphin, precursor for dynorphin, is elevated in the NAc, returning to baseline after 96 h (Przewlocka, Turchan, Lason, & Przewlocki, 1997). Thirty minutes following final ethanol exposure in a chronic binge-like ethanol self-administration paradigm of repeated injections of ethanol for 2 weeks, dynorphin as measured by radioimmunoassay is increased in the NAc and decreased in the cingulate, and the increase in NAc dynorphin is maintained as long as 21 days after final ethanol exposure (Lindholm et al., 2000). Following 6 months of voluntary intake, sensitivity of KORs was observed to be increased in the dorsolateral caudate and the NAc of macaques (Siciliano et al., 2015). Collectively, these studies suggest that chronic ethanol exposure upregulates the dynorphin/KOR system function.

2.4 Ethanol, Kappa Opioid Receptors, and Dopamine

In humans, ethanol withdrawal promotes anhedonia, which often contributes to anxiety, depression, and increased rates of relapse, especially in individuals with alcohol use disorders (Weiss, Griffin, & Mirin, 1992). Similarly, rodents with multiple cycles of ethanol exposure and withdrawal exhibit anxiety- and depression-like behavior outlined earlier (Holleran et al., 2016; Rose et al., 2016) as well as stress-induced relapse and reinstatement (Anderson, Lopez, & Becker, 2016). Evidence from previous studies point to dysfunctional dopamine transmission as a major contributor and promoter of anhedonia (Koob & Volkow, 2010). The following section will examine modulation of dopamine signaling following acute and chronic ethanol, and excitation and inhibition of KORs.

As mentioned in Section 1, among other cell types, KORs are localized on dopamine terminals in the NAc (Spanagel, Herz, & Shippenberg, 1992) and activation of these receptors inhibits dopamine release (Donzanti,

Althaus, Payson, & Von Voigtlander, 1992; Spanagel et al., 1992). Behavioral evidence from the rodent studies suggests that multiple cycles of chronic ethanol exposure and withdrawal result in hyperactive KORs. Indeed, recent literature shows that agonist-stimulated KOR activation concentration dependently decreases stimulated dopamine release in the NAc, and that this inhibition is augmented following chronic exposure to ethanol (Karkhanis, Huggins, et al., 2016; Rose et al., 2016). Additionally, local infusion of KOR agonist results in a greater reduction in extracellular levels of dopamine in animals with chronic ethanol exposure compared to control animals (Karkhanis, Huggins, et al., 2016). This further confirms the observation that KOR function is upregulated following chronic ethanol exposure. In control rodents, exposed to ethanol-free air alone, acute systemic administration of ethanol results in an increase in extracellular levels of dopamine. Following chronic ethanol exposure, however, this same acute systemic administration of ethanol results in a reduction of extracellular levels of dopamine significantly below baseline levels (Karkhanis, Huggins, et al., 2016). This phenomenon is likely driven by hyperfunctioning KORs, since pretreatment with a KOR antagonist reverses the dopamine-decreasing effect of ethanol. This also provides a possible neurobiological mechanism for the opponent process theory.

3. COCAINE AND KAPPA OPIOID RECEPTORS

Chronic stress during development and in adulthood elevates the risk of developing cocaine addiction and relapse, and augments cocaine craving in adulthood (Brady & Sinha, 2005; Elton, Smitherman, Young, & Kilts, 2015; Jacobsen, Southwick, & Kosten, 2001; Rovaris et al., 2015). Both chronic stress and cocaine exposure increase dynorphin/KOR system function (Karkhanis, Rose, et al., 2016; Unterwald, Rubenfeld, & Kreek, 1994), and activation of KORs reinstates cocaine seeking in rodents (e.g., Valdez, Platt, Rowlett, Ruedi-Bettschen, & Spealman, 2007). A recent human study showed a strong correlation between stress and augmented cocaine craving as well as increased risk of relapse, and these addiction vulnerability markers were associated with a variant of the KOR gene, *opkr1* (Xu et al., 2013). These studies indicate that there may be a direct link between KOR function and cocaine addiction. The following section will explore various aspects of the interaction of the dynorphin/KOR system and cocaine (Fig. 4).

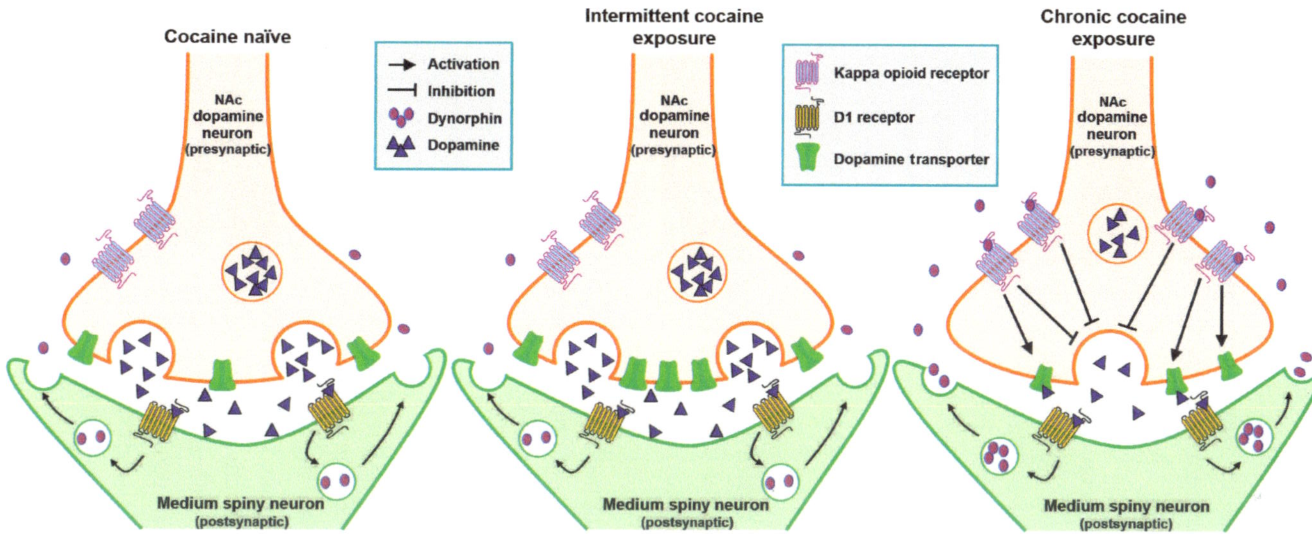

Fig. 4 Effects of chronic cocaine on kappa opioid receptors. In cocaine naïve animals, after dopamine release activates D1-containing medium spiny neurons, dynorphin is released resulting in activation of KORs which further leads to inhibition of dopamine release. Such a feedback maintains homeostatic balance. After noncontingent intermittent cocaine exposure while there is no change in KOR function, DAT function is upregulated. After noncontingent chronic cocaine exposure, however, there is an upregulation of KOR function and down-regulation in DAT function. Thus, once dopamine is released and it activates D1-containing medium spiny neurons dynorphin is released, but it is hypothesized that this dynorphin release is greater. Furthermore, studies have shown that KOR function is upregulated following chronic ethanol exposure. Thus, when KOR is activated by dynorphin, there is an enhanced effect of KORs on the inhibition of dopamine release resulting in low dopamine in the extracellular space. *DAT*, dopamine transporter; *KOR*, kappa opioid receptor.

3.1 Chronic Stress, Kappa Opioid Receptors, and Cocaine

Rodent studies have shown that stress experienced in adolescence or adulthood can increase KOR function can drive subsequent cocaine seeking in adulthood. Chronic stress in adolescence and in adulthood augments cocaine seeking and reinstatement in adulthood (Baarendse, Limpens, & Vanderschuren, 2014; Yajie, Lin, Baoming, & Lan, 2005). The neurobiological mechanisms underlying stress-induced augmentation of cocaine seeking are not fully understood, but there is evidence that altered KOR signaling may be involved. For example, a study recently showed that repeated exposure to forced swim stress in adulthood augments cocaine seeking, which is then reduced by KOR inhibition (Groblewski et al., 2015). Another study showed that KOR inhibition prevented cold-water forced swim stress-induced reinstatement of cocaine-seeking behavior (Polter et al., 2014). Together these studies suggest that chronic stress-induced increases in KOR function are involved in driving cocaine seeking.

3.2 Timing of Kappa Opioid Receptor Activation, Dopamine Deficits, and Cocaine Effects

It is possible that stress-induced increases in KOR function promote cocaine seeking by reducing dopamine transmission. As explained in Section 1, reductions in dopamine transmission are linked to negative affective states including anhedonia. Cocaine produces positive hedonic effects by blocking dopamine transporters (DATs) and slowing reuptake of dopamine from the synapse, resulting in prolonged stimulation of dopamine receptors. Thus, it is hypothesized that stress-exposed animals have low dopamine levels and therefore "crave" cocaine, which restores homeostatic levels of dopamine signaling. This section explains the effects of acute and prolonged activation of KORs and how it affects reward processing.

Because KORs modulate the VTA–NAc dopamine projection, thereby altering dopamine signaling and reward processing, it is crucial to understand how differential temporal activation of KORs affects reward sensitivity. KOR activation time dependently modulates vulnerability to cocaine and reward processing; for example, a recent study showed that administration of cocaine 15 min after KOR activation with salvinorin A decreased ICSS thresholds of the medial forebrain bundle, consistent with increased positive affect, while cocaine administration 24 h after KOR activation increased ICSS thresholds, indicating a negative affective state (Chartoff et al., 2016). This study shows that prior activation of KORs increases negative affect, potentially driving the augmented motivation to self-administer cocaine. Continuous treatment of monkeys with a KOR agonist, that is, intravenous administration of KOR agonist once every 20 min during sequential 3-day blocks increases cocaine choice (Negus, 2004). Studies have shown that activation of KORs at least 60 min before training or testing results in increased cocaine-conditioned place preference (CPP; McLaughlin, Land, Li, Pintar, & Chavkin, 2006; Schindler et al., 2012). Another study has shown that KOR activation by U50,488 15 min before cocaine administration has been shown to block cocaine CPP; however, KOR activation 60 min before cocaine administration significantly potentiated cocaine CPP (Ehrich, Phillips, & Chavkin, 2014). In addition, this study showed that activation of KORs 15 min prior to cocaine self-administration resulted in significant reductions in cocaine effects on electrically evoked dopamine responses. On the other hand, activation of KORs 60 min before cocaine administration resulted in an enhancement of cocaine effects on dopamine responses (Ehrich et al., 2014). Together, these studies resonate well with the chronic stress literature mentioned in Section 3.1. Prolonged activation of KORs, whether pharmacologically or due to chronic stress exposure, results in anhedonia, further leading to potentiated motivation to seek cocaine. Additionally, prolonged activation of KORs increases reinforcing effects of cocaine via potentiated dopamine release. That is, dopamine elevations induced by cocaine are greater following prolonged KOR activation.

3.3 Kappa Opioid Receptors and DATs

Previous literature suggests that KORs and DATs are functionally linked, such that acute activation of KORs increases DAT function, potentially through multiple intracellular mechanisms (Collins, D'Addario, & Izenwasser, 2001; Kivell, Uzelac, et al., 2014; Simonson et al., 2015; Thompson et al., 2000). A recent study shows that Mesyl Sal B, a specific KOR agonist that activates the ERK1/2 intracellular pathway, potentiates DAT function without altering surface expression (Simonson et al., 2015), while salvinorin A, another KOR agonist, increases DAT function by increasing surface level DAT expression in the striatum (Kivell, Uzelac, et al., 2014). Interestingly, this increase in DAT expression is also mediated by KOR-linked ERK1/2 pathway. Together these studies suggest that the KOR mechanism involved in increased DAT function depends on the pharmacological agent used. Chronic KOR activation, however, reduces DAT densities (Collins et al., 2001). Another study showed that repeated activation of KORs resulted in decreased DAT binding but not total DAT protein expression in the NAc of naïve rats (Thompson et al., 2000), suggesting that the efficacy of the transporter was reduced. Coimmunoprecipitation and BRET experiments have shown that KOR and DAT exist in a complex, the formation of which is promoted by KOR activation (Kivell, Uzelac, et al., 2014). Thus, there is strong evidence that KORs and DATs interact to modify DAT function.

3.4 Cocaine, Kappa Opioid Receptors, and DATs

Noncontingent, experimenter-delivered, repeated cocaine administration results in increased DAT function and locomotor activation (Collins et al., 2001; Thompson et al., 2000). Additionally, noncontingent simultaneous administration of cocaine and a KOR agonist attenuated the effects of cocaine on locomotor behavior without any locomotor effects on its own, suggesting that KORs modulate cocaine's actions on the DAT (Collins et al., 2001; Thompson et al., 2000). Noncontingent chronic continuous infusion of cocaine via an osmotic minipump for 14 days increased KOR density (Collins et al., 2002). This change, however, was not observed following other selective DAT inhibitors such as GBR12909 or RTI-117, indicating that the increase in KOR density was likely driven by a mechanism not involving DATs (Collins et al., 2002). Overall these findings suggest that chronic cocaine exposure leads to increased KOR function. Together these studies also indicate that although KORs and DATs are functionally linked, this functional link is relatively understudied and dependent on the length of activation of the KORs. Furthermore, the activity of DATs appears to be dependent on KOR function, but it is unlikely that DAT function reciprocally affects KOR function. Because cocaine binds to DATs, reducing their function, and due to the strong evidence that KORs and DATs interact, it is hypothesized that KORs act through modulation of intracellular modifications of DAT to time dependently change the behavioral and dopaminergic effects of cocaine.

3.5 Reinstatement of Cocaine Self-Administration

There are conflicting reports in the literature of the effects of KOR modulation on cocaine seeking in reinstatement paradigms. For example, KORs in the VTA were observed to be hyperactive, and potentially constitutively active, following a single exposure to cold-water forced swim stress, albeit only at inhibitory synapses (Polter et al., 2017). This hyperactivation of KORs was shown to contribute to prolonged cocaine-seeking observed after cold-water forced swim stress, which occurred 24h before exposure to the reinstatement procedure (Polter et al., 2017). However, there is also evidence that activation of KORs in the VTA blocks cocaine-induced reinstatement of cocaine self-administration; this effect was blocked by KOR inhibition (Sun, Xue, Huang, & Steketee, 2010). In this study, however, cocaine reinstatement was tested only 5min following microinjection of U50,488 into the VTA. As previously discussed, the timing of KOR activation is crucial in determining the effects of cocaine on dopamine signaling. Therefore, it is possible that the short interval between KOR activation and cocaine self-administration results in a blockade of reinstatement. Stress-induced reinstatement of cocaine self-administration is also blocked by KOR inhibition (Polter et al., 2014). It is possible that KOR blockade "normalizes" the hedonic state of the animals reducing the motivation seek cocaine in the event of exposure to stress. In other words, the response to stress is balanced as a result of blocking excessive activation of KORs.

4. OBESITY AND KAPPA OPIOID RECEPTORS

Similar to substance and alcohol use disorder, previous research has shown that opioids are also involved in overconsumption of food, a primary characteristic of binge eating disorder that often leads to the development of obesity (Bodnar & Klein, 2004; Levine & Billington, 2004). Specifically, KOR agonists promote eating and antagonists reduce eating under specific circumstances (Jarosz, 2007). KOR inhibition does not reduce regular food intake, but it significantly attenuates elevation in feeding induced by the following: food deprivation, exposure to high-fat diet, and KOR activation (Arjune & Bodnar, 1990; Levine, Grace, Billington, & Portoghese, 1990). This section will examine the interaction between KORs and food consumption, obesity, and binge eating disorder (Fig. 5).

4.1 The Dynorphin/Kappa Opioid Receptor System and Hereditary Obesity

One of the most common rodent models to study the underlying factors involved in development and maintenance of obesity, the obese Zucker rat, exhibits hyperphagic eating patterns similar to individuals who engage in binge eating. A spontaneous mutation in the Zucker rat strain resulted in the development of obesity and type II diabetes. The mutation was

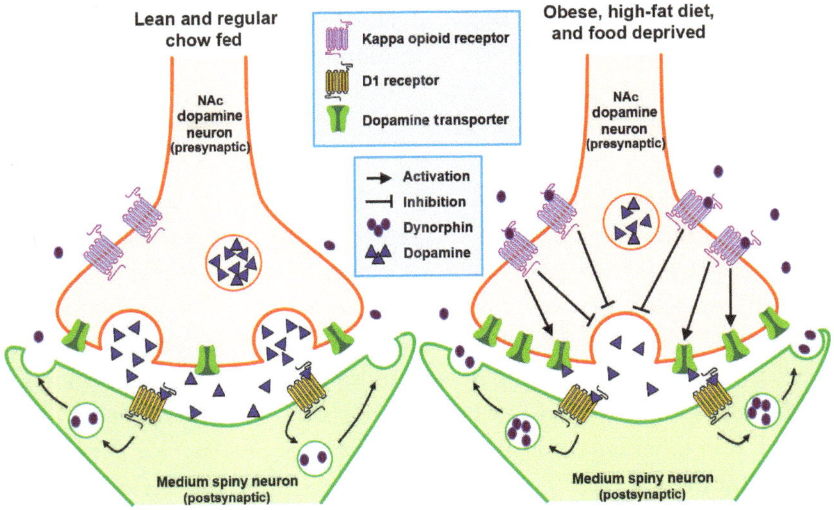

Fig. 5 Effects of high-fat diet/food deprivation on kappa opioid receptors. In lean and regular chow-fed animals, after dopamine release activates D1-containing medium spiny neurons, dynorphin is released resulting in activation of KORs which further leads to inhibition of dopamine release. Such a feedback maintains homeostatic balance. In obese, high-fat diet, and food-deprived animals, similar to lean and regular chow-fed animals, once dopamine is released and it activates D1-containing medium spiny neurons dynorphin is released, but it is hypothesized that this dynorphin release is greater. Furthermore, studies have shown that KOR function is upregulated following high-fat diet consumption and food deprivation. Thus, when KOR is activated by dynorphin, there is an enhanced effect of KORs on the inhibition of dopamine release resulting in low dopamine in the extracellular space.

identified as a missense mutation in the leptin receptor gene (Chua et al., 1996; Phillips et al., 1996; Xie, Sanz-Burgos, Hannon, & Gutierrez, 1996) which results in a lack of response to leptin (Cusin, Rohner-Jeanrenaud, Stricker-Krongrad, & Jeanrenaud, 1996). Leptin exerts a negative control on food consumption and is essential in regulating caloric intake. Inhibition of KOR function has been observed to reduce food intake in obese Zucker rats. KOR antagonists decrease meal size (Jarosz & Metzger, 2002) and duration, while elongating postmeal interval (Kirkham & Blundell, 1987), thus reducing overall food intake. Because of this reduction in feeding, KOR blockade also reduces body weight (Jarosz, 2007; Jarosz & Metzger, 2002). Neurobiologically, these rats show elevated tissue levels of dynorphin in the medial and lateral hypothalamus, cortex, and midbrain (Roane, Iadarola, & Porter, 1988; Tsujii et al., 1987) and a reduction in dynorphin levels in the paraventricular nucleus

(Kim et al., 2000). Although the connection between the leptin mutation and dynorphin changes has not been elucidated, KOR antagonists may be useful in treating leptin-tolerant obesity. In addition to the obese Zucker rat strain, studies examining obesity-driven neurobiological alterations and potential therapeutic targets use genetically obese (ob/ob) and lean mouse models. The ob/ob mice have a mutation in the gene responsible for leptin production. Similar to the obese Zucker rats, ob/ob mice become profoundly obese and develop type II diabetes. Studies have shown alterations in the dynorphin/KOR system in these mice as well. The concentration of KOR binding sites was observed to be 2.7-fold higher in ob/ob compared to lean mice. In addition, dynorphin tissue content was observed to be 12-fold lower in the pituitary of obese mice (Khawaja, Bailey, & Green, 1989). However, another study showed an increase in dynorphin immunoreactivity in the posterior pituitary. In this study, a KOR agonist augmented eating in lean mice and a KOR antagonist attenuated eating in obese mice (Ferguson-Segall, Flynn, Walker, & Margules, 1982). Some of the opposing results observed in these two models may be due to species differences. There are only handful studies that have investigated the effects of genetic obesity on the dynorphin/KOR system. More research needs to be conducted in this field in order to draw general conclusions.

4.2 Dynorphin/Kappa Opioid Receptor System and Palatable Food Consumption

While the genetic makeup of an individual may contribute to the development of obesity, another contributing factor is overconsumption of a high-fat diet or high-caloric intake. High-fat food has been shown to have reinforcing effects. Specifically, high-fat diet increases extracellular levels of dopamine in the NAc shell; this high-fat food-induced elevation in dopamine is not related to the quantity of food consumed nor does it habituate following daily scheduled access. Moreover, KOR inhibition reduced both intake of high-fat food and high-fat diet-induced elevation in dopamine (Sahr et al., 2008). These data suggest that a high-fat diet is consumed due to its reinforcing effects, and this consumption may be reduced by KOR inhibition as a result of reduction in its reinforcing value, i.e., decrease in dopamine elevation.

Studies have also shown that KOR inhibition reduces consumption of fatty foods. KOR inhibition attenuated consumption of high-fat food with no effect on low-fat food in fasted rats. In the same study, activation of KORs in food-sated rats increased consumption of high-fat food with no effect on consumption of low-fat food (Ookuma, Barton, York, & Bray, 1997). Another study showed that KOR inhibition blocked short-term consumption of palatable high-fat diet (Arjune & Bodnar, 1990), and chronic KOR inhibition prolonged the effects of reduction in the consumption of fat but not total intake in rats exposed to a diet of fat, low-fat pellets, milk, and water (Cole, Leventhal, Pasternak, Bowen, & Bodnar, 1995). In addition, KOR inhibition selectively reduced deprivation-induced high-fat food consumption in rats that were genetically susceptible to obesity (Osborne–Mendel rats) while having no effect on effect on low-fat diet intake in these animals. KOR activation markedly increased high-fat food consumption in these rats while increasing both high-fat and low-fat food intake in rats not susceptible to obesity (Ookuma, Barton, York, & Bray, 1998). These findings indicate that modulation of the dynorphin/KOR system selectively influences consumption of foods high in fat. Sucrose intake was also found to be reduced following site-specific injection of nBNI, KOR antagonist, into the paraventricular nucleus of the hypothalamus (Koch, Glass, Cooper, & Bodnar, 1995). Collectively, these studies suggest that the dynorphin/KOR system exerts profound influence on regulating palatable and high-fat food consumption, thereby potentially contributing to the development of obesity.

4.3 Dynorphin/Kappa Opioid Receptor System and Food Deprivation

On the other hand, in studies of the opposite of overeating—food deprivation, the abundance of peptide, and receptor as well as function of the dynorphin/KOR system has been shown to be altered as well. Food deprivation of 72 h increased dynorphin levels in the cortex of lean Zucker rats and in the cortex and hippocampus of obese Zucker rats (Tsujii et al., 1987). However, dynorphin levels were reduced in the striatum of lean Zucker rats and the midbrain of obese Zucker rats (Tsujii et al., 1987). Furthermore, KOR binding was increased in the cortex, midbrain, and striatum of lean rats but only the cortex and midbrain in obese rats following food deprivation (Tsujii et al., 1986). Low dynorphin levels and increased KOR binding in the striatum in lean Zucker rats and low dynorphin and increased KOR binding in the midbrain of obese Zucker rats suggest that the dynorphin/KOR system is upregulated in the striatum of lean and midbrain of obese Zucker rats following food restriction. Chronic food restriction, resulting in body weight equal to around 70% of controls, also altered KOR function in multiple brain areas. For example, KOR binding efficiency was reduced in medial habenula, but increased in BNST, ventral pallidum, and medial preoptic area (Wolinsky, Carr, Hiller, & Simon, 1994). Studies have also shown behavioral effects of systemic and site-specific blockade of KORs. KOR inhibition blocked deprivation-induced elevated feeding in rats and mice food deprived for 12 or 24h (Hadjimarkou et al., 2004; Jewett et al., 2001). A study examining dynorphin/KOR system following 16h of food deprivation in mice showed that KOR blockade markedly reduced food intake, while KOR activation did not elevate intake. Further, mRNA levels of prodynorphin were decreased in the hypothalamus, midbrain, and ventral striatum. Site-specific injection of nBNI, KOR antagonist, into the lateral hypothalamus and VTA, but not the NAc, attenuated food intake (Ikeda et al., 2015). Another study showed that KOR inhibition specifically in the paraventricular nucleus attenuated deprivation-induced food intake following 24h without food (Koch et al., 1995). Together these findings suggest that dynorphin/KOR system function in the lateral hypothalamus, paraventricular nucleus, and the midbrain is implicated in feeding behaviors especially following food deprivation.

It may be surprising to note that alterations in KOR function are remarkably similar between animals consuming copious amounts of high-fat food and those undergoing food restriction/deprivation. The increase in KOR function in high-fat diet-exposed animals is due, in part, to adaptation in areas involved in reward processing, while the increase in KOR function in food-deprived animals is due to stress. Activation of reward circuitry and stress both increase dopamine signaling and the augmented KOR function may be due to neuroadaptations aimed at reducing this signaling.

5. KAPPA OPIOID RECEPTOR-BASED THERAPEUTICS

Somewhat paradoxically, both agonists and antagonists of KORs have been proposed as potential therapeutic drugs to treat addiction and other disorders of excessive consumption, such as obesity. Acute activation of KORs with agonists such as salvinorin A and mesyl salvinorin B has been shown to reduce drug-primed reinstatement of cocaine seeking after self-administration in rats (Kivell, Ewald, et al., 2014; Simonson et al., 2015). This effect may be the result of a KOR-mediated decrease in the reinforcing effects of the priming cocaine injection. As discussed earlier with relation to ethanol reinforcement, KOR agonists have been shown to produce aversion, and thereby adjunctive treatment of KOR agonists in combination with drug (or alcohol) administration in nondependent animals would decrease rewarding effects of the drug through the anxiogenic "punishment" of KOR activation. In addition, delivery of salvinorin A intravenously along with cocaine (or the opiate remifentanil) during self-administration reduced drug choice in a food–drug choice paradigm (Freeman et al., 2014). Similar to cocaine, administration of a KOR agonist in conjunction with ethanol reduces voluntary ethanol intake in non-dependent animals (Lindholm et al., 2001; Nestby et al., 1999). Again, these findings can be attributed to punishment of drug responding due to the dysphoric properties of the KOR agonist (Freeman et al., 2014). While KOR activation may reduce acute drug or ethanol consumption, it is unlikely that individuals with alcohol or substance use disorders would voluntarily comply with this treatment regime. Additionally, a defining feature of drug and alcohol dependence is continued use despite negative consequences, meaning that punishment-based treatment strategies would have particularly poor outcomes. Moreover, prolonged activation of KORs mimics the effects of a chronic stressor and can contribute to facilitation of addiction-related behaviors (Chartoff et al., 2016; Sperling, Gomes, Sypek, Carey, & McLaughlin, 2010), which would clearly be counterproductive for a treatment drug.

Treatment strategies with KOR antagonists appear to show greater therapeutic potential than agonist-based treatments. Inhibition of KORs blocks stress-induced reinstatement of cocaine seeking and reduces chronic stress-induced elevations in cocaine seeking (Graziane et al., 2013; Groblewski et al., 2015). In preclinical studies, KOR antagonists have been successful in reliably reducing voluntary ethanol intake in dependent animals (Anderson et al., 2016; Berger et al., 2013; Rose et al., 2016) and animals exposed to chronic stress (Anderson et al., 2016; Karkhanis, Rose, et al., 2016). Seemingly contradictory, in two studies nBNI, a KOR antagonist, failed to reduce ethanol intake (Lindholm et al., 2001; Mitchell, Liang, & Fields, 2005). However, these findings should be interpreted with caution based on two aspects of the study design: first, the rats did not exhibit ethanol dependence. Earlier in this chapter, the contribution of ethanol dependence on responsivity to KOR agonism vs antagonism was discussed. While non-dependent animals consistently show an absence of responding to KOR antagonists (and usually reduction in drug seeking when given KOR agonists, as shown above), dependent animals, with augmented KOR function, consistently respond well to treatment with KOR antagonists. Second, the Lindholm et al. study failed to take into account a pharmacological characteristic of nBNI itself. nBNI has been shown to inhibit mu opioid receptors initially (within the first 2 h of administration) followed by a reduction in mu antagonism and an increase in KOR blockade, with its peak activity at the KORs at 4 h and lasting for several days (Munro et al., 2012). In the study outlined earlier, nBNI was administered only 30 min before the drinking test, which would affect the outcome through off-target interactions with the mu opioid receptor. Therefore, the results obtained are likely due to inhibition of the mu opioid receptor, not blockade of the KOR. Collectively these studies indicate that KOR inhibition, not activation, would result in a reduction in drug intake especially in individuals suffering from comorbid affective and addictive disorders.

Currently, a number of clinical trials are underway examining the effectiveness of KOR modulation for disorders of affect as well as alcohol and substance use disorders. Early trials were initially disheartening, as a trial investigating the KOR antagonist JDTic (ClinicalTrials.gov ID: NCT01431586) was terminated early due to adverse events experienced by the subjects involved. However, new clinical trials of KOR antagonists for affective and addictive disorders using new, safer, compounds are on the rise. Currently, multiple trials are actively recruiting subjects to investigate the involvement of KORs in these conditions and the therapeutic potential of KOR antagonists.

CERC-501, one of these antagonists, is being tested as a possible therapeutic to combat anhedonia associated with mood and anxiety-spectrum disorders (ClinicalTrials.gov ID: NCT02218736), and primary outcome measures of this trial are scheduled to be completed in August 2017. Given that anhedonia is also a component of drug and alcohol withdrawal which has been theorized to contribute to relapse, positive results of this trial would be encouraging for KOR antagonist treatment of individuals with alcohol and substance use disorders in prolonged withdrawal. Additionally, this trial will also use fMRI to examine reward processing in the ventral striatum via the monetary incentive delay task, which will also be informative in relation to disorders of excessive consumption such as drug and alcohol addiction.

In another trial that is currently recruiting, CERC-501 is being examined as a potential therapeutic to prevent stress-precipitated smoking relapse (ClinicalTrials.gov ID: NCT02800928). Initial results are expected in June 2018. Based on the body of work discussed in this chapter, stress-induced reinstatement of cigarette seeking in individuals dependent on nicotine is likely mediated to some extent by augmented KOR responsiveness. The results of this study will be highly informative for the use of KOR antagonism to attenuate stress-induced reinstatements of other drugs of abuse and alcohol.

Finally, naltrexone—a KOR and mu opioid receptor antagonist—has been successful in reducing relapse in some individuals with alcohol use disorders. An ongoing clinical trial aims to determine if naltrexone's effectiveness in reducing alcohol drinking behavior is correlated to its differential ability to bind KORs in alcohol drinking subjects using PET imaging (ClinicalTrials.gov ID: NCT01625611). Such findings could help prescribers take into consideration individual differences in KOR responsiveness that could give certain therapeutic avenues greater likelihood of success. KORs are crucially important in the regulation of mood and consummatory behaviors. It is likely that KOR augmentation during drug and alcohol abuse, dependency, and withdrawal contributes to anhedonic aspects of the "dark side" of addiction, and may leave individuals vulnerable to relapse, especially following a stressor.

While the field still has much work to do in fully understanding the role of KORs in disorders of excessive consumption, the therapeutic potential of KOR modulation—particularly KOR antagonism—is beginning to be more greatly appreciated. It is an exciting and burgeoning area of research, and KOR-based treatments may soon help to alleviate symptoms in individuals suffering from the negative affect associated with disorders of excessive consumption.

REFERENCES

Abercrombie, E. D., Keefe, K. A., DiFrischia, D. S., & Zigmond, M. J. (1989). Differential effect of stress on in vivo dopamine release in striatum, nucleus accumbens, and medial frontal cortex. *Journal of Neurochemistry*, *52*, 1655–1658.

Al-Hasani, R., McCall, J. G., Shin, G., Gomez, A. M., Schmitz, G. P., Bernardi, J. M., et al. (2015). Distinct subpopulations of nucleus accumbens dynorphin neurons drive aversion and reward. *Neuron*, *87*, 1063–1077.

Anderson, R. I., Lopez, M. F., & Becker, H. C. (2016). Stress-induced enhancement of ethanol intake in C57BL/6J mice with a history of chronic ethanol exposure: Involvement of kappa opioid receptors. *Frontiers in Cellular Neuroscience*, *10*, 45.

Arjune, D., & Bodnar, R. J. (1990). Suppression of nocturnal, palatable and glucoprivic intake in rats by the kappa opioid antagonist, nor-binaltorphamine. *Brain Research*, *534*, 313–316.

Baarendse, P. J., Limpens, J. H., & Vanderschuren, L. J. (2014). Disrupted social development enhances the motivation for cocaine in rats. *Psychopharmacology*, *231*, 1695–1704.

Baimel, C., Lau, B. K., Qiao, M., & Borgland, S. L. (2017). Projection-target-defined effects of orexin and dynorphin on VTA dopamine neurons. *Cell Reports*, *18*, 1346–1355.

Bals-Kubik, R., Ableitner, A., Herz, A., & Shippenberg, T. S. (1993). Neuroanatomical sites mediating the motivational effects of opioids as mapped by the conditioned place preference paradigm in rats. *The Journal of Pharmacology and Experimental Therapeutics*, *264*, 489–495.

Berger, A. L., Williams, A. M., McGinnis, M. M., & Walker, B. M. (2013). Affective cue-induced escalation of alcohol self-administration and increased 22-kHz ultrasonic vocalizations during alcohol withdrawal: Role of kappa-opioid receptors. *Neuropsychopharmacology*, *38*, 647–654.

Bodnar, R. J., & Klein, G. E. (2004). Endogenous opiates and behavior: 2003. *Peptides*, *25*, 2205–2256.

Bohn, L. M., & Aube, J. (2017). Seeking (and finding) biased ligands of the kappa opioid receptor. *ACS Medicinal Chemistry Letters*, *8*, 694–700.

Brady, K. T., & Sinha, R. (2005). Co-occurring mental and substance use disorders: The neurobiological effects of chronic stress. *The American Journal of Psychiatry*, *162*, 1483–1493.

Brodie, M. S., Pesold, C., & Appel, S. B. (1999). Ethanol directly excites dopaminergic ventral tegmental area reward neurons. *Alcoholism, Clinical and Experimental Research*, *23*, 1848–1852.

Bruchas, M. R., & Chavkin, C. (2010). Kinase cascades and ligand-directed signaling at the kappa opioid receptor. *Psychopharmacology*, *210*, 137–147.

Brust, T. F., Morgenweck, J., Kim, S. A., Rose, J. H., Locke, J. L., Schmid, C. L., et al. (2016). Biased agonists of the kappa opioid receptor suppress pain and itch without causing sedation or dysphoria. *Science Signaling*, *9*, ra117.

Burke, A. R., & Miczek, K. A. (2014). Stress in adolescence and drugs of abuse in rodent models: Role of dopamine, CRF, and HPA axis. *Psychopharmacology*, *231*, 1557–1580.

Butler, T. R., Karkhanis, A. N., Jones, S. R., & Weiner, J. L. (2016). Adolescent social isolation as a model of heightened vulnerability to comorbid alcoholism and anxiety disorders. *Alcoholism, Clinical and Experimental Research*, *40*, 1202–1214.

Cashman, J. R., & Azar, M. R. (2014). Potent inhibition of alcohol self-administration in alcohol-preferring rats by a kappa-opioid receptor antagonist. *The Journal of Pharmacology and Experimental Therapeutics*, *350*, 171–180.

Chartoff, E. H., Ebner, S. R., Sparrow, A., Potter, D., Baker, P. M., Ragozzino, M. E., et al. (2016). Relative timing between kappa opioid receptor activation and cocaine determines the impact on reward and dopamine release. *Neuropsychopharmacology*, *41*, 989–1002.

Chartoff, E. H., & Mavrikaki, M. (2015). Sex differences in kappa opioid receptor function and their potential impact on addiction. *Frontiers in Neuroscience*, *9*, 466.

Chartoff, E. H., Papadopoulou, M., MacDonald, M. L., Parsegian, A., Potter, D., Konradi, C., et al. (2009). Desipramine reduces stress-activated dynorphin expression and CREB phosphorylation in NAc tissue. *Molecular Pharmacology*, *75*, 704–712.

Chavkin, C. (2011). The therapeutic potential of kappa-opioids for treatment of pain and addiction. *Neuropsychopharmacology*, *36*, 369–370.

Chavkin, C. (2013). Dynorphin—Still an extraordinarily potent opioid peptide. *Molecular Pharmacology*, *83*, 729–736.

Chavkin, C., James, I. F., & Goldstein, A. (1982). Dynorphin is a specific endogenous ligand of the kappa opioid receptor. *Science*, *215*, 413–415.

Chua, S. C., Jr., Chung, W. K., Wu-Peng, X. S., Zhang, Y., Liu, S. M., Tartaglia, L., et al. (1996). Phenotypes of mouse diabetes and rat fatty due to mutations in the OB (leptin) receptor. *Science*, *271*, 994–996.

Cole, J. L., Leventhal, L., Pasternak, G. W., Bowen, W. D., & Bodnar, R. J. (1995). Reductions in body weight following chronic central opioid receptor subtype antagonists during development of dietary obesity in rats. *Brain Research*, *678*, 168–176.

Collins, S. L., D'Addario, C., & Izenwasser, S. (2001). Effects of kappa-opioid receptor agonists on long-term cocaine use and dopamine neurotransmission. *European Journal of Pharmacology*, *426*, 25–34.

Collins, S. L., Kunko, P. M., Ladenheim, B., Cadet, J. L., Carroll, F. I., & Izenwasser, S. (2002). Chronic cocaine increases kappa-opioid receptor density: Lack of effect by selective dopamine uptake inhibitors. *Synapse, 45*, 153–158.

Craft, R. M., Kruzich, P. J., Boyer, J. S., Harding, J. W., & Hanesworth, J. M. (1998). Sex differences in discriminative stimulus and diuretic effects of the kappa opioid agonist U69,593 in the rat. *Pharmacology, Biochemistry, and Behavior, 61*, 395–403.

Crowley, N. A., & Kash, T. L. (2015). Kappa opioid receptor signaling in the brain: Circuitry and implications for treatment. *Progress in Neuro-Psychopharmacology & Biological Psychiatry, 62*, 51–60.

Cusin, I., Rohner-Jeanrenaud, F., Stricker-Krongrad, A., & Jeanrenaud, B. (1996). The weight-reducing effect of an intracerebroventricular bolus injection of leptin in genetically obese fa/fa rats. Reduced sensitivity compared with lean animals. *Diabetes, 45*, 1446–1450.

Dogra, S., & Yadav, P. N. (2015). Biased agonism at kappa opioid receptors: Implication in pain and mood disorders. *European Journal of Pharmacology, 763*, 184–190.

Donzanti, B. A., Althaus, J. S., Payson, M. M., & Von Voigtlander, P. F. (1992). Kappa agonist-induced reduction in dopamine release: Site of action and tolerance. *Research Communications in Chemical Pathology and Pharmacology, 78*, 193–210.

Doyon, W. M., Howard, E. C., Shippenberg, T. S., & Gonzales, R. A. (2006). Kappa-opioid receptor modulation of accumbal dopamine concentration during operant ethanol self-administration. *Neuropharmacology, 51*, 487–496.

Dube, S. R., Anda, R. F., Felitti, V. J., Chapman, D. P., Williamson, D. F., & Giles, W. H. (2001). Childhood abuse, household dysfunction, and the risk of attempted suicide throughout the life span: Findings from the adverse childhood experiences study. *JAMA, 286*, 3089–3096.

Dube, S. R., Felitti, V. J., Dong, M., Chapman, D. P., Giles, W. H., & Anda, R. F. (2003). Childhood abuse, neglect, and household dysfunction and the risk of illicit drug use: The adverse childhood experiences study. *Pediatrics, 111*, 564–572.

Ehrich, J. M., Phillips, P. E., & Chavkin, C. (2014). Kappa opioid receptor activation potentiates the cocaine-induced increase in evoked dopamine release recorded in vivo in the mouse nucleus accumbens. *Neuropsychopharmacology, 39*, 3036–3048.

Elton, A., Smitherman, S., Young, J., & Kilts, C. D. (2015). Effects of childhood maltreatment on the neural correlates of stress- and drug cue-induced cocaine craving. *Addiction Biology, 20*, 820–831.

Ferguson-Segall, M., Flynn, J. J., Walker, J., & Margules, D. L. (1982). Increased immunoreactive dynorphin and leu-enkephalin in posterior pituitary of obese mice (ob/ob) and super-sensitivity to drugs that act at kappa receptors. *Life Sciences, 31*, 2233–2236.

Ford, C. P., Mark, G. P., & Williams, J. T. (2006). Properties and opioid inhibition of mesolimbic dopamine neurons vary according to target location. *The Journal of Neuroscience, 26*, 2788–2797.

Fowler, C. J., & Fraser, G. L. (1994). Mu-, delta-, kappa-opioid receptors and their subtypes. A critical review with emphasis on radioligand binding experiments. *Neurochemistry International, 24*, 401–426.

Freeman, K. B., Naylor, J. E., Prisinzano, T. E., & Woolverton, W. L. (2014). Assessment of the kappa opioid agonist, salvinorin A, as a punisher of drug self-administration in monkeys. *Psychopharmacology, 231*, 2751–2758.

Funk, D., Coen, K., & Le, A. D. (2014). The role of kappa opioid receptors in stress-induced reinstatement of alcohol seeking in rats. *Brain and Behavior, 4*, 356–367.

Graziane, N. M., Polter, A. M., Briand, L. A., Pierce, R. C., & Kauer, J. A. (2013). Kappa opioid receptors regulate stress-induced cocaine seeking and synaptic plasticity. *Neuron, 77*, 942–954.

Groblewski, P. A., Zietz, C., Willuhn, I., Phillips, P. E., & Chavkin, C. (2015). Repeated stress exposure causes strain-dependent shifts in the behavioral economics of cocaine in rats. *Addiction Biology*, *20*, 297–301.

Hadjimarkou, M. M., Singh, A., Kandov, Y., Israel, Y., Pan, Y. X., Rossi, G. C., et al. (2004). Opioid receptor involvement in food deprivation-induced feeding: Evaluation of selective antagonist and antisense oligodeoxynucleotide probe effects in mice and rats. *The Journal of Pharmacology and Experimental Therapeutics*, *311*, 1188–1202.

Henderson-Redmond, A., & Czachowski, C. (2014). Effects of systemic opioid receptor ligands on ethanol- and sucrose seeking and drinking in alcohol-preferring (P) and Long Evans rats. *Psychopharmacology*, *231*, 4309–4321.

Holleran, K. M., Wilson, H. H., Fetterly, T. L., Bluett, R. J., Centanni, S. W., Gilfarb, R. A., et al. (2016). Ketamine and MAG lipase inhibitor-dependent reversal of evolving depressive-like behavior during forced abstinence from alcohol drinking. *Neuropsychopharmacology*, *41*, 2062–2071.

Holter, S. M., Henniger, M. S., Lipkowski, A. W., & Spanagel, R. (2000). Kappa-opioid receptors and relapse-like drinking in long-term ethanol-experienced rats. *Psychopharmacology*, *153*, 93–102.

Ikeda, H., Ardianto, C., Yonemochi, N., Yang, L., Ohashi, T., Ikegami, M., et al. (2015). Inhibition of opioid systems in the hypothalamus as well as the mesolimbic area suppresses feeding behavior of mice. *Neuroscience*, *311*, 9–21.

Jacobsen, L. K., Southwick, S. M., & Kosten, T. R. (2001). Substance use disorders in patients with posttraumatic stress disorder: A review of the literature. *The American Journal of Psychiatry*, *158*, 1184–1190.

Jarosz, P. A. (2007). The effect of kappa opioid receptor antagonism on energy expenditure in the obese Zucker rat. *Biological Research for Nursing*, *8*, 294–299.

Jarosz, P. A., & Metzger, B. L. (2002). The effect of opioid antagonism on food intake behavior and body weight in a biobehavioral model of obese binge eating. *Biological Research for Nursing*, *3*, 198–209.

Jewett, D. C., Grace, M. K., Jones, R. M., Billington, C. J., Portoghese, P. S., & Levine, A. S. (2001). The kappa-opioid antagonist GNTI reduces U50,488-, DAMGO-, and deprivation-induced feeding, but not butorphanol- and neuropeptide Y-induced feeding in rats. *Brain Research*, *909*, 75–80.

Karkhanis, A. N., Huggins, K. N., Rose, J. H., & Jones, S. R. (2016). Switch from excitatory to inhibitory actions of ethanol on dopamine levels after chronic exposure: Role of kappa opioid receptors. *Neuropharmacology*, *110*, 190–197.

Karkhanis, A. N., Locke, J. L., McCool, B. A., Weiner, J. L., & Jones, S. R. (2014). Social isolation rearing increases nucleus accumbens dopamine and norepinephrine responses to acute ethanol in adulthood. *Alcoholism, Clinical and Experimental Research*, *38*, 2770–2779.

Karkhanis, A. N., Rose, J. H., Weiner, J. L., & Jones, S. R. (2016). Early-life social isolation stress increases kappa opioid receptor responsiveness and downregulates the dopamine system. *Neuropsychopharmacology*, *41*, 2263–2274.

Khawaja, X. Z., Bailey, C. J., & Green, I. C. (1989). Central mu, delta, and kappa opioid binding sites, and brain and pituitary beta-endorphin and met-enkephalin in genetically obese (ob/ob) and lean mice. *Life Sciences*, *44*, 1097–1105.

Kim, E. M., O'Hare, E., Grace, M. K., Welch, C. C., Billington, C. J., & Levine, A. S. (2000). ARC POMC mRNA and PVN alpha-MSH are lower in obese relative to lean zucker rats. *Brain Research*, *862*, 11–16.

Kirkham, T. C., & Blundell, J. E. (1987). Effects of naloxone and naltrexone on meal patterns of freely-feeding rats. *Pharmacology, Biochemistry, and Behavior*, *26*, 515–520.

Kivell, B. M., Ewald, A. W., & Prisinzano, T. E. (2014). Salvinorin A analogs and other kappa-opioid receptor compounds as treatments for cocaine abuse. *Advances in Pharmacology*, *69*, 481–511.

Kivell, B., Uzelac, Z., Sundaramurthy, S., Rajamanickam, J., Ewald, A., Chefer, V., et al. (2014). Salvinorin A regulates dopamine transporter function via a kappa opioid receptor and ERK1/2-dependent mechanism. *Neuropharmacology*, *86*, 228–240.
Koch, J. E., Glass, M. J., Cooper, M. L., & Bodnar, R. J. (1995). Alterations in deprivation, glucoprivic and sucrose intake following general, mu and kappa opioid antagonists in the hypothalamic paraventricular nucleus of rats. *Neuroscience*, *66*, 951–957.
Koob, G. F., & Le Moal, M. (2001). Drug addiction, dysregulation of reward, and allostasis. *Neuropsychopharmacology*, *24*, 97–129.
Koob, G. F., & Volkow, N. D. (2010). Neurocircuitry of addiction. *Neuropsychopharmacology*, *35*, 217–238.
Lam, M. P., Marinelli, P. W., Bai, L., & Gianoulakis, C. (2008). Effects of acute ethanol on opioid peptide release in the central amygdala: An in vivo microdialysis study. *Psychopharmacology*, *201*, 261–271.
Laman-Maharg, A. R., Copeland, T., Sanchez, E. O., Campi, K. L., & Trainor, B. C. (2017). The long-term effects of stress and kappa opioid receptor activation on conditioned place aversion in male and female California mice. *Behavioural Brain Research*, *332*, 299–307.
Lammel, S., Hetzel, A., Hackel, O., Jones, I., Liss, B., & Roeper, J. (2008). Unique properties of mesoprefrontal neurons within a dual mesocorticolimbic dopamine system. *Neuron*, *57*, 760–773.
Levine, A. S., & Billington, C. J. (2004). Opioids as agents of reward-related feeding: A consideration of the evidence. *Physiology & Behavior*, *82*, 57–61.
Levine, A. S., Grace, M., Billington, C. J., & Portoghese, P. S. (1990). Nor-binaltorphimine decreases deprivation and opioid-induced feeding. *Brain Research*, *534*, 60–64.
Lindholm, S., Ploj, K., Franck, J., & Nylander, I. (2000). Repeated ethanol administration induces short- and long-term changes in enkephalin and dynorphin tissue concentrations in rat brain. *Alcohol*, *22*, 165–171.
Lindholm, S., Werme, M., Brene, S., & Franck, J. (2001). The selective kappa-opioid receptor agonist U50,488H attenuates voluntary ethanol intake in the rat. *Behavioural Brain Research*, *120*, 137–146.
Logrip, M. L., Janak, P. H., & Ron, D. (2009). Blockade of ethanol reward by the kappa opioid receptor agonist U50,488H. *Alcohol*, *43*, 359–365.
Mantsch, J. R., Baker, D. A., Funk, D., Le, A. D., & Shaham, Y. (2016). Stress-induced reinstatement of drug seeking: 20 Years of progress. *Neuropsychopharmacology*, *41*, 335–356.
Margolis, E. B., Hjelmstad, G. O., Bonci, A., & Fields, H. L. (2003). Kappa-opioid agonists directly inhibit midbrain dopaminergic neurons. *The Journal of Neuroscience*, *23*, 9981–9986.
Margolis, E. B., Hjelmstad, G. O., Bonci, A., & Fields, H. L. (2005). Both kappa and mu opioid agonists inhibit glutamatergic input to ventral tegmental area neurons. *Journal of Neurophysiology*, *93*, 3086–3093.
Margolis, E. B., Lock, H., Chefer, V. I., Shippenberg, T. S., Hjelmstad, G. O., & Fields, H. L. (2006). Kappa opioids selectively control dopaminergic neurons projecting to the prefrontal cortex. *Proceedings of the National Academy of Sciences of the United States of America*, *103*, 2938–2942.
Margolis, E. B., Mitchell, J. M., Ishikawa, J., Hjelmstad, G. O., & Fields, H. L. (2008). Midbrain dopamine neurons: Projection target determines action potential duration and dopamine D(2) receptor inhibition. *The Journal of Neuroscience*, *28*, 8908–8913.
Marinelli, P. W., Lam, M., Bai, L., Quirion, R., & Gianoulakis, C. (2006). A microdialysis profile of dynorphin A(1-8) release in the rat nucleus accumbens following alcohol administration. *Alcoholism, Clinical and Experimental Research*, *30*, 982–990.
McLaughlin, J. P., Land, B. B., Li, S., Pintar, J. E., & Chavkin, C. (2006). Prior activation of kappa opioid receptors by U50,488 mimics repeated forced swim stress to potentiate cocaine place preference conditioning. *Neuropsychopharmacology*, *31*, 787–794.

Mitchell, J. M., Liang, M. T., & Fields, H. L. (2005). A single injection of the kappa opioid antagonist norbinaltorphimine increases ethanol consumption in rats. *Psychopharmacology, 182*, 384–392.

Morales, M., Anderson, R. I., Spear, L. P., & Varlinskaya, E. I. (2014). Effects of the kappa opioid receptor antagonist, nor-binaltorphimine, on ethanol intake: Impact of age and sex. *Developmental Psychobiology, 56*, 700–712.

Morikawa, H., & Morrisett, R. A. (2010). Ethanol action on dopaminergic neurons in the ventral tegmental area: Interaction with intrinsic ion channels and neurotransmitter inputs. *International Review of Neurobiology, 91*, 235–288.

Munro, T. A., Berry, L. M., Van't Veer, A., Beguin, C., Carroll, F. I., Zhao, Z., et al. (2012). Long-acting kappa opioid antagonists nor-BNI, GNTI and JDTic: Pharmacokinetics in mice and lipophilicity. *BMC Pharmacology, 12*, 5.

Negus, S. S. (2004). Effects of the kappa opioid agonist U50,488 and the kappa opioid antagonist nor-binaltorphimine on choice between cocaine and food in rhesus monkeys. *Psychopharmacology, 176*, 204–213.

Nestby, P., Schoffelmeer, A. N., Homberg, J. R., Wardeh, G., De Vries, T. J., Mulder, A. H., et al. (1999). Bremazocine reduces unrestricted free-choice ethanol self-administration in rats without affecting sucrose preference. *Psychopharmacology, 142*, 309–317.

Ookuma, K., Barton, C., York, D. A., & Bray, G. A. (1997). Effect of enterostatin and kappa-opioids on macronutrient selection and consumption. *Peptides, 18*, 785–791.

Ookuma, K., Barton, C., York, D. A., & Bray, G. A. (1998). Differential response to kappa-opioidergic agents in dietary fat selection between Osborne-Mendel and S5B/P1 rats. *Peptides, 19*, 141–147.

Perez, E. E., & De Biasi, M. (2015). Assessment of affective and somatic signs of ethanol withdrawal in C57BL/6J mice using a short-term ethanol treatment. *Alcohol, 49*, 237–243.

Phillips, M. S., Liu, Q., Hammond, H. A., Dugan, V., Hey, P. J., Caskey, C. J., et al. (1996). Leptin receptor missense mutation in the fatty Zucker rat. *Nature Genetics, 13*, 18–19.

Polter, A. M., Barcomb, K., Chen, R. W., Dingess, P. M., Graziane, N. M., Brown, T. E., et al. (2017). Constitutive activation of kappa opioid receptors at ventral tegmental area inhibitory synapses following acute stress. *eLife, 6*. pii: e23785.

Polter, A. M., Bishop, R. A., Briand, L. A., Graziane, N. M., Pierce, R. C., & Kauer, J. A. (2014). Poststress block of kappa opioid receptors rescues long-term potentiation of inhibitory synapses and prevents reinstatement of cocaine seeking. *Biological Psychiatry, 76*, 785–793.

Prisinzano, T. E., Tidgewell, K., & Harding, W. W. (2005). Kappa opioids as potential treatments for stimulant dependence. *The AAPS Journal, 7*, E592–599.

Przewlocka, B., Turchan, J., Lason, W., & Przewlocki, R. (1997). Ethanol withdrawal enhances the prodynorphin system activity in the rat nucleus accumbens. *Neuroscience Letters, 238*, 13–16.

Racz, I., Markert, A., Mauer, D., Stoffel-Wagner, B., & Zimmer, A. (2013). Long-term ethanol effects on acute stress responses: Modulation by dynorphin. *Addiction Biology, 18*, 678–688.

Radke, A. K., & Gewirtz, J. C. (2012). Increased dopamine receptor activity in the nucleus accumbens shell ameliorates anxiety during drug withdrawal. *Neuropsychopharmacology, 37*, 2405–2415.

Roane, D. S., Iadarola, M. J., & Porter, J. R. (1988). Decreased [3H]-naloxone binding and elevated dynorphin-A(1-8) content in Zucker rat brain. *Physiology & Behavior, 43*, 371–374.

Robles, C. F., McMackin, M. Z., Campi, K. L., Doig, I. E., Takahashi, E. Y., Pride, M. C., et al. (2014). Effects of kappa opioid receptors on conditioned place aversion and social interaction in males and females. *Behavioural Brain Research, 262*, 84–93.

Rose, J. H., Karkhanis, A. N., Chen, R., Gioia, D., Lopez, M. F., Becker, H. C., et al. (2016). Supersensitive kappa opioid receptors promotes ethanol withdrawal-related behaviors and reduce dopamine signaling in the nucleus accumbens. *International Journal of Neuropsychopharmacology*, *19*. pii: pyv127.

Rovaris, D. L., Mota, N. R., Bertuzzi, G. P., Aroche, A. P., Callegari-Jacques, S. M., Guimaraes, L. S., et al. (2015). Corticosteroid receptor genes and childhood neglect influence susceptibility to crack/cocaine addiction and response to detoxification treatment. *Journal of Psychiatric Research*, *68*, 83–90.

Russell, S. E., Rachlin, A. B., Smith, K. L., Muschamp, J., Berry, L., Zhao, Z., et al. (2014). Sex differences in sensitivity to the depressive-like effects of the kappa opioid receptor agonist U-50488 in rats. *Biological Psychiatry*, *76*, 213–222.

Saddoris, M. P., Sugam, J. A., Cacciapaglia, F., & Carelli, R. M. (2013). Rapid dopamine dynamics in the accumbens core and shell: Learning and action. *Frontiers in Bioscience (Elite Edition)*, *5*, 273–288.

Saddoris, M. P., Sugam, J. A., & Carelli, R. M. (2017). Prior cocaine experience impairs normal phasic dopamine signals of reward value in accumbens shell. *Neuropsychopharmacology*, *42*, 766–773.

Sahr, A. E., Sindelar, D. K., Alexander-Chacko, J. T., Eastwood, B. J., Mitch, C. H., & Statnick, M. A. (2008). Activation of mesolimbic dopamine neurons during novel and daily limited access to palatable food is blocked by the opioid antagonist LY255582. *American Journal of Physiology Regulatory, Integrative and Comparative Physiology*, *295*, R463–R471.

Schindler, A. G., Messinger, D. I., Smith, J. S., Shankar, H., Gustin, R. M., Schattauer, S. S., et al. (2012). Stress produces aversion and potentiates cocaine reward by releasing endogenous dynorphins in the ventral striatum to locally stimulate serotonin reuptake. *The Journal of Neuroscience*, *32*, 17582–17596.

Selye, H. (1998). A syndrome produced by diverse nocuous agents. 1936. *The Journal of Neuropsychiatry and Clinical Neurosciences*, *10*, 230–231.

Sershen, H., Hashim, A., & Lajtha, A. (1998). Gender differences in kappa-opioid modulation of cocaine-induced behavior and NMDA-evoked dopamine release. *Brain Research*, *801*, 67–71.

Shippenberg, T. S., Chefer, V. I., Zapata, A., & Heidbreder, C. A. (2001). Modulation of the behavioral and neurochemical effects of psychostimulants by kappa-opioid receptor systems. *Annals of the New York Academy of Sciences*, *937*, 50–73.

Shippenberg, T. S., LeFevour, A., & Heidbreder, C. (1996). Kappa-opioid receptor agonists prevent sensitization to the conditioned rewarding effects of cocaine. *The Journal of Pharmacology and Experimental Therapeutics*, *276*, 545–554.

Shirayama, Y., Ishida, H., Iwata, M., Hazama, G. I., Kawahara, R., & Duman, R. S. (2004). Stress increases dynorphin immunoreactivity in limbic brain regions and dynorphin antagonism produces antidepressant-like effects. *Journal of Neurochemistry*, *90*, 1258–1268.

Siciliano, C. A., Calipari, E. S., Cuzon Carlson, V. C., Helms, C. M., Lovinger, D. M., Grant, K. A., et al. (2015). Voluntary ethanol intake predicts kappa-opioid receptor supersensitivity and regionally distinct dopaminergic adaptations in macaques. *The Journal of Neuroscience*, *35*, 5959–5968.

Simonson, B., Morani, A. S., Ewald, A. W., Walker, L., Kumar, N., Simpson, D., et al. (2015). Pharmacology and anti-addiction effects of the novel kappa opioid receptor agonist Mesyl Sal B, a potent and long-acting analogue of salvinorin A. *British Journal of Pharmacology*, *172*, 515–531.

Sinha, R., Fox, H. C., Hong, K. A., Bergquist, K., Bhagwagar, Z., & Siedlarz, K. M. (2009). Enhanced negative emotion and alcohol craving, and altered physiological responses following stress and cue exposure in alcohol dependent individuals. *Neuropsychopharmacology*, *34*, 1198–1208.

Sinha, R., Fuse, T., Aubin, L. R., & O'Malley, S. S. (2000). Psychological stress, drug-related cues and cocaine craving. *Psychopharmacology, 152*, 140–148.

Sirohi, S., Bakalkin, G., & Walker, B. M. (2012). Alcohol-induced plasticity in the dynorphin/kappa-opioid receptor system. *Frontiers in Molecular Neuroscience, 5*, 95.

Spanagel, R., Herz, A., & Shippenberg, T. S. (1992). Opposing tonically active endogenous opioid systems modulate the mesolimbic dopaminergic pathway. *Proceedings of the National Academy of Sciences of the United States of America, 89*, 2046–2050.

Sperling, R. E., Gomes, S. M., Sypek, E. I., Carey, A. N., & McLaughlin, J. P. (2010). Endogenous kappa-opioid mediation of stress-induced potentiation of ethanol-conditioned place preference and self-administration. *Psychopharmacology, 210*, 199–209.

Sun, W., Xue, Y., Huang, Z., & Steketee, J. D. (2010). Regulation of cocaine-reinstated drug-seeking behavior by kappa-opioid receptors in the ventral tegmental area of rats. *Psychopharmacology, 210*, 179–188.

Svingos, A. L., Chavkin, C., Colago, E. E., & Pickel, V. M. (2001). Major coexpression of kappa-opioid receptors and the dopamine transporter in nucleus accumbens axonal profiles. *Synapse, 42*, 185–192.

Tejeda, H. A., Wu, J., Kornspun, A. R., Pignatelli, M., Kashtelyan, V., Krashes, M. J., et al. (2017). Pathway- and cell-specific kappa-opioid receptor modulation of excitation-inhibition balance differentially gates D1 and D2 accumbens neuron activity. *Neuron, 93*, 147–163.

Theile, J. W., Morikawa, H., Gonzales, R. A., & Morrisett, R. A. (2011). GABAergic transmission modulates ethanol excitation of ventral tegmental area dopamine neurons. *Neuroscience, 172*, 94–103.

Thompson, A. C., Zapata, A., Justice, J. B., Jr., Vaughan, R. A., Sharpe, L. G., & Shippenberg, T. S. (2000). Kappa-opioid receptor activation modifies dopamine uptake in the nucleus accumbens and opposes the effects of cocaine. *The Journal of Neuroscience, 20*, 9333–9340.

Threlfell, S., & Cragg, S. J. (2011). Dopamine signaling in dorsal versus ventral striatum: The dynamic role of cholinergic interneurons. *Frontiers in Systems Neuroscience, 5*, 11.

Tsujii, S., Nakai, Y., Fukata, J., Nakaishi, S., Takahashi, H., Usui, T., et al. (1987). Effects of food deprivation and high fat diet on immunoreactive dynorphin A(1-8) levels in brain regions of Zucker rats. *Peptides, 8*, 1075–1078.

Tsujii, S., Nakai, Y., Koh, T., Takahashi, H., Usui, T., Ikeda, H., et al. (1986). Effect of food deprivation on opioid receptor binding in the brain of lean and fatty Zucker rats. *Brain Research, 399*, 200–203.

Umathe, S., Bhutada, P., Dixit, P., & Shende, V. (2008). Increased marble-burying behavior in ethanol-withdrawal state: Modulation by gonadotropin-releasing hormone agonist. *European Journal of Pharmacology, 587*, 175–180.

Unterwald, E. M., Rubenfeld, J. M., & Kreek, M. J. (1994). Repeated cocaine administration upregulates kappa and mu, but not delta, opioid receptors. *Neuroreport, 5*, 1613–1616.

Valdez, G. R., & Harshberger, E. (2012). Kappa opioid regulation of anxiety-like behavior during acute ethanol withdrawal. *Pharmacology, Biochemistry, and Behavior, 102*, 44–47.

Valdez, G. R., Platt, D. M., Rowlett, J. K., Ruedi-Bettschen, D., & Spealman, R. D. (2007). Kappa agonist-induced reinstatement of cocaine seeking in squirrel monkeys: A role for opioid and stress-related mechanisms. *The Journal of Pharmacology and Experimental Therapeutics, 323*, 525–533.

Van't Veer, A., & Carlezon, W. A., Jr. (2013). Role of kappa-opioid receptors in stress and anxiety-related behavior. *Psychopharmacology, 229*, 435–452.

Vendruscolo, L. F., Barbier, E., Schlosburg, J. E., Misra, K. K., Whitfield, T. W., Jr., Logrip, M. L., et al. (2012). Corticosteroid-dependent plasticity mediates compulsive alcohol drinking in rats. *The Journal of Neuroscience, 32*, 7563–7571.

Vijay, A., Wang, S., Worhunsky, P., Zheng, M. Q., Nabulsi, N., Ropchan, J., et al. (2016). PET imaging reveals sex differences in kappa opioid receptor availability in humans, in vivo. *American Journal of Nuclear Medicine and Molecular Imaging*, *6*, 205–214.

Walker, B. M., & Koob, G. F. (2008). Pharmacological evidence for a motivational role of kappa-opioid systems in ethanol dependence. *Neuropsychopharmacology*, *33*, 643–652.

Wee, S., & Koob, G. F. (2010). The role of the dynorphin-kappa opioid system in the reinforcing effects of drugs of abuse. *Psychopharmacology*, *210*, 121–135.

Weiss, R. D., Griffin, M. L., & Mirin, S. M. (1992). Drug abuse as self-medication for depression: An empirical study. *The American Journal of Drug and Alcohol Abuse*, *18*, 121–129.

Werling, L. L., Frattali, A., Portoghese, P. S., Takemori, A. E., & Cox, B. M. (1988). Kappa receptor regulation of dopamine release from striatum and cortex of rats and guinea pigs. *The Journal of Pharmacology and Experimental Therapeutics*, *246*, 282–286.

Wolinsky, T. D., Carr, K. D., Hiller, J. M., & Simon, E. J. (1994). Effects of chronic food restriction on mu and kappa opioid binding in rat forebrain: A quantitative autoradiographic study. *Brain Research*, *656*, 274–280.

Xie, Q., Sanz-Burgos, A. P., Hannon, G. J., & Gutierrez, C. (1996). Plant cells contain a novel member of the retinoblastoma family of growth regulatory proteins. *The EMBO Journal*, *15*, 4900–4908.

Xu, K., Seo, D., Hodgkinson, C., Hu, Y., Goldman, D., & Sinha, R. (2013). A variant on the kappa opioid receptor gene (OPRK1) is associated with stress response and related drug craving, limbic brain activation and cocaine relapse risk. *Translational Psychiatry*, *3*. e292.

Yajie, D., Lin, K., Baoming, L., & Lan, M. (2005). Enhanced cocaine self-administration in adult rats with adolescent isolation experience. *Pharmacology, Biochemistry, and Behavior*, *82*, 673–677.

Yim, H. J., & Gonzales, R. A. (2000). Ethanol-induced increases in dopamine extracellular concentration in rat nucleus accumbens are accounted for by increased release and not uptake inhibition. *Alcohol*, *22*, 107–115.

Yim, H. J., Schallert, T., Randall, P. K., & Gonzales, R. A. (1998). Comparison of local and systemic ethanol effects on extracellular dopamine concentration in rat nucleus accumbens by microdialysis. *Alcoholism, Clinical and Experimental Research*, *22*, 367–374.

Zhou, Y., Colombo, G., Gessa, G. L., & Kreek, M. J. (2013). Effects of voluntary alcohol drinking on corticotropin-releasing factor and preprodynorphin mRNA levels in the central amygdala of Sardinian alcohol-preferring rats. *Neuroscience Letters*, *554*, 110–114.

Zhou, Y., Mabrouk, O. S., & Kennedy, R. T. (2013). Rapid preconcentration for liquid chromatography-mass spectrometry assay of trace level neuropeptides. *Journal of the American Society for Mass Spectrometry*, *24*, 1700–1709.

Zhou, Y., Rubinstein, M., Low, M. J., & Kreek, M. J. (2017). Hypothalamic-specific proopiomelanocortin deficiency reduces alcohol drinking in male and female mice. *Genes, Brain, and Behavior*, *16*, 449–461.

Zhou, L., Stahl, E. L., Lovell, K. M., Frankowski, K. J., Prisinzano, T. E., Aube, J., et al. (2015). Characterization of kappa opioid receptor mediated, dynorphin-stimulated [35S]GTPgammaS binding in mouse striatum for the evaluation of selective KOR ligands in an endogenous setting. *Neuropharmacology*, *99*, 131–141.

FURTHER READING

Calipari, E. S., Ferris, M. J., & Jones, S. R. (2014). Extended access of cocaine self-administration results in tolerance to the dopamine-elevating and locomotor-stimulating effects of cocaine. *Journal of Neurochemistry*, *128*, 224–232.

Ferris, M. J., Calipari, E. S., Mateo, Y., Melchior, J. R., Roberts, D. C., & Jones, S. R. (2012). Cocaine self-administration produces pharmacodynamic tolerance: Differential effects on the potency of dopamine transporter blockers, releasers, and methylphenidate. *Neuropsychopharmacology*, *37*, 1708–1716.

CHAPTER FOUR

The Role of the Ghrelin System in Drug Addiction

Lia J. Zallar*[,†,2], Mehdi Farokhnia*[,2], Brendan J. Tunstall[†], Leandro F. Vendruscolo[†], Lorenzo Leggio*[,‡,1]

*Section on Clinical Psychoneuroendocrinology and Neuropsychopharmacology, National Institute on Alcohol Abuse and Alcoholism and National Institute on Drug Abuse, National Institutes of Health, Bethesda, MD, United States

[†]Neurobiology of Addiction Section, Intramural Research Program, National Institute on Drug Abuse, National Institutes of Health, Baltimore, MD, United States

[‡]Center for Alcohol and Addiction Studies, Brown University, Providence, RI, United States

[1]Corresponding author: e-mail address: lorenzo.leggio@nih.gov

Contents

Abstract

In the past years, a significant volume of research has implicated the appetitive hormone ghrelin in the mechanisms underlying drug use and addiction. From a neuroscientific standpoint, ghrelin modulates both reward and stress pathways, two key drivers of substance use behaviors. Previous investigations support a connection between the

[2] The first two authors contributed equally to this work.

International Review of Neurobiology, Volume 136
ISSN 0074-7742
http://dx.doi.org/10.1016/bs.irn.2017.08.002

ghrelin system and alcohol, stimulants, and tobacco use in both animals and humans, while the research on opioids and cannabis is scarce. In general, upregulation of the ghrelin system seems to enhance craving for drugs as well as substances use. On the other hand, acute and chronic exposure to drugs of abuse influences the ghrelin system at different levels. This chapter summarizes the literature on the relationship between the ghrelin system and substance-related behaviors. We also review recent work investigating the ghrelin system as a potential pharmacological target for treating substance use disorders and discuss the need for additional research.

1. INTRODUCTION

Substance use disorder, commonly known as drug addiction, is a public health challenge which imposes enormous burden upon economic, psychosocial, and health care systems (Grant et al., 2016). From a medical standpoint, addiction is a chronic relapsing brain disease with a complex and multifaceted etiology and natural history. Determining the main biological mechanisms involved in the pathophysiology of drug addiction can substantially facilitate the efforts to develop better preventative and therapeutic interventions (Koob, 2005; Wise, 2000). One promising approach to scientifically decompose and better understand the nature of this disease is to study the biological overlap between substance use behavior and other—sometimes better understood—behaviors. For example, the neurobiology of food-related appetitive and consummatory behaviors shares several fundamental features with the neurobiology of drug craving and use (Avena, Bocarsly, Hoebel, & Gold, 2011; Volkow, Wang, Fowler, Tomasi, & Baler, 2012). This overlap provides a valid platform rooted in decades of research to study the biobehavioral correlates of drug addiction.

Ghrelin, commonly referred to as the "hunger hormone," is a 28-amino acid peptide known primarily for its role in increasing appetite and food intake. Additionally, ghrelin has been implicated in a variety of physiological and behavioral processes such as energy homeostasis, glucose metabolism, gastrointestinal motility, learning/memory, sexual behavior, stress, and reward processing (Bali & Jaggi, 2016; Castaneda, Tong, Datta, Culler, & Tschop, 2010; Dickson et al., 2011; Edwards & Abizaid, 2016; Egecioglu, Prieto-Garcia, Studer, Westberg, & Jerlhag, 2016; Howick, Griffin, Cryan, & Schellekens, 2017; Hsu, Suarez, & Kanoski, 2016; Inui et al., 2004; Kojima & Kangawa, 2005; Kunath & Dresler, 2014; Leggio et al., 2011; Meyer, Burgos-Robles, Liu, Correia, & Goosens, 2014;

Panagopoulos & Ralevski, 2014; Perboni & Inui, 2010; Revitsky & Klein, 2013; Schellekens, Dinan, & Cryan, 2013; Spencer, Emmerzaal, Kozicz, & Andrews, 2015; Stievenard et al., 2017; Tack et al., 2006). Ghrelin is synthesized primarily by endocrine cells in the stomach and, in smaller amounts, by other organs such as the pancreas, salivary glands, and placenta (Kojima et al., 1999). There are reports suggesting that ghrelin may also be produced in the brain, but this topic remains controversial (Cabral, Lopez Soto, Epelbaum, & Perello, 2017; Ferrini, Salio, Lossi, & Merighi, 2009). Nonetheless, it is well known that peripherally produced ghrelin communicates with the central nervous system and affects the brain directly by crossing the blood–brain barrier and indirectly by stimulating the vagus nerve (Banks, Tschop, Robinson, & Heiman, 2002; Cabral et al., 2017; Dockray, 2013; Edwards & Abizaid, 2017; Howick et al., 2017; Tamboli et al., 2017).

The *GHRL* gene on chromosome 3 encodes the preproghrelin peptide, which is cleaved off to form the ghrelin peptide (Ariyasu et al., 2001; Gualillo, Lago, Casanueva, & Dieguez, 2006; Kojima et al., 1999). Ghrelin is then modified by the ghrelin-*O*-acyltransferase (GOAT) enzyme to form acyl-ghrelin (Al Massadi, Tschop, & Tong, 2011; Gahete et al., 2014; Gutierrez et al., 2008; Yang, Brown, Liang, Grishin, & Goldstein, 2008). Acyl-ghrelin has traditionally been referred to as "active ghrelin," and most research efforts have been focused on this form. On the other hand, the unacylated form of ghrelin is often referred to as "inactive ghrelin," given its low affinity for the ghrelin receptor (for more details about the ghrelin receptor, see later). However, the designations "active" vs "inactive" ghrelin may be misleading considering recent works that suggest a number of physiological functions for unacylated ghrelin, such as anorectic effects, adipogenesis, and insulin regulation (Delhanty, Neggers, & van der Lely, 2013, 2014; Hopkins et al., 2017).

The physiological functions of acyl-ghrelin are executed through binding to the growth hormone secretagogue receptor 1a (GHS-R1a), which is encoded by the *GHSR* gene on chromosome 3 (Howard et al., 1996). GHS-R1a is expressed widely throughout both the peripheral organs and the brain, and is notable for its high level of constitutive activity, i.e., intrinsic activity of the receptor independent of ligand binding (Guan et al., 1997; Mason, Wang, & Zigman, 2014; Mear, Enjalbert, & Thirion, 2013; Zigman, Jones, Lee, Saper, & Elmquist, 2006). Another variant of the ghrelin receptor, named GHS-R1b, forms heterodimers with GHS-R1a and modulates the receptor's activity and cell trafficking, although the functional role of this so-called inactive variant has been less explored (Albarran-Zeckler & Smith, 2013; Laviano, Molfino, Rianda, & Rossi Fanelli, 2012; Muccioli, Baragli, Granata, Papotti, & Ghigo, 2007).

While implicated in numerous physiological processes, the orexigenic effect of acyl-ghrelin and its role in metabolism and food consumption are the most extensively documented (Cummings, Frayo, Marmonier, Aubert, & Chapelot, 2004; Currie, Mirza, Fuld, Park, & Vasselli, 2005; Melis et al., 2002; Schele, Bake, Rabasa, & Dickson, 2016; Tschop, Smiley, & Heiman, 2000; Wren et al., 2001, 2000). In addition to the homeostatic functions, ghrelin has been shown to modulate the hedonic mechanisms of feeding behavior as well (Davis, Perello, et al., 2012; Naleid, Grace, Cummings, & Levine, 2005; Olszewski, Schioth, & Levine, 2008; Perello & Zigman, 2012; Saper, Chou, & Elmquist, 2002; Schellekens et al., 2013; Wei et al., 2015). Acyl-ghrelin is also being increasingly recognized for its key role in processing general reward from reinforcers other than food (Mason et al., 2014; Revitsky & Klein, 2013; Stievenard et al., 2017; Vengeliene, 2013). For example, ghrelin administration modulates dopamine transmission in the brain and stimulates behavioral measures of reward processing (e.g., conditioned place preference). GHS-R1a is expressed throughout the mesolimbic reward pathway and forms heterodimers with dopamine D1 and D2 receptors (Abizaid et al., 2006; Guan et al., 1997; Jerlhag, 2008; Jerlhag et al., 2006, 2007; Jerlhag, Janson, Waters, & Engel, 2012; Jiang, Betancourt, & Smith, 2006; Quarta et al., 2009; Schellekens et al., 2013). These findings suggest that ghrelin may play a central role in reward processing. Moreover, acyl-ghrelin interacts with the hypothalamic–pituitary–adrenal (HPA) axis and has been shown to modulate stress-related behaviors (Bali & Jaggi, 2016; Currie et al., 2012; Labarthe et al., 2014; Meyer et al., 2014; Schellekens et al., 2013; Spencer et al., 2015, 2012). In rodent models, centrally administered ghrelin increases anxiety, and ghrelin receptor knockout mice have decreased corticosterone release in response to stressors (Chuang et al., 2011; Currie et al., 2012). Collectively, ghrelin's involvement in both reward and stress pathways suggests a potential role for the ghrelin signaling in addiction, as drug-seeking behavior is linked to positive reinforcement (drug reward) and negative reinforcement (stress relief) mechanisms (Koob & Volkow, 2016; Panagopoulos & Ralevski, 2014).

In this chapter, we will review preclinical and clinical studies on the interaction between the ghrelin system and drugs of abuse. We will summarize our current knowledge and will highlight the remaining gaps to be addressed in future research.

2. GHRELIN AND ALCOHOL

Ghrelin's role in substance use has been most extensively explored with alcohol. Alcohol is a unique drug as it is not only addictive, but also has palatable properties and is a direct source of calorie. Therefore, among all substances, the overlap between food- and drug-seeking is most conceivable with alcohol, from both behavioral and biological standpoints.

2.1 Preclinical Evidence

A significant number of preclinical studies have investigated the relationship between ghrelin signaling and alcohol intake using a variety of methods and procedures. Measurements of endogenous ghrelin in rodents show that exposure to alcohol modulates blood ghrelin concentrations. In two independent set of experiments, prolonged alcohol intake resulted in decreased blood ghrelin levels; this reduction was smaller in high alcohol preferring rats compared with low alcohol preferring rats (Landgren, Engel, et al., 2011; Szulc et al., 2013). Szulc and colleagues also examined the effects of acute alcohol exposure and found that a single dose of experimenter-administered alcohol increased blood ghrelin concentrations in high alcohol preferring rats, but an opposite effect was observed in low alcohol preferring rats (Szulc et al., 2013). These studies indicate a close relationship between circulating ghrelin concentrations and alcohol exposure, and also suggest that alcohol consumption may differentially change ghrelin concentrations depending on the amount of consumption, length of exposure, and alcohol preference.

Exogenous ghrelin administration has been used to further examine the role of ghrelin in regulating alcohol consumption. A number of studies with different experimental procedures show that both systemic and central administration of ghrelin increase alcohol intake in rodent models (Cepko et al., 2014; Davis, Schurdak, et al., 2012; Jerlhag et al., 2009; Jerlhag, Landgren, Egecioglu, Dickson, & Engel, 2011). Two studies, however, found no effect of exogenous ghrelin on alcohol intake (Lyons, Lowery, Sparta, & Thiele, 2008; Schneider, Rada, Darby, Leibowitz, & Hoebel, 2007).

Given the fact that ghrelin is a peripherally produced peptide with central actions, it is important to investigate the potential brain regions involved in ghrelin's alcohol-related effects. Central administration of ghrelin into the ventral tegmental area, laterodorsal tegmental nucleus, and cerebral ventricles potentiated alcohol consumption in two rodent studies (Fig. 1)

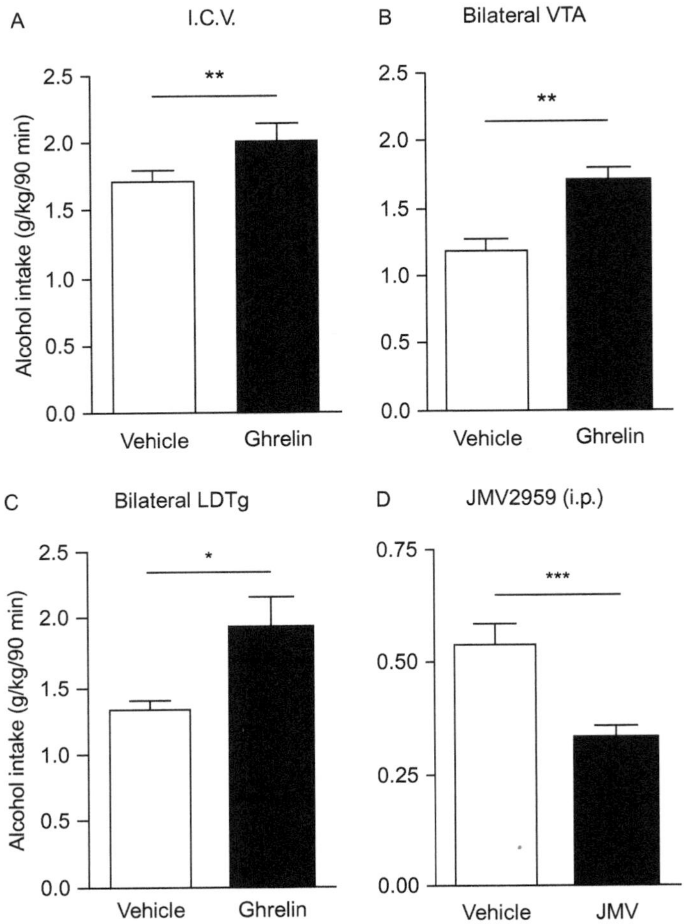

Fig. 1 Compared to vehicle, ghrelin administration into the third ventricle (A), the ventral tegmental area (B), and the laterodorsal tegmental nucleus (C) significantly increased alcohol intake in mice in a 2-bottle (alcohol/water) free choice limited access paradigm. Using the same paradigm, systemic injection of a GHS-R1a antagonist (JMV2959), compared to vehicle, significantly decreased alcohol intake in mice (D). $*P < 0.05$, $**P < 0.01$, $***P < 0.001$. *ICV*, intracerebroventricular; *LDTg*, laterodorsal tegmental nucleus; *VTA*, ventral tegmental area. *Copied with permission from Jerlhag, E., Egecioglu, E., Landgren, S., Salomé, N., Heilig, M, Moechars, D., et al. (2009). Requirement of central ghrelin signaling for alcohol reward.* PNAS, 106*(27), 11318–11323.*

(Cepko et al., 2014; Jerlhag et al., 2009). Landgren and colleagues reported that after prolonged exposure to alcohol, ghrelin receptor gene expression was increased in high preferring vs low preferring rats in the prefrontal cortex, nucleus accumbens, hippocampus, amygdala, and ventral tegmental area; the alcohol access paradigm (continuous vs limited) also contributed to the ghrelin receptor gene expression pattern in this study (Landgren, Engel, et al., 2011). In addition to reward-related brain regions, different studies suggest that the amygdala is one of ghrelin's key targets (Alvarez-Crespo et al., 2012; Song et al., 2013) through which ghrelin modulates alcohol signaling, mainly via GABAergic (Cruz, Herman, Cote, Ryabinin, & Roberto, 2013) and serotonergic (Hansson et al., 2014; Yoshimoto et al., 2017) neurotransmission. It is not well established how peripheral ghrelin may affect central signaling related to alcohol use. In a rodent study, blocking the action of ghrelin in the periphery by administering an antighrelin L-oligonucleotide (Spiegelmer NOX-B11-2) did not affect alcohol intake, alcohol-induced conditioned place preference, locomotor activity, or accumbal dopamine release (Jerlhag, Ivanoff, Vater, & Engel, 2014), suggesting that central, but not peripheral, ghrelin signaling contributes to alcohol-seeking behavior. However, additional studies are required to obtain a definite answer in this regard.

Pharmacological or genetic suppression of the ghrelin system and studying the subsequent biobehavioral changes in animal models provide a valuable framework to understand how the ghrelin system interacts with alcohol-related behaviors. To date, three GHS-R1a antagonist compounds have been tested in preclinical studies: JMV2959 (Bahi et al., 2013; Davis, Schurdak, et al., 2012; Gomez & Ryabinin, 2014; Jerlhag et al., 2009; Landgren et al., 2012; Stevenson et al., 2015, 2016; Suchankova, Engel, & Jerlhag, 2016; Suchankova, Steensland, Fredriksson, Engel, & Jerlhag, 2013), D-Lys3-GHRP-6 (Gomez et al., 2015; Gomez & Ryabinin, 2014; Hajnal et al., 2012; Kaur & Ryabinin, 2010), and BIM28163. Experiments in prairie voles, mice, and rats indicate that systemic administration of these GHS-R1a antagonists reduces voluntary alcohol intake and operant alcohol self-administration. Ghrelin antagonism was also found to suppress alcohol-induced conditioned place preference, hyperlocomotion, and dopamine release in the nucleus accumbens (Fig. 1) (Bahi et al., 2013; Davis, Schurdak, et al., 2012; Gomez et al., 2015; Gomez & Ryabinin, 2014; Hajnal et al., 2012; Jerlhag et al., 2009; Kaur & Ryabinin, 2010; Landgren et al., 2012; Stevenson et al., 2015, 2016; Suchankova, Engel, et al., 2016; Suchankova, Steensland, et al., 2013). Converging results from these studies suggest a key role for the ghrelin system in modulating the rewarding effects of alcohol.

Inhibitory genetic manipulation of the ghrelin system in knockout models has yielded results similar to GHS-R1a pharmacological blockade.

Compared to wild-type mice, ghrelin peptide gene (*GHRL*) knockout mice exhibited lower alcohol intake and alcohol-induced conditioned place preference in one study (Bahi et al., 2013), and reduced alcohol-induced hyperlocomotion and accumbal dopamine release in another study (Jerlhag et al., 2011). However, in the latter experiment, neither spontaneous nor ghrelin-induced alcohol intake was affected in the knockout model (Jerlhag et al., 2011). As part of a comprehensive set of experiments, Jerlhag and colleagues found that the ghrelin receptor gene (*GHSR*) knockout mice, compared to the wild-type, demonstrated lower locomotor activity and accumbal dopamine release in response to intraperitoneal injection of ethanol (Jerlhag et al., 2009).

Taken together, the aforementioned preclinical evidence suggests a mutual interaction between the ghrelin system and alcohol-related pathways. On one hand, alcohol modulates the ghrelin system and on the other hand, ghrelin modulates alcohol consummatory behaviors. However, the ghrelin-alcohol interplay is clearly complex, and it is not surprising that experiments from different laboratories under different experimental conditions yield somewhat varying results. Proceeding from bench to bedside, we next summarize the clinical studies looking at the cross-talk between the ghrelin system and alcohol-related behaviors in humans.

2.2 Clinical Evidence

Clinical investigations have extensively studied the relationship between endogenous ghrelin concentrations and alcohol consumption in both alcohol-dependent and nondependent populations. Oral alcohol administration in healthy individuals results in an acute decrease in blood ghrelin levels, as observed in several studies (Calissendorff, Danielsson, Brismar, & Rojdmark, 2005, 2006; Calissendorff, Gustafsson, Holst, Brismar, & Rojdmark, 2012; Zimmermann, Buchmann, Steffin, Dieterle, & Uhr, 2007). Similarly, two studies have shown that intravenous alcohol administration suppresses the ghrelin system in nondependent social drinkers (Leggio, Schwandt, Oot, Dias, & Ramchandani, 2013; Ralevski et al., 2017). One of these studies suggests that fasting ghrelin levels may predict the intensity of subjective responses to alcohol (Ralevski et al., 2017).

Several observational studies have compared blood ghrelin concentrations between alcohol-dependent patients and healthy controls, and correlated ghrelin measurements with alcohol-related parameters such as craving, withdrawal, and abstinence. A number of studies have detected higher concentrations of total and/or acyl-ghrelin in alcohol-dependent vs healthy individuals (Kim et al., 2005; Kraus et al., 2005; Wurst et al., 2007), while others report an opposite relationship, i.e., decreased ghrelin levels in alcohol-dependent individuals compared to controls (Addolorato et al., 2006; Badaoui et al., 2008; de Timary et al., 2012).

There is agreement in several reports that alcohol abstinence compared to active drinking leads to increased blood ghrelin concentrations in the short-term (Kim et al., 2013; Kraus et al., 2005; Leggio et al., 2012). However, the literature is not conclusive regarding protracted abstinence: some studies report that blood ghrelin increases (Kim et al., 2013; Koopmann et al., 2012; Leggio et al., 2012), whereas others report it decreases over time during abstinence (Akkisi Kumsar & Dilbaz, 2015; Wurst et al., 2007). Various factors may contribute to these inconsistent findings, such as differences in the study designs, participants' characteristics, timing and methodology of sample collection, etc. Future studies should work to parse out the dynamics of the relationship between alcohol use and the ghrelin system, with a particular focus on changes over time during active dependence and abstinence.

In addition to alcohol consumption, it is crucial to understand how the ghrelin system may relate to other biobehavioral elements of alcohol-seeking behavior. Craving, or intense desire, is a key correlate of excessive alcohol use, which has been shown to predict the likelihood of relapse (Monti, Rohsenow, & Hutchison, 2000). In fact, craving was added to the Diagnostic and Statistical Manual of Mental Disorders 5 (DSM-5) as a diagnostic criterion for substance use disorders (Hasin et al., 2013). Numerous studies using different techniques and designs indicate that ghrelin concentration in the blood is positively associated with the intensity of self-reported craving for alcohol, as assessed through valid measures such as Obsessive–Compulsive Drinking Scale and Penn Alcohol Craving Scale (Addolorato et al., 2006; Akkisi Kumsar & Dilbaz, 2015; Koopmann et al., 2012; Leggio et al., 2012; Wurst et al., 2007). However, this positive association was not found in two studies, where alcohol craving was assessed as a secondary outcome (Hillemacher et al., 2007; Kraus et al., 2005).

Although the clinical studies presented so far suggest a close relationship between the ghrelin system and alcohol craving, they do not establish causality. In other words, these observational studies do not verify whether ghrelin signaling directly modulates alcohol craving or changes in the ghrelin concentration are epiphenomenal. To address this key question, Leggio and colleagues conducted a double-blind, placebo-controlled, human laboratory experiment in which 45 heavy drinking alcohol-dependent individuals were randomized to receive a 10-min intravenous infusion of exogenous ghrelin (0, 1, or 3 μg/kg), and thereafter underwent a cue-reactivity session. Cue-reactivity is a well validated and laboratory procedure during which participants rate the intensity of their craving after being exposed to visual, tactile, olfactory, and proprioceptive stimuli associated with nonalcoholic and alcoholic beverages (Courtney, Schacht, Hutchison, Roche, & Ray, 2016; Monti et al., 1987; O'Malley, Krishnan-Sarin, Farren, Sinha, & Kreek, 2002). Given the known orexigenic effects of ghrelin, fruit punch was used as the nonalcoholic cue in this study in order to specify the effects of ghrelin on craving for alcohol vs general calorie in heavy drinking alcohol-dependent individuals. Results showed that exogenous ghrelin administration (3 μg/kg), compared to placebo, significantly increased cue-induced alcohol craving, with no significant effect on urge to drink the juice (Fig. 2). Moreover, postinfusion ghrelin concentrations in the blood were positively correlated with the increase in alcohol craving (Leggio et al., 2014).

The aforementioned study provided the first direct translational evidence from a ghrelin pharmacological challenge in alcohol-dependent individuals, highlighting the central action of peripherally administered ghrelin (a debatable topic questioned in a rodent experiment, Jerlhag et al., 2014), and showing ghrelin's specific role in the context of cue-elicited alcohol craving (Kiefer, 2014). No significant effect of ghrelin, compared to placebo, was found on urge to drink juice, which proposes a partially independent role for ghrelin in regulating alcohol craving, regardless of its caloric content and palatable properties, at least in the population of this study (i.e., nonobese alcohol-dependent individuals). This notion is further corroborated by rodent studies which indicate that the link between ghrelin and alcohol-seeking behaviors extends beyond orexigenic effects, as manipulation of the ghrelin system also modulates (a) the consumption of saccharin, a highly rewarding noncaloric sweetener (Disse et al., 2010; Landgren, Simms, et al., 2011); (b) conditioned place preference, a measure of reward in absence of alcohol (Bahi et al., 2013; Jerlhag, 2008; Jerlhag et al., 2009); and (c) the rewarding effects of other addictive drugs with no caloric content, e.g., stimulants, nicotine, and opioids (Engel, Nylander, & Jerlhag, 2015; Jerlhag & Engel, 2011; Suchankova, Engel, et al., 2016), as detailed later in upcoming sections of this chapter.

While direct translation of gene knockout models into human research is not feasible, some preliminary work has investigated the potential relationship between alcohol-related outcomes and genetic variation of the ghrelin

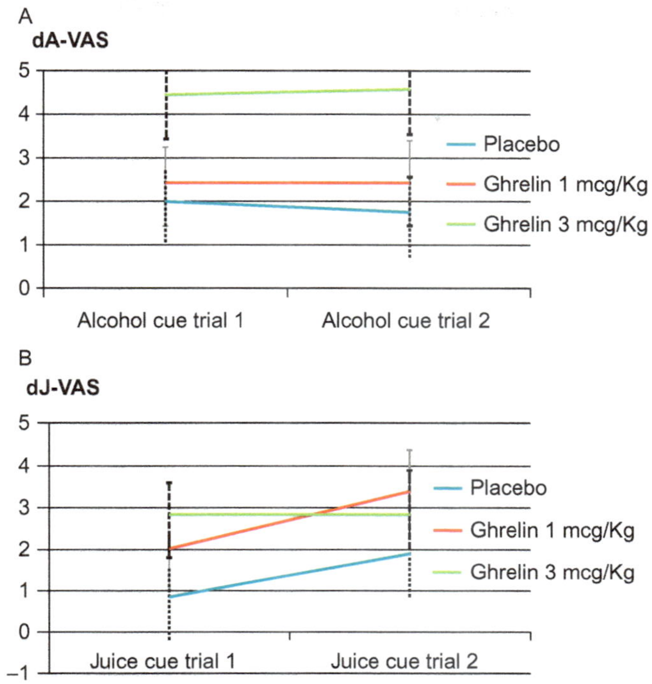

Fig. 2 Intravenous ghrelin administration dose-dependently increased alcohol cue-induced craving (A), but did not significantly change juice cue-induced craving (B) in alcohol-dependent, heavy drinking individuals. Post hoc analysis revealed a significant effect of ghrelin 3 mcg/kg compared to placebo in increasing alcohol craving with a large effect size. *dA-VAS*, delta alcohol-visual analogue scale; *dJ-VAS*, delta juice-visual analogue scale. *Copied with permission from Leggio, L., Zywiak, W. H., Fricchione, S. R., Edwards, S. M., de la Monte, S. M., Swift, R. M., et al. (2014). Intravenous ghrelin administration increases alcohol craving in alcohol-dependent heavy drinkers: A preliminary investigation.* Biological Psychiatry, 76*(9): 734–741.*

system, mainly through analyzing the frequency of single-nucleotide polymorphisms (SNPs). The majority of previous studies report no association between variants of the ghrelin peptide gene (*GHRL*) and risk or severity of alcohol dependence (Landgren, Berglund, et al., 2011; Landgren et al., 2010, 2008; Suchankova, Nilsson et al., 2016). Nevertheless, one of these studies reported a weak association between the *GHRL* gene variation and (a) paternal alcohol dependence, (b) alcohol withdrawal symptoms, in females with severe alcohol dependence (Landgren et al., 2010). Notably, a more recent case–control analysis found that a *GHRL* gene missense polymorphism (Leu72Met) may be associated with the likelihood of having an alcohol use disorder diagnosis. Significant associations were also found between this polymorphism and alcohol-related endophenotypes including subjective responses to intravenous alcohol as well as self-reported amount of alcohol drinking (Suchankova et al., 2017). Another study found significant associations between personality traits that are linked to pathological alcohol use (i.e., self-directedness and self-transcendence) and variations in the ghrelin peptide and receptor genes (Landgren, Berglund, et al., 2011). Compared to the *GHRL* gene, studies on the ghrelin receptor gene (*GHSR*) have yielded more consistent results; significant associations between the *GHSR* gene variation and both risk and severity of alcohol dependence have been reported (Landgren, Berglund, et al., 2011; Landgren et al., 2008; Suchankova, Nilsson, et al., 2016). These findings along with the previously described high constitutive activity of GHS-R1a (Mear et al., 2013) underscore the key role of this receptor in the ghrelin signaling pathway. Therefore, blocking the GHS-R1a or targeting other components of the ghrelin system may represent potential pharmacological approaches to treat alcohol use disorder.

To summarize, both preclinical and clinical literature indicate a significant role for the ghrelin system in alcohol use disorder. In most rodent and human research, increased ghrelin signaling appears to be associated with increased alcohol craving and consumption. However, from a mechanistic point of view, more studies are required to deconstruct the complex interaction between alcohol and ghrelin signaling, and to determine how manipulations of the ghrelin system may impact alcohol-related behaviors.

3. GHRELIN AND STIMULANTS

Stimulants include a wide range of synthetic and natural compounds. The most commonly abused stimulant drugs include amphetamines, methamphetamine, and cocaine. A considerable amount of work has explored the role of the ghrelin system in stimulants use and stimulants-related effects.

3.1 Preclinical Evidence

Exposure to stimulants alters peripheral endogenous ghrelin concentrations in rodent experiments. Intraperitoneal injection of a single dose of methamphetamine or 3,4-methylenedioxy-methamphetamine (MDMA, also known as ecstasy) increased blood ghrelin concentrations in rats (Crowley et al., 2005; Kobeissy, Jeung, Warren, Geier, & Gold, 2008). In another study with rats, the blood ghrelin concentration measured prior to the test session was positively correlated with cue-triggered reinstatement of cocaine seeking after a period of extinction (Tessari et al., 2007).

Manipulation of the ghrelin system, either through exogenous ghrelin administration or GHS-R1a antagonism, further supports a role for ghrelin in stimulants use. Rats intraperitoneally injected with ghrelin exhibited enhanced cocaine-induced motor hyperactivity compared with animals injected with saline; ghrelin alone, without cocaine, appeared to have no effect on locomotion (Wellman, Davis, & Nation, 2005). Potentiation of cocaine-induced hyperlocomotion was also found with repeated systemic administration of ghrelin prior to acute cocaine exposure (Wellman, Hollas, & Elliott, 2008), as well as with microinjection of ghrelin into the nucleus accumbens core (Jang, Kim, Cho, Lee, & Kim, 2013). Pretreatment with systemic ghrelin (Davis, Wellman, & Clifford, 2007) or microinjection of ghrelin into the ventral tegmental area (Schuette, Gray, & Currie, 2013) also increased rats' preference for cocaine as measured by conditioned place preference. Additionally, ghrelin may modulate cocaine's interactions with the rewarding effects of other drugs of abuse, as both central (ventral tegmental area) and peripheral administration of ghrelin increased cocaine-induced potentiation of alcohol consumption in rats (Cepko et al., 2014).

Taken together, these findings suggest that ghrelin signaling may play a role in motivation for stimulant drugs. Further supporting this conclusion, a number of studies show that pharmacological blockade of GHS-R1a attenuates the biobehavioral outcomes associated with stimulants use. Indeed, injection of a GHS-R1a antagonist (JMV2959), compared with placebo, prior to intraperitoneal administration of cocaine or methamphetamine, led to decreased drug-induced conditioned place preference, hyperlocomotion, and accumbal dopamine release in rats (Jerlhag, Egecioglu, Dickson, & Engel, 2010). In two other studies, GHS-R1a antagonism reduced the stimulant-induced hyperlocomotion and also attenuated the development of stimulant-induced locomotor sensitization, a phenomenon in which repeated exposure to a drug leads to progressive amplification of the rodent's motor response (Clifford et al., 2012; Suchankova, Engel, et al., 2016).

Genetic manipulation of the ghrelin system in rodents has yielded mixed results regarding stimulants-related outcomes. In one study, ghrelin peptide gene (*GHRL*) knockout mice showed decreased cocaine-induced hyperlocomotion, as well as reduced dopamine release in the striatum (Abizaid et al., 2011), while another study found no difference in locomotor effects of cocaine between *GHRL* knockout and wild-type mice

(Clifford et al., 2011). Furthermore, while ghrelin receptor gene (*GHSR*) knockout mice compared to wild-type showed no significant change in cocaine-induced hyperlocomotion (Clifford et al., 2011), attenuated locomotor sensitization was observed in rats knockout for the *GHSR* compared to the wild-type (Clifford et al., 2012). Altogether, these data point to a potential relationship between the ghrelin system and the neurocircutries involved in using stimulant substances.

3.2 Clinical Evidence

The human research on the role of the ghrelin system in stimulant use and dependence is limited to two genetic analyses comparing dependent individuals with healthy controls. The first study showed no association between variation of the *GHRL* gene and likelihood of methamphetamine dependence in a Korean sample, but found a significant correlation between carrying the *GHRL* Leu72Met polymorphism and emotional problems associated with drug abuse (Yoon et al., 2005). The second study found no significant differences in the *GHRL* gene polymorphisms between Swedish amphetamine-dependent individuals and healthy controls, but did show a significant association between a SNP on *GHRL* and severity of amphetamine use (Suchankova, Jerlhag, et al., 2013). Notably, Suchankova and colleagues also looked at the ghrelin receptor gene in the same study and found a significant relationship between one SNP on the *GHSR* gene and risk of amphetamine dependence, as the minor allele of this SNP was more common among amphetamine-dependent individuals compared to healthy controls (Suchankova, Jerlhag, et al., 2013). Albeit preliminary, these findings suggest that variation of the *GHSR* gene, rather than *GHRL*, and the subsequent downstream effects may contribute to the development of dependence on stimulant drugs.

To summarize, preclinical experiments suggest that ghrelin enhances the effects of stimulant drugs and may modulate the biobehavioral correlates of stimulants abuse; however, human data are still scarce. Additional basic and clinical research is required to elucidate more detailed aspects of the relationship between the ghrelin system and stimulants.

4. GHRELIN AND TOBACCO/NICOTINE

Nicotine is the main psychoactive/addictive ingredient of tobacco, a plant that has been used (smoked, chewed, or sniffed) for centuries. Several lines of research have studied the effects of nicotine use on appetitive signals as well as the potential involvement of the ghrelin system in nicotine-seeking behavior.

4.1 Preclinical Evidence

In two separate in vitro experiments on rat brain slices, Palotai and colleagues showed that ghrelin acts similar to nicotine in stimulating dopamine release in the amygdala (Palotai et al., 2013b) and increases nicotine-induced dopamine release in the striatum (Palotai et al., 2013a).

Rodent experiments also indicate that nicotine may influence endogenous ghrelin concentrations. In one study, 4-week repeated exposure to cigarette smoke significantly increased blood ghrelin concentrations in rats (Tomoda et al., 2012). Another study assessed the effects of both oral and smoked nicotine in rats and found that nicotine administration by either route resulted in increased peripheral ghrelin concentrations (Ali, Hamed, Ayuob, Shaker Ali, & Suliman, 2015). Ypsilantis and colleagues, however, did not find differences in blood ghrelin levels between rats passively exposed to cigarette smoke for 8 weeks and the nonexposed control group, although this study measured total ghrelin, as opposed to acyl-ghrelin which is commonly measured in most investigations (Ypsilantis et al., 2013).

The role of ghrelin signaling in mediating the rewarding properties of nicotine has been further investigated in rodent models through ghrelin receptor antagonism. In two separate studies, systemic administration of a GHS-R1a antagonist (JMV2959) resulted in decreased nicotine-induced hyperlocomotion, conditioned place preference, and accumbal dopamine release (Jerlhag & Engel, 2011; Wellman et al., 2011), suggesting a potential link between ghrelin and nicotine-related effects.

4.2 Clinical Evidence

Research in humans has shown that cigarette smokers and nonsmokers do not appear to differ in regard to their fasting blood ghrelin concentrations (Bouhours-Nouet et al., 2006; Kokkinos et al., 2007), but there is controversy in the literature about the acute effects of cigarette smoking on endogenous ghrelin. Kokkinos and colleagues reported that smoking two consecutive cigarettes had no immediate effect on ghrelin levels in habitual cigarette smokers, but decreased the blood ghrelin concentrations in nonsmokers (Kokkinos et al., 2007). In contrast, another study found a significant increase in blood ghrelin concentrations in both smokers and nonsmokers after smoking one filtered cigarette (Bouros et al., 2006). Other results come from a number of sporadic studies looking at the relationship between nicotine exposure and feeding outcomes in nonsmoking individuals: smoking one cigarette reduced ghrelin levels in saliva, but not in blood (Kaabi & Khalifa, 2014), while chewing nicotine gum had no effect on blood ghrelin concentrations (Kroemer, Wuttig, Bidlingmaier, Zimmermann, & Smolka, 2015; Pilhatsch et al., 2014). Variations in the study designs and the enrolled samples may at least partially explain these inconsistent results.

Studies in habitual smokers suggest that endogenous ghrelin levels positively correlate with years of smoking and may predict the risk of smoking relapse (al'Absi, Lemieux, & Nakajima, 2014; Kokkinos et al., 2007; Lemieux & al'Absi, 2017). However, no association has been found between ghrelin concentrations and the severity of nicotine smoking, craving, or withdrawal (al'Absi et al., 2014; Kokkinos et al., 2007; Mutschler et al., 2012). Another study suggested that smoking cessation may lead to decreased blood ghrelin levels (Lee et al., 2006). Finally, genetic variation of the *GHSR* gene, but not *GHRL*, appears to be associated with increased severity of cigarette smoking (Suchankova, Nilsson, et al., 2016). Interestingly, findings related to the ghrelin receptor gene (*GHSR*) again appear more robust compared to the ghrelin peptide gene (*GHRL*); this seems to be a consistent pattern across different drugs of abuse including nicotine, alcohol, and stimulants.

To summarize, the ghrelin system seems to interact with nicotine-related pathways, but the findings to date are quite inconsistent to draw any conclusions. Future studies should standardize their methodology and outcome measures to allow for comparison across their findings and to provide conclusive information about a potential link between nicotine and ghrelin.

5. GHRELIN AND OPIOIDS

Both illegal opioid substances (e.g., heroin) and prescription opioids (e.g., pain relievers such as oxycodone and morphine) carry high liability for abuse and dependence. The literature on ghrelin signaling in opioid use is scarce and limited to preclinical investigations.

5.1 Preclinical Evidence

In one study, intracerebroventricular administration of ghrelin in rats increased intravenous heroin consumption and motivation for heroin, as shown by increased breakpoints on a progressive ratio schedule, but intracerebroventricular administration of a GHS-R1a antagonist (D-Lys3-GHRP-6) had no effect on heroin self-administration or food deprivation-induced reinstatement of heroin seeking (Maric, Sedki, Ronfard, Chafetz, & Shalev, 2012).

In a comprehensive set of experiments in mice, Engel and colleagues studied the effects of repeated intraperitoneal administration of ghrelin and a GHS-R1a antagonist (JMV2959) on opioid peptides [Met-enkephalin-Arg6Phe7 (MEAP), dynorphin B (DynB), and Leu-enkephalin-Arg6 (LeuArg)] in reward- and stress-related brain regions, such as amygdala, striatum, prefrontal cortex, ventral tegmental area, and hippocampus (Engel et al., 2015). Ghrelin receptor antagonism increased the opioid peptides levels in the brain tissue, whereas no change was observed in response to ghrelin administration. Systemic administration of JMV2959 also resulted in decreased morphine-induced hyperlocomotion, conditioned place preference, and dopamine release in the nucleus accumbens (Engel et al., 2015).

Rat studies with JMV2959 have found similar reductions of morphine-induced behavioral stimulation and dopamine release in the nucleus accumbens (Sustkova-Fiserova, Jerabek, Havlickova, Kacer, & Krsiak, 2014). Based on the known interactions between the endocannabinoid system and both opioid and ghrelin signaling (Navarro et al., 2001; Storr & Sharkey, 2007), Sustkova-Fiserova and colleagues designed a study to investigate the potential role of ghrelin in modulating the cross-talk between opioid and endocannabinoid pathways. Results showed that pretreatment with JMV2959 reversed both short-term and long-term morphine-induced increases in anandamide levels in the nucleus accumbens, intensified acute morphine-induced decreases in 2-arachidonoylglycerol levels, and attenuated long-term morphine-induced decreases in 2-arachidonoylglycerol levels (Sustkova-Fiserova, Jerabek, Havlickova, Syslova, & Kacer, 2016).

To summarize, the literature points to a possible connection between ghrelin and the opioid system, but more work is required. GHS-R1a antagonism seems to affect behavioral and biological correlates of opioid use; therefore, future animal and human research should probe the safety and efficacy of this approach as a novel pharmacological treatment for opioid addiction.

6. GHRELIN AND CANNABIS

Cannabis is a flowering plant used for both recreational and medicinal purposes. Three main types (species, subspecies, or varieties) of the cannabis genus have been recognized: *Cannabis sativa*, *Cannabis indica*, and *Cannabis ruderalis*. Δ(9)-Tetrahydrocannabinol is the main psychoactive/addictive ingredient of cannabis; the effects of this compound are most commonly associated with binding to the cannabinoid receptor 1 (Hoffman & Lupica, 2013).

6.1 Preclinical Evidence

To our knowledge, only one rodent study has investigated the relationship between ghrelin and cannabis. Mazidi and colleagues administered *C. sativa* via oral gavage and found that *C. sativa*, compared to a distilled water control, increased blood ghrelin concentrations and food consumption in rats (Mazidi et al., 2014).

6.2 Clinical Evidence

In a study with HIV-positive adult men, blood concentrations of ghrelin and other appetitive hormones (leptin, PYY, and insulin) were tested after having received smoked medicinal cannabis or matched placebo for HIV-associated neuropathic pain. Cannabis administration, as compared to placebo, significantly increased ghrelin concentrations in this study. In addition, leptin and PYY levels were, respectively, increased and decreased, but no impact on insulin levels was found (Riggs et al., 2012).

While these two small studies in rats and humans suggest that cannabis exposure may affect endogenous ghrelin concentrations, more research is needed to understand the potential connection between ghrelin and cannabis use.

7. CONCLUSIONS

Preclinical and clinical evidence supports the role of ghrelin in mediating several behavioral measures related to alcohol and, to a lesser degree, stimulants, nicotine, opioids, and cannabis. To our knowledge, there is no published data on the ghrelin system in the context of other abused drug categories including hallucinogens, inhalants, sedatives, hypnotics, anxiolytics, and caffeine.

Although the research on ghrelin in the context of drug addiction is somewhat in infancy, both preclinical and clinical studies indicate a key role for different components of the ghrelin system in the neurobiology of drug addiction. The human genetic work is not conclusive, possibly because a few genetic variants with unclear functionality have been studied in small sample sizes. Future studies should apply genome-wide association and next-generation sequencing approaches in larger samples to provide more information on the potential contribution of the ghrelin system. It is also plausible that rapid adaptive and compensatory changes occur in rodents when a single gene is knocked out, therefore, reducing the ability of knockout models to be fully informative. The role of these models, however, remains important in better understanding how ghrelin may affect addictive behaviors.

Studying the endogenous ghrelin system in humans supports the notion that peripheral ghrelin levels correlate with different substance-related behaviors. Given both ethical and technical limitations of human research, most studies do not establish a causal role for ghrelin in relation to substance use. However, a human laboratory study (Leggio et al., 2014) did support such a causality link in heavy drinking individuals with alcohol dependence. The rodent experiments are more conclusive in detecting a causal relationship between the ghrelin system and substance-related behaviors. Specifically, research generated by independent laboratories and using different animal models is largely consistent in indicating that ghrelin administration increases, whereas ghrelin receptor blockade decreases substance use and related behaviors. Since most preclinical studies were conducted in nondependent rodents, future studies should investigate the role of ghrelin in animal models of drug dependence.

A number of other feeding-related neuropeptides (e.g., leptin, insulin, glucagon-like peptide-1) have close interactions with the ghrelin system (Haass-Koffler et al., 2015, 2016; Suchankova et al., 2015). Ghrelin's effects on substance-related outcomes may be directly driven by ghrelin, mediated through the cross-talk with other neuropeptides, or both; a set of questions which requires further investigation. Future studies, especially in humans, are needed to investigate whether GHS-R1a may represent a safe and effective pharmacological target to treat patients with drug addiction. Recent advances in developing new radioligands for GHS-R1a (Kawamura et al., 2017) may facilitate the medication development efforts by providing information on receptor occupancy as a direct measure of target engagement. Given the complexity of the ghrelin system, translational studies are also needed to explore alternative and additional ways to manipulate the ghrelin system, therefore, informing the neurobiology of drug addiction and contributing to the medication development efforts for substance use disorders.

ACKNOWLEDGMENTS AND DISCLOSURES

This work was supported by the Division of Intramural Clinical and Biological Research of the National Institute on Alcohol Abuse and Alcoholism (L.J.Z., M.F., L.L.) and the Intramural Research Program of the National Institute on Drug Abuse (L.J.Z., M.F., B.J.T., L.F.V., L.L.). We thank Karen Smith, National Institutes of Health Library, for bibliographic assistance. The content of this chapter is solely the responsibility of the authors and does not necessarily represent the official views of the National Institutes of Health. The authors report no biomedical financial interests or potential conflicts of interest.

REFERENCES

Abizaid, A., Liu, Z. W., Andrews, Z. B., Shanabrough, M., Borok, E., Elsworth, J. D., et al. (2006). Ghrelin modulates the activity and synaptic input organization of midbrain dopamine neurons while promoting appetite. *Journal of Clinical Investigation, 116*(12), 3229–3239. http://dx.doi.org/10.1172/JCI29867.

Abizaid, A., Mineur, Y. S., Roth, R. H., Elsworth, J. D., Sleeman, M. W., Picciotto, M. R., et al. (2011). Reduced locomotor responses to cocaine in ghrelin-deficient mice. *Neuroscience, 192*, 500–506.

Addolorato, G., Capristo, E., Leggio, L., Ferrulli, A., Abenavoli, L., Malandrino, N., et al. (2006). Relationship between ghrelin levels, alcohol craving, and nutritional status in current alcoholic patients. *Alcoholism: Clinical and Experimental Research, 30*(11), 1933–1937. http://dx.doi.org/10.1111/j.1530-0277.2006.00238.x.

Akkisi Kumsar, N., & Dilbaz, N. (2015). Relationship between craving and ghrelin, adiponectin, and resistin levels in patients with alcoholism. *Alcoholism, Clinical and Experimental Research, 39*(4), 702–709. http://dx.doi.org/10.1111/acer.12689.

Al Massadi, O., Tschop, M. H., & Tong, J. (2011). Ghrelin acylation and metabolic control. *Peptides, 32*(11), 2301–2308. http://dx.doi.org/10.1016/j.peptides.2011.08.020.

al'Absi, M., Lemieux, A., & Nakajima, M. (2014). Peptide yy and ghrelin predict craving and risk for relapse in abstinent smokers. *Psychoneuroendocrinology, 49*, 253–259. http://dx.doi.org/10.1016/j.psyneuen.2014.07.018.

Albarran-Zeckler, R. G., & Smith, R. G. (2013). The ghrelin receptors (GHS-R1a and GHS-R1b). *Endocrine Development, 25*, 5–15. http://dx.doi.org/10.1159/000346042.

Ali, S. S., Hamed, E. A., Ayuob, N. N., Shaker Ali, A., & Suliman, M. I. (2015). Effects of different routes of nicotine administration on gastric morphology and hormonal secretion in rats. *Experimental Physiology, 100*(8), 881–895. http://dx.doi.org/10.1113/ep085015.

Alvarez-Crespo, M., Skibicka, K. P., Farkas, I., Molnar, C. S., Egecioglu, E., Hrabovszky, E., et al. (2012). The amygdala as a neurobiological target for ghrelin in rats: Neuroanatomical, electrophysiological and behavioral evidence. *PLoS One*, 7(10), e46321. http://dx.doi.org/10.1371/journal.pone.0046321.

Ariyasu, H., Takaya, K., Tagami, T., Ogawa, Y., Hosoda, K., Akamizu, T., et al. (2001). Stomach is a major source of circulating ghrelin, and feeding state determines plasma ghrelin-like immunoreactivity levels in humans. *The Journal of Clinical Endocrinology and Metabolism, 86*(10), 4753–4758. http://dx.doi.org/10.1210/jcem.86.10.7885.

Avena, N. M., Bocarsly, M. E., Hoebel, B. G., & Gold, M. S. (2011). Overlaps in the nosology of substance abuse and overeating: The translational implications of "food addiction". *Current Drug Abuse Reviews*, *4*(3), 133–139.

Badaoui, A., De Saeger, C., Duchemin, J., Gihousse, D., de Timary, P., & Starkel, P. (2008). Alcohol dependence is associated with reduced plasma and fundic ghrelin levels. *European Journal of Clinical Investigation*, *38*(6), 397–403. http://dx.doi.org/10.1111/j.1365-2362.2008.01947.x.

Bahi, A., Tolle, V., Fehrentz, J. A., Brunel, L., Martinez, J., Tomasetto, C. L., et al. (2013). Ghrelin knockout mice show decreased voluntary alcohol consumption and reduced ethanol-induced conditioned place preference. *Peptides*, *43*, 48–55. http://dx.doi.org/10.1016/j.peptides.2013.02.008.

Bali, A., & Jaggi, A. S. (2016). An integrative review on role and mechanisms of ghrelin in stress, anxiety and depression. *Current Drug Targets*, *17*(5), 495–507.

Banks, W. A., Tschop, M., Robinson, S. M., & Heiman, M. L. (2002). Extent and direction of ghrelin transport across the blood-brain barrier is determined by its unique primary structure. *Journal of Pharmacology and Experimental Therapeutics*, *302*(2), 822–827. http://dx.doi.org/10.1124/jpet.102.034827.

Bouhours-Nouet, N., Boux de Casson, F., Rouleau, S., Douay, O., Mathieu, E., Bouderlique, C., et al. (2006). Maternal and cord blood ghrelin in the pregnancies of smoking mothers: Possible markers of nutrient availability for the fetus. *Hormone Research*, *66*(1), 6–12. http://dx.doi.org/10.1159/000092807.

Bouros, D., Tzouvelekis, A., Anevlavis, S., Doris, M., Tryfon, S., Froudarakis, M., et al. (2006). Smoking acutely increases plasma ghrelin concentrations. *Clinical Chemistry*, *52*(4), 777–778. http://dx.doi.org/10.1373/clinchem.2005.065243.

Cabral, A., Lopez Soto, E. J., Epelbaum, J., & Perello, M. (2017). Is ghrelin synthesized in the central nervous system? *International Journal of Molecular Sciences*, *18*(3), 638. http://dx.doi.org/10.3390/ijms18030638.

Calissendorff, J., Danielsson, O., Brismar, K., & Rojdmark, S. (2005). Inhibitory effect of alcohol on ghrelin secretion in normal man. *European Journal of Endocrinology*, *152*(5), 743–747. http://dx.doi.org/10.1530/eje.1.01905.

Calissendorff, J., Danielsson, O., Brismar, K., & Rojdmark, S. (2006). Alcohol ingestion does not affect serum levels of peptide yy but decreases both total and octanoylated ghrelin levels in healthy subjects. *Metabolism*, *55*(12), 1625–1629. http://dx.doi.org/10.1016/j.metabol.2006.08.003.

Calissendorff, J., Gustafsson, T., Holst, J. J., Brismar, K., & Rojdmark, S. (2012). Alcohol intake and its effect on some appetite-regulating hormones in man: Influence of gastroprotection with sucralfate. *Endocrine Research*, *37*(3), 154–162. http://dx.doi.org/10.3109/07435800.2012.662662.

Castaneda, T. R., Tong, J., Datta, R., Culler, M., & Tschop, M. H. (2010). Ghrelin in the regulation of body weight and metabolism. *Frontiers in Neuroendocrinology*, *31*(1), 44–60. http://dx.doi.org/10.1016/j.yfrne.2009.10.008.

Cepko, L. C., Selva, J. A., Merfeld, E. B., Fimmel, A. I., Goldberg, S. A., & Currie, P. J. (2014). Ghrelin alters the stimulatory effect of cocaine on ethanol intake following mesolimbic or systemic administration. *Neuropharmacology*, *85*, 224–231. http://dx.doi.org/10.1016/j.neuropharm.2014.05.030.

Chuang, J. C., Perello, M., Sakata, I., Osborne-Lawrence, S., Savitt, J. M., Lutter, M., et al. (2011). Ghrelin mediates stress-induced food-reward behavior in mice. *The Journal of Clinical Investigation*, *121*(7), 2684–2692. http://dx.doi.org/10.1172/jci57660.

Clifford, P. S., Rodriguez, J., Schul, D., Hughes, S., Kniffin, T., Hart, N., et al. (2012). Attenuation of cocaine-induced locomotor sensitization in rats sustaining genetic or pharmacologic antagonism of ghrelin receptors. *Addiction Biology*, *17*(6), 956–963. http://dx.doi.org/10.1111/j.1369-1600.2011.00339.x.

Clifford, S., Zeckler, R. A., Buckman, S., Thompson, J., Hart, N., Wellman, P. J., et al. (2011). Impact of food restriction and cocaine on locomotion in ghrelin- and ghrelin-receptor knockout mice. *Addiction Biology*, *16*(3), 386–392. http://dx.doi.org/10.1111/j.1369-1600.2010.00253.x.

Courtney, K. E., Schacht, J. P., Hutchison, K., Roche, D. J., & Ray, L. A. (2016). Neural substrates of cue reactivity: Association with treatment outcomes and relapse. *Addiction Biology*, *21*(1), 3–22. http://dx.doi.org/10.1111/adb.12314.

Crowley, W. R., Ramoz, G., Keefe, K. A., Torto, R., Kalra, S. P., & Hanson, G. R. (2005). Differential effects of methamphetamine on expression of neuropeptide y mrna in hypothalamus and on serum leptin and ghrelin concentrations in ad libitum-fed and schedule-fed rats. *Neuroscience*, *132*(1), 167–173. http://dx.doi.org/10.1016/j.neuroscience.2004.11.037.

Cruz, M. T., Herman, M. A., Cote, D. M., Ryabinin, A. E., & Roberto, M. (2013). Ghrelin increases gabaergic transmission and interacts with ethanol actions in the rat central nucleus of the amygdala. *Neuropsychopharmacology*, *38*(2), 364–375. http://dx.doi.org/10.1038/npp.2012.190.

Cummings, D. E., Frayo, R. S., Marmonier, C., Aubert, R., & Chapelot, D. (2004). Plasma ghrelin levels and hunger scores in humans initiating meals voluntarily without time- and food-related cues. *American Journal of Physiology. Endocrinology and Metabolism*, *287*(2), E297–304. http://dx.doi.org/10.1152/ajpendo.00582.2003.

Currie, P. J., Khelemsky, R., Rigsbee, E. M., Dono, L. M., Coiro, C. D., Chapman, C. D., et al. (2012). Ghrelin is an orexigenic peptide and elicits anxiety-like behaviors following administration into discrete regions of the hypothalamus. *Behavioural Brain Research*, *226*(1), 96–105. http://dx.doi.org/10.1016/j.bbr.2011.08.037.

Currie, P. J., Mirza, A., Fuld, R., Park, D., & Vasselli, J. R. (2005). Ghrelin is an orexigenic and metabolic signaling peptide in the arcuate and paraventricular nuclei. *American Journal of Physiology. Regulatory, Integrative and Comparative Physiology*, *289*(2), R353–R358. http://dx.doi.org/10.1152/ajpregu.00756.2004.

Davis, J. F., Perello, M., Choi, D. L., Magrisso, I. J., Kirchner, H., Pfluger, P. T., et al. (2012). Goat induced ghrelin acylation regulates hedonic feeding. *Hormones and Behavior*, *62*(5), 598–604. http://dx.doi.org/10.1016/j.yhbeh.2012.08.009.

Davis, J. F., Schurdak, J. D., Magrisso, I. J., Mul, J. D., Grayson, B. E., Pfluger, P. T., et al. (2012). Gastric bypass surgery attenuates ethanol consumption in ethanol-preferring rats. *Biological Psychiatry*, *72*(5), 354–360. http://dx.doi.org/10.1016/j.biopsych.2012.01.035.

Davis, K. W., Wellman, P. J., & Clifford, P. S. (2007). Augmented cocaine conditioned place preference in rats pretreated with systemic ghrelin. *Regulatory Peptides*, *140*(3), 148–152. http://dx.doi.org/10.1016/j.regpep.2006.12.003.

de Timary, P., Cani, P. D., Duchemin, J., Neyrinck, A. M., Gihousse, D., Laterre, P. F., et al. (2012). The loss of metabolic control on alcohol drinking in heavy drinking alcohol-dependent subjects. *PLoS One*, *7*(7), e38682. http://dx.doi.org/10.1371/journal.pone.0038682.

Delhanty, P. J., Neggers, S. J., & van der Lely, A. J. (2013). Des-acyl ghrelin: A metabolically active peptide. *Endocrine Development*, *25*, 112–121. http://dx.doi.org/10.1159/000346059.

Delhanty, P. J., Neggers, S. J., & van der Lely, A. J. (2014). Should we consider des-acyl ghrelin as a separate hormone and if so, what does it do? *Frontiers of Hormone Research*, *42*, 163–174. http://dx.doi.org/10.1159/000358345.

Dickson, S. L., Egecioglu, E., Landgren, S., Skibicka, K. P., Engel, J. A., & Jerlhag, E. (2011). The role of the central ghrelin system in reward from food and chemical drugs. *Molecular and Cellular Endocrinology*, *340*(1), 80–87. http://dx.doi.org/10.1016/j.mce.2011.02.017.

Disse, E., Bussier, A. L., Veyrat-Durebex, C., Deblon, N., Pfluger, P. T., Tschop, M. H., et al. (2010). Peripheral ghrelin enhances sweet taste food consumption and preference, regardless of its caloric content. *Physiology and Behavior*, *101*(2), 277–281. http://dx.doi.org/10.1016/j.physbeh.2010.05.017.

Dockray, G. J. (2013). Enteroendocrine cell signalling via the vagus nerve. *Current Opinion in Pharmacology*, *13*(6), 954–958. http://dx.doi.org/10.1016/j.coph.2013.09.007.

Edwards, A., & Abizaid, A. (2016). Driving the need to feed: Insight into the collaborative interaction between ghrelin and endocannabinoid systems in modulating brain reward systems. *Neuroscience and Biobehavioral Reviews*, *66*, 33–53. http://dx.doi.org/10.1016/j.neubiorev.2016.03.032.

Edwards, A., & Abizaid, A. (2017). Clarifying the ghrelin system's ability to regulate feeding behaviours despite enigmatic spatial separation of the GHSR and its endogenous ligand. *International Journal of Molecular Sciences*, *18*(4), 859. http://dx.doi.org/10.3390/ijms18040859.

Egecioglu, E., Prieto-Garcia, L., Studer, E., Westberg, L., & Jerlhag, E. (2016). The role of ghrelin signalling for sexual behaviour in male mice. *Addiction Biology*, *21*(2), 348–359. http://dx.doi.org/10.1111/adb.12202.

Engel, J. A., Nylander, I., & Jerlhag, E. (2015). A ghrelin receptor (GHS-R1a) antagonist attenuates the rewarding properties of morphine and increases opioid peptide levels in reward areas in mice. *European Neuropsychopharmacology*, *25*(12), 2364–2371. http://dx.doi.org/10.1016/j.euroneuro.2015.10.004.

Ferrini, F., Salio, C., Lossi, L., & Merighi, A. (2009). Ghrelin in central neurons. *Current Neuropharmacology*, 7(1), 37–49. http://dx.doi.org/10.2174/157015909787602779.

Gahete, M. D., Rincon-Fernandez, D., Villa-Osaba, A., Hormaechea-Agulla, D., Ibanez-Costa, A., Martinez-Fuentes, A. J., et al. (2014). Ghrelin gene products, receptors, and goat enzyme: Biological and pathophysiological insight. *The Journal of Endocrinology*, *220*(1), R1–24. http://dx.doi.org/10.1530/joe-13-0391.

Gomez, J. L., Cunningham, C. L., Finn, D. A., Young, E. A., Helpenstell, L. K., Schuette, L. M., et al. (2015). Differential effects of ghrelin antagonists on alcohol drinking and reinforcement in mouse and rat models of alcohol dependence. *Neuropharmacology*, *97*, 182–193. http://dx.doi.org/10.1016/j.neuropharm.2015.05.026.

Gomez, J. L., & Ryabinin, A. E. (2014). The effects of ghrelin antagonists [D-Lys(3)]-GHRP-6 or JMV2959 on ethanol, water, and food intake in c57bl/6j mice. *Alcoholism, Clinical and Experimental Research*, *38*(9), 2436–2444. http://dx.doi.org/10.1111/acer.12499.

Grant, B. F., Saha, T. D., Ruan, W. J., Goldstein, R. B., Chou, S. P., Jung, J., et al. (2016). Epidemiology of DSM-5 drug use disorder: Results from the national epidemiologic survey on alcohol and related conditions-III. *JAMA Psychiatry*, *73*(1), 39–47. http://dx.doi.org/10.1001/jamapsychiatry.2015.2132.

Gualillo, O., Lago, F., Casanueva, F. F., & Dieguez, C. (2006). One ancestor, several peptides post-translational modifications of preproghrelin generate several peptides with antithetical effects. *Molecular and Cellular Endocrinology*, *256*(1–2), 1–8. http://dx.doi.org/10.1016/j.mce.2006.05.007.

Guan, X. M., Yu, H., Palyha, O. C., McKee, K. K., Feighner, S. D., Sirinathsinghji, D. J., et al. (1997). Distribution of mrna encoding the growth hormone secretagogue receptor in brain and peripheral tissues. *Brain Research. Molecular Brain Research*, *48*(1), 23–29.

Gutierrez, J. A., Solenberg, P. J., Perkins, D. R., Willency, J. A., Knierman, M. D., Jin, Z., et al. (2008). Ghrelin octanoylation mediated by an orphan lipid transferase. *Proceedings of the National Academy of Sciences of the United States of America*, *105*(17), 6320–6325. http://dx.doi.org/10.1073/pnas.0800708105.

Haass-Koffler, C. L., Aoun, E. G., Swift, R. M., de la Monte, S. M., Kenna, G. A., & Leggio, L. (2015). Leptin levels are reduced by intravenous ghrelin administration and correlated with cue-induced alcohol craving. *Translational Psychiatry, 5*, e646. http://dx.doi.org/10.1038/tp.2015.140.

Haass-Koffler, C. L., Giovenco, D. E., Lee, M. R., Zywiak, W. H., de la Monte, S. M., Kenna, G. A., et al. (2016). Serum insulin levels are reduced by intravenous ghrelin administration but do not correlate with alcohol craving in alcohol-dependent individuals. *The International Journal of Neuropsychopharmacology, 19*(10), pyw048. http://dx.doi.org/10.1093/ijnp/pyw048.

Hajnal, A., Zharikov, A., Polston, J. E., Fields, M. R., Tomasko, J., Rogers, A. M., et al. (2012). Alcohol reward is increased after Roux-en-Y gastric bypass in dietary obese rats with differential effects following ghrelin antagonism. *PLoS One*, 7(11), e49121. http://dx.doi.org/10.1371/journal.pone.0049121.

Hansson, C., Alvarez-Crespo, M., Taube, M., Skibicka, K. P., Schmidt, L., Karlsson-Lindahl, L., et al. (2014). Influence of ghrelin on the central serotonergic signaling system in mice. *Neuropharmacology, 79*, 498–505. http://dx.doi.org/10.1016/j.neuropharm.2013.12.012.

Hasin, D. S., O'Brien, C. P., Auriacombe, M., Borges, G., Bucholz, K., Budney, A., et al. (2013). DSM-5 criteria for substance use disorders: Recommendations and rationale. *The American Journal of Psychiatry, 170*(8), 834–851. http://dx.doi.org/10.1176/appi.ajp.2013.12060782.

Hillemacher, T., Kraus, T., Rauh, J., Weiß, J., Schanze, A., Frieling, H., et al. (2007). Role of appetite-regulating peptides in alcohol craving: An analysis in respect to subtypes and different consumption patterns in alcoholism. *Alcoholism: Clinical and Experimental Research, 31*(6), 950–954. http://dx.doi.org/10.1111/j.1530-0277.2007.00388.x.

Hoffman, A. F., & Lupica, C. R. (2013). Synaptic targets of delta9-tetrahydrocannabinol in the central nervous system. *Cold Spring Harbor Perspectives in Medicine, 3*(8), a012237. http://dx.doi.org/10.1101/cshperspect.a012237.

Hopkins, A. L., Nelson, T. A., Guschina, I. A., Parsons, L. C., Lewis, C. L., Brown, R. C., et al. (2017). Unacylated ghrelin promotes adipogenesis in rodent bone marrow via ghrelin *O*-acyl transferase and GHS-R1a activity: Evidence for target cell-induced acylation. *Scientific Reports*, 7, 45541. http://dx.doi.org/10.1038/srep45541.

Howard, A. D., Feighner, S. D., Cully, D. F., Arena, J. P., Liberator, P. A., Rosenblum, C. I., et al. (1996). A receptor in pituitary and hypothalamus that functions in growth hormone release. *Science, 273*(5277), 974–977.

Howick, K., Griffin, B. T., Cryan, J. F., & Schellekens, H. (2017). From belly to brain: Targeting the ghrelin receptor in appetite and food intake regulation. *International Journal of Molecular Sciences, 18*(2), E273. http://dx.doi.org/10.3390/ijms18020273.

Hsu, T. M., Suarez, A. N., & Kanoski, S. E. (2016). Ghrelin: A link between memory and ingestive behavior. *Physiology & Behavior, 162*, 10–17. http://dx.doi.org/10.1016/j.physbeh.2016.03.039.

Inui, A., Asakawa, A., Bowers, C. Y., Mantovani, G., Laviano, A., Meguid, M. M., et al. (2004). Ghrelin, appetite, and gastric motility: The emerging role of the stomach as an endocrine organ. *FASEB Journal, 18*(3), 439–456. http://dx.doi.org/10.1096/fj.03-0641rev.

Jang, J. K., Kim, W. Y., Cho, B. R., Lee, J. W., & Kim, J. H. (2013). Microinjection of ghrelin in the nucleus accumbens core enhances locomotor activity induced by cocaine. *Behavioural Brain Research, 248*, 7–11. http://dx.doi.org/10.1016/j.bbr.2013.03.049.

Jerlhag, E. (2008). Systemic administration of ghrelin induces conditioned place preference and stimulates accumbal dopamine. *Addiction Biology, 13*(3–4), 358–363. http://dx.doi.org/10.1111/j.1369-1600.2008.00125.x.

Jerlhag, E., Egecioglu, E., Dickson, S. L., Andersson, M., Svensson, L., & Engel, J. A. (2006). Ghrelin stimulates locomotor activity and accumbal dopamine-overflow via central cholinergic systems in mice: Implications for its involvement in brain reward. *Addiction Biology*, *11*(1), 45–54. http://dx.doi.org/10.1111/j.1369-1600.2006.00002.x.

Jerlhag, E., Egecioglu, E., Dickson, S. L., Douhan, A., Svensson, L., & Engel, J. A. (2007). Ghrelin administration into tegmental areas stimulates locomotor activity and increases extracellular concentration of dopamine in the nucleus accumbens. *Addiction Biology*, *12*(1), 6–16. http://dx.doi.org/10.1111/j.1369-1600.2006.00041.x.

Jerlhag, E., Egecioglu, E., Dickson, S. L., & Engel, J. A. (2010). Ghrelin receptor antagonism attenuates cocaine- and amphetamine-induced locomotor stimulation, accumbal dopamine release, and conditioned place preference. *Psychopharmacology*, *211*(4), 415–422. http://dx.doi.org/10.1007/s00213-010-1907-7.

Jerlhag, E., Egecioglu, E., Landgren, S., Salome, N., Heilig, M., Moechars, D., et al. (2009). Requirement of central ghrelin signaling for alcohol reward. *Proceedings of the National Academy of Sciences of the United States of America*, *106*(27), 11318–11323. http://dx.doi.org/10.1073/pnas.0812809106.

Jerlhag, E., & Engel, J. A. (2011). Ghrelin receptor antagonism attenuates nicotine-induced locomotor stimulation, accumbal dopamine release and conditioned place preference in mice. *Drug and Alcohol Dependence*, *117*(2–3), 126–131. http://dx.doi.org/10.1016/j.drugalcdep.2011.01.010.

Jerlhag, E., Ivanoff, L., Vater, A., & Engel, J. A. (2014). Peripherally circulating ghrelin does not mediate alcohol-induced reward and alcohol intake in rodents. *Alcoholism: Clinical and Experimental Research*, *38*(4), 959–968. http://dx.doi.org/10.1111/acer.12337.

Jerlhag, E., Janson, A. C., Waters, S., & Engel, J. A. (2012). Concomitant release of ventral tegmental acetylcholine and accumbal dopamine by ghrelin in rats. *PLoS One*, 7(11), e49557. http://dx.doi.org/10.1371/journal.pone.0049557.

Jerlhag, E., Landgren, S., Egecioglu, E., Dickson, S. L., & Engel, J. A. (2011). The alcohol-induced locomotor stimulation and accumbal dopamine release is suppressed in ghrelin knockout mice. *Alcohol*, *45*(4), 341–347. http://dx.doi.org/10.1016/j.alcohol.2010.10.002.

Jiang, H., Betancourt, L., & Smith, R. G. (2006). Ghrelin amplifies dopamine signaling by cross talk involving formation of growth hormone secretagogue receptor/dopamine receptor subtype 1 heterodimers. *Molecular Endocrinology*, *20*(8), 1772–1785. http://dx.doi.org/10.1210/me.2005-0084.

Kaabi, Y. A., & Khalifa, M. A. (2014). Acute one-cigarette smoking decreases ghrelin hormone in saliva: A pilot study. *International Journal of Endocrinology*, *2014*, 575671. http://dx.doi.org/10.1155/2014/575671.

Kaur, S., & Ryabinin, A. E. (2010). Ghrelin receptor antagonism decreases alcohol consumption and activation of perioculomotor urocortin-containing neurons. *Alcoholism, Clinical and Experimental Research*, *34*(9), 1525–1534. http://dx.doi.org/10.1111/j.1530-0277.2010.01237.x.

Kawamura, K., Fujinaga, M., Shimoda, Y., Yamasaki, T., Zhang, Y., Hatori, A., et al. (2017). Developing new pet tracers to image the growth hormone secretagogue receptor 1a (GHS-R1a). *Nuclear Medicine and Biology*, *52*, 49–56. http://dx.doi.org/10.1016/j.nucmedbio.2017.06.002.

Kiefer, F. (2014). Ghrelin in addictive behaviors: Plenus venter non studet libenter. *Biological Psychiatry*, *76*(9), 676–677. http://dx.doi.org/10.1016/j.biopsych.2014.08.014.

Kim, J. H., Kim, S. J., Lee, W. Y., Cheon, Y. H., Lee, S. S., Ju, A., et al. (2013). The effects of alcohol abstinence on bdnf, ghrelin, and leptin secretions in alcohol-dependent patients with glucose intolerance. *Alcoholism, Clinical and Experimental Research*, *37*(Suppl. 1), E52–58. http://dx.doi.org/10.1111/j.1530-0277.2012.01921.x.

Kim, D. J., Yoon, S. J., Choi, B., Kim, T. S., Woo, Y. S., Kim, W., et al. (2005). Increased fasting plasma ghrelin levels during alcohol abstinence. *Alcohol and Alcoholism*, *40*(1), 76–79. http://dx.doi.org/10.1093/alcalc/agh108.

Kobeissy, F. H., Jeung, J. A., Warren, M. W., Geier, J. E., & Gold, M. S. (2008). Changes in leptin, ghrelin, growth hormone and neuropeptide-y after an acute model of mdma and methamphetamine exposure in rats. *Addiction Biology*, *13*(1), 15–25. http://dx.doi.org/10.1111/j.1369-1600.2007.00083.x.

Kojima, M., Hosoda, H., Date, Y., Nakazato, M., Matsuo, H., & Kangawa, K. (1999). Ghrelin is a growth-hormone-releasing acylated peptide from stomach. *Nature*, *402*(6762), 656–660. http://dx.doi.org/10.1038/45230.

Kojima, M., & Kangawa, K. (2005). Ghrelin: Structure and function. *Physiological Reviews*, *85*(2), 495–522. http://dx.doi.org/10.1152/physrev.00012.2004.

Kokkinos, A., Tentolouris, N., Kyriakaki, E., Argyrakopoulou, G., Doupis, J., Psallas, M., et al. (2007). Differentiation in the short- and long-term effects of smoking on plasma total ghrelin concentrations between male nonsmokers and habitual smokers. *Metabolism*, *56*(4), 523–527. http://dx.doi.org/10.1016/j.metabol.2006.11.012.

Koob, G. F. (2005). The neurocircuitry of addiction: Implications for treatment. *Clinical Neuroscience Research*, *5*(2), 89–101. http://dx.doi.org/10.1016/j.cnr.2005.08.005.

Koob, G. F., & Volkow, N. D. (2016). Neurobiology of addiction: A neurocircuitry analysis. *The Lancet Psychiatry*, *3*(8), 760–773. http://dx.doi.org/10.1016/s2215-0366(16)00104-8.

Koopmann, A., von der Goltz, C., Grosshans, M., Dinter, C., Vitale, M., Wiedemann, K., et al. (2012). The association of the appetitive peptide acetylated ghrelin with alcohol craving in early abstinent alcohol dependent individuals. *Psychoneuroendocrinology*, *37*(7), 980–986. http://dx.doi.org/10.1016/j.psyneuen.2011.11.005.

Kraus, T., Schanze, A., Groschl, M., Bayerlein, K., Hillemacher, T., Reulbach, U., et al. (2005). Ghrelin levels are increased in alcoholism. *Alcoholism: Clinical and Experimental Research*, *29*(12), 2154–2157.

Kroemer, N. B., Wuttig, F., Bidlingmaier, M., Zimmermann, U. S., & Smolka, M. N. (2015). Nicotine enhances modulation of food-cue reactivity by leptin and ghrelin in the ventromedial prefrontal cortex. *Addiction Biology*, *20*(4), 832–844. http://dx.doi.org/10.1111/adb.12167.

Kunath, N., & Dresler, M. (2014). Ghrelin and memory. In J. Portelli & I. Smolders (Eds.), *Central functions of the ghrelin receptor* (pp. 167–175). New York, NY: Springer.

Labarthe, A., Fiquet, O., Hassouna, R., Zizzari, P., Lanfumey, L., Ramoz, N., et al. (2014). Ghrelin-derived peptides: A link between appetite/reward, gh axis, and psychiatric disorders? *Frontiers in Endocrinology (Lausanne)*, *5*, 163. http://dx.doi.org/10.3389/fendo.2014.00163.

Landgren, S., Berglund, K., Jerlhag, E., Fahlke, C., Balldin, J., Berggren, U., et al. (2011). Reward-related genes and personality traits in alcohol-dependent individuals: A pilot case control study. *Neuropsychobiology*, *64*(1), 38–46. http://dx.doi.org/10.1159/000324992.

Landgren, S., Engel, J. A., Hyytia, P., Zetterberg, H., Blennow, K., & Jerlhag, E. (2011). Expression of the gene encoding the ghrelin receptor in rats selected for differential alcohol preference. *Behavioural Brain Research*, *221*(1), 182–188. http://dx.doi.org/10.1016/j.bbr.2011.03.003.

Landgren, S., Jerlhag, E., Hallman, J., Oreland, L., Lissner, L., Strandhagen, E., et al. (2010). Genetic variation of the ghrelin signaling system in females with severe alcohol dependence. *Alcoholism: Clinical and Experimental Research*, *34*(9), 1519–1524. http://dx.doi.org/10.1111/j.1530-0277.2010.01236.x.

Landgren, S., Jerlhag, E., Zetterberg, H., Gonzalez-Quintela, A., Campos, J., Olofsson, U., et al. (2008). Association of pro-ghrelin and GHS-R1a gene polymorphisms and haplotypes with heavy alcohol use and body mass. *Alcoholism, Clinical and Experimental Research*, *32*(12), 2054–2061. http://dx.doi.org/10.1111/j.1530-0277.2008.00793.x.

Landgren, S., Simms, J. A., Hyytia, P., Engel, J. A., Bartlett, S. E., & Jerlhag, E. (2012). Ghrelin receptor (GHS-R1a) antagonism suppresses both operant alcohol self-administration and high alcohol consumption in rats. *Addiction Biology*, *17*(1), 86–94. http://dx.doi.org/10.1111/j.1369-1600.2010.00280.x.

Landgren, S., Simms, J. A., Thelle, D. S., Strandhagen, E., Bartlett, S. E., Engel, J. A., et al. (2011). The ghrelin signalling system is involved in the consumption of sweets. *PLoS One*, *6*(3), e18170. http://dx.doi.org/10.1371/journal.pone.0018170.

Laviano, A., Molfino, A., Rianda, S., & Rossi Fanelli, F. (2012). The growth hormone secretagogue receptor (GHS-R). *Current Pharmaceutical Design*, *18*(31), 4749–4754.

Lee, H., Joe, K. H., Kim, W., Park, J., Lee, D. H., Sung, K. W., et al. (2006). Increased leptin and decreased ghrelin level after smoking cessation. *Neuroscience Letters*, *409*(1), 47–51. http://dx.doi.org/10.1016/j.neulet.2006.09.013.

Leggio, L., Addolorato, G., Cippitelli, A., Jerlhag, E., Kampov-Polevoy, A. B., & Swift, R. M. (2011). Role of feeding-related pathways in alcohol dependence: A focus on sweet preference, npy, and ghrelin. *Alcoholism: Clinical and Experimental Research*, *35*(2), 194–202. http://dx.doi.org/10.1111/j.1530-0277.2010.01334.x.

Leggio, L., Ferrulli, A., Cardone, S., Nesci, A., Miceli, A., Malandrino, N., et al. (2012). Ghrelin system in alcohol-dependent subjects: Role of plasma ghrelin levels in alcohol drinking and craving. *Addiction Biology*, *17*(2), 452–464. http://dx.doi.org/10.1111/j.1369-1600.2010.00308.x.

Leggio, L., Schwandt, M. L., Oot, E. N., Dias, A. A., & Ramchandani, V. A. (2013). Fasting-induced increase in plasma ghrelin is blunted by intravenous alcohol administration: A within-subject placebo-controlled study. *Psychoneuroendocrinology*, *38*(12), 3085–3091. http://dx.doi.org/10.1016/j.psyneuen.2013.09.005.

Leggio, L., Zywiak, W. H., Fricchione, S. R., Edwards, S. M., de la Monte, S. M., Swift, R. M., et al. (2014). Intravenous ghrelin administration increases alcohol craving in alcohol-dependent heavy drinkers: A preliminary investigation. *Biological Psychiatry*, *76*(9), 734–741. http://dx.doi.org/10.1016/j.biopsych.2014.03.019.

Lemieux, A. M., & al'Absi, M. (2017). Changes in circulating peptide YY and ghrelin are associated with early smoking relapse. *Biological Psychology*. http://dx.doi.org/10.1016/j.biopsycho.2017.03.007. (in press).

Lyons, A. M., Lowery, E. G., Sparta, D. R., & Thiele, T. E. (2008). Effects of food availability and administration of orexigenic and anorectic agents on elevated ethanol drinking associated with drinking in the dark procedures. *Alcoholism, Clinical and Experimental Research*, *32*(11), 1962–1968. http://dx.doi.org/10.1111/j.1530-0277.2008.00784.x.

Maric, T., Sedki, F., Ronfard, B., Chafetz, D., & Shalev, U. (2012). A limited role for ghrelin in heroin self-administration and food deprivation-induced reinstatement of heroin seeking in rats. *Addiction Biology*, *17*(3), 613–622. http://dx.doi.org/10.1111/j.1369-1600.2011.00396.x.

Mason, B. L., Wang, Q., & Zigman, J. M. (2014). The central nervous system sites mediating the orexigenic actions of ghrelin. *Annual Review of Physiology*, *76*, 519–533. http://dx.doi.org/10.1146/annurev-physiol-021113-170310.

Mazidi, M., Baghban Taraghdari, S., Rezaee, P., Kamgar, M., Jomezadeh, M. R., Akbarieh Hasani, O., et al. (2014). The effect of hydroalcoholic extract of *Cannabis sativa* on appetite hormone in rat. *Journal of Complementary & Integrative Medicine*, *11*(4), 253–257. http://dx.doi.org/10.1515/jcim-2014-0006.

Mear, Y., Enjalbert, A., & Thirion, S. (2013). GHS-R1a constitutive activity and its physiological relevance. *Frontiers in Neuroscience*, *7*, 87. http://dx.doi.org/10.3389/fnins.2013.00087.

Melis, M. R., Mascia, M. S., Succu, S., Torsello, A., Muller, E. E., Deghenghi, R., et al. (2002). Ghrelin injected into the paraventricular nucleus of the hypothalamus of male rats induces feeding but not penile erection. *Neuroscience Letters*, *329*(3), 339–343.

Meyer, R. M., Burgos-Robles, A., Liu, E., Correia, S. S., & Goosens, K. A. (2014). A ghrelin-growth hormone axis drives stress-induced vulnerability to enhanced fear. *Molecular Psychiatry*, *19*(12), 1284–1294. http://dx.doi.org/10.1038/mp.2013.135.

Monti, P. M., Binkoff, J. A., Abrams, D. B., Zwick, W. R., Nirenberg, T. D., & Liepman, M. R. (1987). Reactivity of alcoholics and nonalcoholics to drinking cues. *Journal of Abnormal Psychology*, *96*(2), 122–126.

Monti, P. M., Rohsenow, D. J., & Hutchison, K. E. (2000). Toward bridging the gap between biological, psychobiological and psychosocial models of alcohol craving. *Addiction*, *95*(Suppl. 2), S229–236.

Muccioli, G., Baragli, A., Granata, R., Papotti, M., & Ghigo, E. (2007). Heterogeneity of ghrelin/growth hormone secretagogue receptors. Toward the understanding of the molecular identity of novel ghrelin/GHS receptors. *Neuroendocrinology*, *86*(3), 147–164. http://dx.doi.org/10.1159/000105141.

Mutschler, J., Graf, N., Spanaus, K. S., Rössler, W., Hergan, K., Binkert, C. A., et al. (2012). Circulating ghrelin levels are not associated with craving and withdrawal symptoms in acute nicotine withdrawal. *Psychiatria Danubina*, *24*(2), 229–230.

Naleid, A. M., Grace, M. K., Cummings, D. E., & Levine, A. S. (2005). Ghrelin induces feeding in the mesolimbic reward pathway between the ventral tegmental area and the nucleus accumbens. *Peptides*, *26*(11), 2274–2279. http://dx.doi.org/10.1016/j.peptides.2005.04.025.

Navarro, M., Carrera, M. R., Fratta, W., Valverde, O., Cossu, G., Fattore, L., et al. (2001). Functional interaction between opioid and cannabinoid receptors in drug self-administration. *The Journal of Neuroscience*, *21*(14), 5344–5350.

Olszewski, P. K., Schioth, H. B., & Levine, A. S. (2008). Ghrelin in the CNS: From hunger to a rewarding and memorable meal? *Brain Research Reviews*, *58*(1), 160–170. http://dx.doi.org/10.1016/j.brainresrev.2008.01.004.

O'Malley, S. S., Krishnan-Sarin, S., Farren, C., Sinha, R., & Kreek, M. J. (2002). Naltrexone decreases craving and alcohol self-administration in alcohol-dependent subjects and activates the hypothalamo-pituitary-adrenocortical axis. *Psychopharmacology*, *160*(1), 19–29. http://dx.doi.org/10.1007/s002130100919.

Palotai, M., Bagosi, Z., Jaszberenyi, M., Csabafi, K., Dochnal, R., Manczinger, M., et al. (2013a). Ghrelin amplifies the nicotine-induced dopamine release in the rat striatum. *Neurochemistry International*, *63*(4), 239–243. http://dx.doi.org/10.1016/j.neuint.2013.06.014.

Palotai, M., Bagosi, Z., Jaszberenyi, M., Csabafi, K., Dochnal, R., Manczinger, M., et al. (2013b). Ghrelin and nicotine stimulate equally the dopamine release in the rat amygdala. *Neurochemical Research*, *38*(10), 1989–1995. http://dx.doi.org/10.1007/s11064-013-1105-1.

Panagopoulos, V. N., & Ralevski, E. (2014). The role of ghrelin in addiction: A review. *Psychopharmacology*, *231*(14), 2725–2740. http://dx.doi.org/10.1007/s00213-014-3640-0.

Perboni, S., & Inui, A. (2010). Appetite and gastrointestinal motility: Role of ghrelin-family peptides. *Clinical Nutrition*, *29*(2), 227–234. http://dx.doi.org/10.1016/j.clnu.2008.10.016.

Perello, M., & Zigman, J. M. (2012). The role of ghrelin in reward-based eating. *Biological Psychiatry*, *72*(5), 347–353. http://dx.doi.org/10.1016/j.biopsych.2012.02.016.

Pilhatsch, M., Scheuing, H., Kroemer, N., Kobiella, A., Bidlingmaier, M., Farger, G., et al. (2014). Nicotine administration in healthy non-smokers reduces appetite but does not alter plasma ghrelin. *Human Psychopharmacology*, *29*(4), 384–387. http://dx.doi.org/10.1002/hup.2405.

Quarta, D., Di Francesco, C., Melotto, S., Mangiarini, L., Heidbreder, C., & Hedou, G. (2009). Systemic administration of ghrelin increases extracellular dopamine in the shell but not the core subdivision of the nucleus accumbens. *Neurochemistry International*, *54*(2), 89–94. http://dx.doi.org/10.1016/j.neuint.2008.12.006.

Ralevski, E., Horvath, T. L., Shanabrough, M., Hayden, R., Newcomb, J., & Petrakis, I. (2017). Ghrelin is supressed by intravenous alcohol and is related to stimulant and sedative effects of alcohol. *Alcohol and Alcoholism*, *52*(4), 431–438. http://dx.doi.org/10.1093/alcalc/agx022.

Revitsky, A. R., & Klein, L. C. (2013). Role of ghrelin in drug abuse and reward-relevant behaviors: A burgeoning field and gaps in the literature. *Current Drug Abuse Reviews*, *6*(3), 231–244.

Riggs, P. K., Vaida, F., Rossi, S. S., Sorkin, L. S., Gouaux, B., Grant, I., et al. (2012). A pilot study of the effects of cannabis on appetite hormones in HIV-infected adult men. *Brain Research*, *1431*, 46–52. http://dx.doi.org/10.1016/j.brainres.2011.11.001.

Saper, C. B., Chou, T. C., & Elmquist, J. K. (2002). The need to feed: Homeostatic and hedonic control of eating. *Neuron*, *36*(2), 199–211.

Schele, E., Bake, T., Rabasa, C., & Dickson, S. L. (2016). Centrally administered ghrelin acutely influences food choice in rodents. *PLoS One*, *11*(2), e0149456. http://dx.doi.org/10.1371/journal.pone.0149456.

Schellekens, H., Dinan, T. G., & Cryan, J. F. (2013). Taking two to tango: A role for ghrelin receptor heterodimerization in stress and reward. *Frontiers in Neuroscience*, 7, 148. http://dx.doi.org/10.3389/fnins.2013.00148.

Schneider, E. R., Rada, P., Darby, R. D., Leibowitz, S. F., & Hoebel, B. G. (2007). Orexigenic peptides and alcohol intake: Differential effects of orexin, galanin, and ghrelin. *Alcoholism, Clinical and Experimental Research*, *31*(11), 1858–1865. http://dx.doi.org/10.1111/j.1530-0277.2007.00510.x.

Schuette, L. M., Gray, C. C., & Currie, P. J. (2013). Microinjection of ghrelin into the ventral tegmental area potentiates cocaine-induced conditioned place preference. *Journal of Behavioral and Brain Science*, *3*(8), 576–580.

Song, L., Zhu, Q., Liu, T., Yu, M., Xiao, K., Kong, Q., et al. (2013). Ghrelin modulates lateral amygdala neuronal firing and blocks acquisition for conditioned taste aversion. *PLoS One*, *8*(6), e65422. http://dx.doi.org/10.1371/journal.pone.0065422.

Spencer, S. J., Emmerzaal, T. L., Kozicz, T., & Andrews, Z. B. (2015). Ghrelin's role in the hypothalamic-pituitary-adrenal axis stress response: Implications for mood disorders. *Biological Psychiatry*, *78*(1), 19–27. http://dx.doi.org/10.1016/j.biopsych.2014.10.021.

Spencer, S. J., Xu, L., Clarke, M. A., Lemus, M., Reichenbach, A., Geenen, B., et al. (2012). Ghrelin regulates the hypothalamic-pituitary-adrenal axis and restricts anxiety after acute stress. *Biological Psychiatry*, *72*(6), 457–465. http://dx.doi.org/10.1016/j.biopsych.2012.03.010.

Stevenson, J. R., Buirkle, J. M., Buckley, L. E., Young, K. A., Albertini, K. M., & Bohidar, A. E. (2015). GHS-R1a antagonism reduces alcohol but not sucrose preference in prairie voles. *Physiology and Behavior*, *147*, 23–29. http://dx.doi.org/10.1016/j.physbeh.2015.04.001.

Stevenson, J. R., Francomacaro, L. M., Bohidar, A. E., Young, K. A., Pesarchick, B. F., Buirkle, J. M., et al. (2016). Ghrelin receptor (GHS-R1a) antagonism alters preference for ethanol and sucrose in a concentration-dependent manner in prairie voles. *Physiology and Behavior*, *155*, 231–236.

Stievenard, A., Mequinion, M., Andrews, Z. B., Destee, A., Chartier-Harlin, M. C., Viltart, O., et al. (2017). Is there a role for ghrelin in central dopaminergic systems? Focus on nigrostriatal and mesocorticolimbic pathways. *Neuroscience and Biobehavioral Reviews*, *73*, 255–275. http://dx.doi.org/10.1016/j.neubiorev.2016.11.021.

Storr, M. A., & Sharkey, K. A. (2007). The endocannabinoid system and gut-brain signalling. *Current Opinion in Pharmacology*, 7(6), 575–582. http://dx.doi.org/10.1016/j.coph.2007.08.008.

Suchankova, P., Engel, J. A., & Jerlhag, E. (2016). Sub-chronic ghrelin receptor blockade attenuates alcohol- and amphetamine-induced locomotor stimulation in mice. *Alcohol and Alcoholism*, *51*(2), 121–127. http://dx.doi.org/10.1093/alcalc/agv100.

Suchankova, P., Jerlhag, E., Jayaram-Lindström, N., Nilsson, S., Toren, K., Rosengren, A., et al. (2013). Genetic variation of the ghrelin signalling system in individuals with amphetamine dependence. *PLoS One*, *8*(4), e61242.

Suchankova, P., Nilsson, S., von der Pahlen, B., Santtila, P., Sandnabba, K., Johansson, A., et al. (2016). Genetic variation of the growth hormone secretagogue receptor gene is associated with alcohol use disorders identification test scores and smoking. *Addiction Biology*, *21*(2), 481–488. http://dx.doi.org/10.1111/adb.12277.

Suchankova, P., Steensland, P., Fredriksson, I., Engel, J. A., & Jerlhag, E. (2013). Ghrelin receptor (GHS-R1a) antagonism suppresses both alcohol consumption and the alcohol deprivation effect in rats following long-term voluntary alcohol consumption. *PLoS One*, *8*(8), e71284. http://dx.doi.org/10.1371/journal.pone.0071284.

Suchankova, P., Yan, J., Schwandt, M. L., Stangl, B. L., Caparelli, E. C., Momenan, R., et al. (2015). The glucagon-like peptide-1 receptor as a potential treatment target in alcohol use disorder: Evidence from human genetic association studies and a mouse model of alcohol dependence. *Translational Psychiatry*, *5*, e583. http://dx.doi.org/10.1038/tp.2015.68.

Suchankova, P., Yan, J., Schwandt, M. L., Stangl, B. L., Jerlhag, E., Engel, J. A., et al. (2017). The Leu72Met polymorphism of the prepro-ghrelin gene is associated with alcohol consumption and subjective responses to alcohol: Preliminary findings. *Alcohol and Alcoholism*, *52*(4), 425–430. http://dx.doi.org/10.1093/alcalc/agx021.

Sustkova-Fiserova, M., Jerabek, P., Havlickova, T., Kacer, P., & Krsiak, M. (2014). Ghrelin receptor antagonism of morphine-induced accumbens dopamine release and behavioral stimulation in rats. *Psychopharmacology*, *231*(14), 2899–2908. http://dx.doi.org/10.1007/s00213-014-3466-9.

Sustkova-Fiserova, M., Jerabek, P., Havlickova, T., Syslova, K., & Kacer, P. (2016). Ghrelin and endocannabinoids participation in morphine-induced effects in the rat nucleus accumbens. *Psychopharmacology*, *233*(3), 469–484. http://dx.doi.org/10.1007/s00213-015-4119-3.

Szulc, M., Mikolajczak, P. L., Geppert, B., Wachowiak, R., Dyr, W., & Bobkiewicz-Kozlowska, T. (2013). Ethanol affects acylated and total ghrelin levels in peripheral blood of alcohol-dependent rats. *Addiction Biology*, *18*(4), 689–701. http://dx.doi.org/10.1111/adb.12025.

Tack, J., Depoortere, I., Bisschops, R., Delporte, C., Coulie, B., Meulemans, A., et al. (2006). Influence of ghrelin on interdigestive gastrointestinal motility in humans. *Gut*, *55*(3), 327–333. http://dx.doi.org/10.1136/gut.2004.060426.

Tamboli, R. A., Antoun, J., Sidani, R. M., Clements, A., Harmata, E. E., Marks-Shulman, P., et al. (2017). Metabolic responses to exogenous ghrelin in obesity and early after Roux-en-Y gastric bypass in humans. *Diabetes, Obesity & Metabolism*, *19*(9), 267–1275. http://dx.doi.org/10.1111/dom.12952.

Tessari, M., Catalano, A., Pellitteri, M., Di Francesco, C., Marini, F., Gerrard, P. A., et al. (2007). Correlation between serum ghrelin levels and cocaine-seeking behaviour triggered by cocaine-associated conditioned stimuli in rats. *Addiction Biology*, *12*(1), 22–29.

Tomoda, K., Kubo, K., Nishii, Y., Yamamoto, Y., Yoshikawa, M., & Kimura, H. (2012). Changes of ghrelin and leptin levels in plasma by cigarette smoke in rats. *The Journal of Toxicological Sciences*, *37*(1), 131–138.

Tschop, M., Smiley, D. L., & Heiman, M. L. (2000). Ghrelin induces adiposity in rodents. *Nature*, *407*(6806), 908–913. http://dx.doi.org/10.1038/35038090.

Vengeliene, V. (2013). The role of ghrelin in drug and natural reward. *Addiction Biology*, *18*(6), 897–900. http://dx.doi.org/10.1111/adb.12114.

Volkow, N. D., Wang, G. J., Fowler, J. S., Tomasi, D., & Baler, R. (2012). Food and drug reward: Overlapping circuits in human obesity and addiction. *Current Topics in Behavioral Neurosciences*, *11*, 1–24. http://dx.doi.org/10.1007/7854_2011_169.

Wei, X. J., Sun, B., Chen, K., Lv, B., Luo, X., & Yan, J. Q. (2015). Ghrelin signaling in the ventral tegmental area mediates both reward-based feeding and fasting-induced hyperphagia on high-fat diet. *Neuroscience, 300*, 53–62. http://dx.doi.org/10.1016/j.neuroscience.2015.05.001.

Wellman, P. J., Clifford, P. S., Rodriguez, J., Hughes, S., Eitan, S., Brunel, L., et al. (2011). Pharmacologic antagonism of ghrelin receptors attenuates development of nicotine induced locomotor sensitization in rats. *Regulatory Peptides, 172*(1–3), 77–80. http://dx.doi.org/10.1016/j.regpep.2011.08.014.

Wellman, P. J., Davis, K. W., & Nation, J. R. (2005). Augmentation of cocaine hyperactivity in rats by systemic ghrelin. *Regulatory Peptides, 125*(1–3), 151–154. http://dx.doi.org/10.1016/j.regpep.2004.08.013.

Wellman, P. J., Hollas, C. N., & Elliott, A. E. (2008). Systemic ghrelin sensitizes cocaine-induced hyperlocomotion in rats. *Regulatory Peptides, 146*(1–3), 33–37. http://dx.doi.org/10.1016/j.regpep.2007.07.007.

Wise, R. A. (2000). Addiction becomes a brain disease. *Neuron, 26*(1), 27–33.

Wren, A. M., Seal, L. J., Cohen, M. A., Brynes, A. E., Frost, G. S., Murphy, K. G., et al. (2001). Ghrelin enhances appetite and increases food intake in humans. *The Journal of Clinical Endocrinology and Metabolism, 86*(12), 5992. http://dx.doi.org/10.1210/jcem.86.12.8111.

Wren, A. M., Small, C. J., Ward, H. L., Murphy, K. G., Dakin, C. L., Taheri, S., et al. (2000). The novel hypothalamic peptide ghrelin stimulates food intake and growth hormone secretion. *Endocrinology, 141*(11), 4325–4328. http://dx.doi.org/10.1210/endo.141.11.7873.

Wurst, F. M., Graf, I., Ehrenthal, H. D., Klein, S., Backhaus, J., Blank, S., et al. (2007). Gender differences for ghrelin levels in alcohol-dependent patients and differences between alcoholics and healthy controls. *Alcoholism: Clinical and Experimental Research, 31*(12), 2006–2011. http://dx.doi.org/10.1111/j.1530-0277.2007.00527.x.

Yang, J., Brown, M. S., Liang, G., Grishin, N. V., & Goldstein, J. L. (2008). Identification of the acyltransferase that octanoylates ghrelin, an appetite-stimulating peptide hormone. *Cell, 132*(3), 387–396. http://dx.doi.org/10.1016/j.cell.2008.01.017.

Yoon, S. J., Pae, C. U., Lee, H., Choi, B., Kim, T. S., Lyoo, I. K., et al. (2005). Ghrelin precursor gene polymorphism and methamphetamine dependence in the korean population. *Neuroscience Research, 53*(4), 391–395.

Yoshimoto, K., Nagao, M., Watanabe, Y., Yamaguchi, T., Ueda, S., Kitamura, Y., et al. (2017). Enhanced alcohol-drinking behavior associated with active ghrelinergic and serotoninergic neurons in the lateral hypothalamus and amygdala. *Pharmacology, Biochemistry and Behavior, 153*, 1–11. http://dx.doi.org/10.1016/j.pbb.2016.12.001.

Ypsilantis, P., Politou, M., Anagnostopoulos, C., Tsigalou, C., Kambouromiti, G., Kortsaris, A., et al. (2013). Effects of cigarette smoke exposure and its cessation on body weight, food intake and circulating leptin, and ghrelin levels in the rat. *Nicotine & Tobacco Research, 15*(1), 206–212. http://dx.doi.org/10.1093/ntr/nts113.

Zigman, J. M., Jones, J. E., Lee, C. E., Saper, C. B., & Elmquist, J. K. (2006). Expression of ghrelin receptor mrna in the rat and the mouse brain. *The Journal of Comparative Neurology, 494*(3), 528–548. http://dx.doi.org/10.1002/cne.20823.

Zimmermann, U. S., Buchmann, A., Steffin, B., Dieterle, C., & Uhr, M. (2007). Alcohol administration acutely inhibits ghrelin secretion in an experiment involving psychosocial stress. *Addiction Biology, 12*(1), 17–21. http://dx.doi.org/10.1111/j.1369-1600.2006.00026.x.

CHAPTER FIVE

The Role of the Melanocortin System in Drug and Alcohol Abuse

Montserrat Navarro[1]
University of North Carolina, Chapel Hill, NC, United States
[1]Corresponding author: e-mail address: mthiele@unc.edu

Contents

Abstract

The melanocortin (MC) system is composed of different peptides centrally produced by proopiomelanocortin neurons in the arcuate nucleus of the hypothalamus (N.Arc) and the medulla. These peptides act through five subtypes of melanocortin receptors (MCRs) that belong to the family of seven-transmembrane G-protein-coupled receptors that are coupled through $G\alpha_s$ signaling pathways. MC3R and MC4R are the predominant MCR subtypes in the brain, and they are widely expressed in brain regions thought to modulate drug self-administration, including the nucleus accumbens, the hypothalamus, and the ventral tegmental area. The MC system is unique in terms of having an endogenous antagonist/inverse agonist, agouti-related protein (AgRP), also produces in the N.Arc and secreted in terminals expressing MCRs. There is a large body of research showing the role of the MC and AgRP systems in neurobiological responses to drugs of abuse, in particular, neurobiological responses to ethanol. In this chapter, we discuss the most recent evidence that supports the role of the MC/AgRP systems in modulating neurobiological responses to drugs of abuse, with a focus on ethanol consumption.

International Review of Neurobiology, Volume 136
ISSN 0074-7742
http://dx.doi.org/10.1016/bs.irn.2017.06.009

1. INTRODUCTION

The melanocortin (MC) system is composed by peptides that are cleaved from the polypeptide precursor proopiomelanocortin (POMC) (Castro & Morrison, 1997; Eipper & Mains, 1978; Mains & Eipper, 1976). These peptides include adrenocorticotropic hormone (ACTH) alpha-melanocyte-stimulating hormone (α-MSH), β-MSH, and γ-MSH (Hadley & Haskell-Luevano, 1999). The POMC gene is located on chromosome 2p23.3. POMC is a 31-kDa protein that is synthesized in the corticotrophs and melanotrophs cells of the anterior (PA) and intermediate (PI) lobes of the pituitary, respectively (Akil, Shiomi, & Matthews, 1985; Browne, Bennett, & Solomon, 1981; Castro & Morrison, 1997; Mains & Eipper, 1981), as well as in peptidergic neurons in the arcuate nucleus of the hypothalamus (N.Arc) and the nucleus of the solitary tract (NTS) (Akil et al., 1985). Peripheral expression of POMC has also been observed in the skin, spleen, thyroid, and the gastrointestinal tract (Smith & Funder, 1988; Yeo, Farroqi, Challis, Jackson, & O'Rahilly, 2000).

As most prohormones, POMC-derived peptides are generated by extensive posttranslational processing that involves several enzymes, including prohormone convertases (PCs) 1/3 (PC1) and 2 (PC2) (Benjannet, Rondeau, Day, Chretien, & Seidah, 1991). The differential expressions of these PCs in various tissues make this posttranslational processing tissue specific (Castro & Morrison, 1997; Pritchard, Turnbull, & White, 2002; Smith & Funder, 1988), which allows a multitude of peptides with different physiological functions. In the PA, POMC is expressed in the corticotroph cells and is posttranslationally cleaved by PC1 into peptide hormones that include ACTH, β-lipotropin (β-LPH), and N-terminal POMC fragment (N-POC). In the PI and the hypothalamus (HT), the melanotrophs cells expressed PC2. PC2 further processes ACTH to generate ACTH 1–17, corticotrophin-like intermediate lobe peptide (CLIP), and γ-MSH from N-terminal POM (Emerson & Eipper, 1986; Pritchard & White, 2007). In addition to PCs, other processing enzymes such as carboxypeptidase E, peptidyl α-ami-dating monooxygenase, and *n*-acetyltransferase are necessary to generate α-MSH (Wilkinson, 2006). PC2 also cleaved β-LPH to γ-LPH and beta endorphin (β-endorphin) (Akil et al., 1985; Browne et al., 1981; Castro & Morrison, 1997; Eipper & Mains, 1981; Mains & Eipper, 1981; Pritchard & White, 2007; Wardlaw, 2011) (Fig. 1).

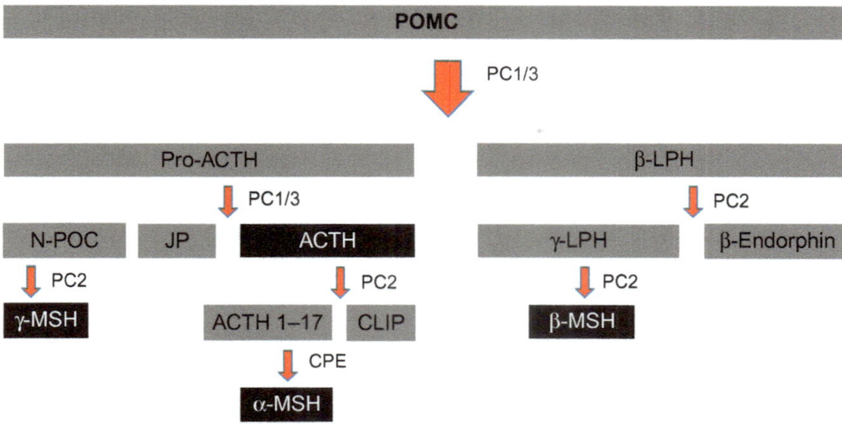

Fig. 1 Simplified model of posttranslational processing of POMC. *ACTH*, adrenocorticotropin hormone; *CLIP*, corticotrophin-like intermediate lobe peptide; *JP*, junctional peptide; *LPH*, lipotropic hormone; *MSH*, melanocyte-stimulating hormone; *N-POC*, *N*-pro-opiocortin; *PC*, prohormone convertase; *POMC*, proopiomelanocortin.

Two main discoveries changed the view of the MC system. The first one was in the 1990s, when the existence of melanocortin receptors (MCRs) in the brain was discovered (Chhajlani & Wikberg, 1992; Chhajlani, Muceniece, & Wikberg, 1993; Mountjoy, Mortrud, Low, Simerly, & Cone, 1994; Roselli-Rehfuss et al., 1993; Tatro, 1990). And the second one in 1994, with the demonstration that the agouti protein was an antagonist for the melanocyte-stimulating hormone receptor (Lu et al., 1994). These two findings have been some of the most important discovers in the field of the MC system and have subsequently helped stimulate an extensive amount of research.

2. MELANOCORTIN RECEPTORS

The MC peptides exert their actions through at least five MCRs subtypes, namely, MC1 receptor (MC1R), MC2R, MC3R, MC4R, and MC5R, all of which couple to heterotrimeric G-proteins that stimulate adenylyl cyclase activity and induce cyclic 3′,5′-adenosine monophosphate (Voisey, Carrol, & van Daal, 2003; for review, see Cone, 2006; Dores, Liang, Davis, Thomas, & Petko, 2016; Dores et al., 2014). In addition to the MCR, a unique characteristic of the MC system is the existence of its own endogenous antagonists/inverse agonist, *agouti-signaling protein* (ASP) and *agouti-related protein* (AgRP) (Adan & Kas, 2003; Fong et al., 1997; Lu et al., 1994; Quillan et al., 1998).

All the MCRs exhibit different expression patterns and have different affinity for the MC peptides. MC1R is a 317-amino acid protein (García-Borrón, Sánchez-Laorden, & Jiménez-Cervantes, 2005) whose gene is localized in the 16q24.3 chromosome (Gantz et al., 1994). It was cloned for the first time in 1992 from mouse and human melanoma cells and melanocytes (Chhajlani & Wikberg, 1992; Mountjoy, Robbins, Mortrud, & Cone, 1992). This receptor is widely expressed in skin (keratinocytes, melanocytes, and fibroblasts), melanoma cells, immune cells (monocytes and neutrophils cells, dendritic cells, B-lymphocytes), endothelial cells and glial cells (Böhm et al., 1999; Catania et al., 1996; Chhajlani, 1996; Loir et al., 1999; Mountjoy et al., 1992; Stander et al., 2002). MC1R has been implicated in different biological functions, such as antiinflammatory and immunomodulatory responses, and it is well known its role in pigmentation (Rees, 2003; Wikberg et al., 2000). Briefly, α-MSH binds to this receptor to stimulate the synthesis of eumelamin (brown-black pigments). On the other hand, ASP binds to MC1R, inhibiting the binding of α-MSH and stimulating the synthesis of pheomelanin (yellow–red pigments) (for review, see Wolf Horrell, Boulanger, & D'Orazio, 2016). In order of ligand potency, α-MSH and ACTH are equipotent, followed by β-MSH and γ-MSH, the least potent of the MC peptides for MC1R (Suzuki, Cone, Im, Nordlund, & Abdel-Malek, 1996) (Table 1). As mentioned earlier, MC1R is widely expressed

Table 1 MC Peptides Act Through 5G-Coupled Subtypes of MCRs, Differentially Expressed Throughout the Body, With Varying Degrees of Potency

MCR	Potency of MC Ligand	Antagonist	Primary Expression Site
MC1R	α-MSH = ACTH > β-MSH > γ-MSK	ASP	Melanocytes
MC2R	ACTH		Adrenal cortex
MC3R	α-MSH = ACTH = β-MSH = γ-MSH	AgRP	Hypothalamus, brainstem
MC4R	α-MSH = ACTH = β-MSH > γ-MSH	AgRP/ASP	Hippocampus, cortex, amygdala, hypothalamus, striatum, thalamus, brainstem
MC5R	α-MSK > ACTH > β-MSH > γ-MSH	AgRP	Exocrine glands

The MC system has its own endogenous antagonists.

and has additional functions other than its role in pigmentation. Immunohistochemistry studies (IHC) and in situ hybridization studies in both humans and rodents have shown the expression of MCR1 in the central nervous system, specifically in the periaqueductal gray (PAG) region (Xia, Wikberg, & Chhajlani, 1995). This area is recognized as an important control center for pain modulation, and the presence of MC1R suggests a functional implication of the MC1R in the modulation of pain and analgesia (Mogil et al., 2005, 2003).

MC2R was cloned from the adrenal gland in 1992 (Chhajlani & Wikberg, 1992). It is a 297-amino acid protein whose gene is localized in the 18p11.2 chromosome (Mountjoy et al., 1992). MC2R is the classical adrenocortical ACTH receptor, and ACTH is the only MC peptide that binds to MC2R (Schioth, Chhajlani, Muceniece, Klusa, & Wikberg, 1996) (Table 1). In situ hybridization studies have shown that MC2R is expressed in the adrenal cortex, zone reticularis, and zona fasciculate, where it mediates the production of glucocorticoids and mineralocorticoids, respectively (Mountjoy et al., 1992). Rodents studies have shown the presence of MC2R in adipocytes where it mediates stress-induced lipolysis via central ACTH (Boston & Cone, 1996; Møller et al., 2011). In humans, there is no evidence of MC2R on adipocytes (Bousquet-Melou, Galizky, La Fontan, & Berlan, 1995; Chhajlani, 1996; Hoch et al., 2007).

The MC3R is a 361-amino acid peptide whose gene is localized in the 20q13.2–13.3 chromosome, and it was first cloned from humans and rats in 1993 (Gantz, Konda, et al., 1993; Roselli-Rehfuss et al., 1993). All the MC peptides are approximately equipotent at the MC3R (Roselli-Rehfuss et al., 1993), though γ-MSH is reported to possess much greater selectivity and potency at the MC3R than at any other MCR subtype (Adan & Gispen, 2000; Renquist, Lippert, Sebag, Ellacott, & Cone, 2011) (Table 1). MC3R is expressed in many areas in the central nervous system and in peripheral tissues such as the gastrointestinal tract and placenta (Chhajlani, 1996). In the brain, Northern blot hybridization studies demonstrate that the highest densities of MC3R gene are found in the HT and limbic system (Gantz, Konda, et al., 1993; Lindblom, Schiot, Larsson, Wikberg, & Bergstrom, 1998; Roselli-Rehfuss et al., 1993). In situ hybridization studies have shown that in the HT, the greatest density of MC3R containing neurons is found in the N.Arc, ventromedial HT, lateral and posterior areas of the HT and preoptic area. Outside the HT, MC3R is found in the medial habenula nucleus of the thalamus, nucleus accumbens (NAc), ventral tegmental area (VTA), and raphe nucleus (Roselli-Rehfuss et al., 1993;

Xia & Wikberg, 1997). In the central nervous system, AgRP is the endogenous antagonist for MC3R (Ollmann et al., 1997). Double labeling in situ hybridization has been used to show that POMC, AgRP, and NPY neurons in the N.Arc are positive for MC3R mRNA (Bagnol et al., 1999; Mounien, Bizet, Boutelet, Vaudry, & Jegou, 2005). The presence of MC3R on POMC neurons suggests that it functions as an autoreceptor to control the release of POMC-derived peptides, and pharmacological and genetic studies support this idea (Lee et al., 2008; Marks, Hruby, Brookhart, & Cone, 2006; Olney, Navarro, & Thiele, 2014). MC3R is known to be involved in the regulation of various body functions such as circadian rhythm and immune responses (antiinflammatory). However, data from genetic studies using MC3R-deficient mice and pharmacological studies highlight the implication of this receptor in the regulation of energy homeostasis and food intake (Begriche et al., 2011; Butler et al., 2000; Chen, Marsh, et al., 2000; Girardet, Begriche, Ptitsyn, & Butler, 2014; Lee et al., 2008).

The MC4R is a 333-amino acid protein whose gene is localized in the 18q21.3 chromosome. It was first cloned by two independent laboratories in 1993 from mice (Gantz, Miwa, et al., 1993) and rats (Mountjoy et al., 1994). In order of ligand potency, α-MSH, β-MSH, and ACTH show similar potency for the MC4R, while γ-MSH is the least potent (Gantz, Miwa, et al., 1993) (Table 1). MC4R is widely expressed in the central nervous system including the cortex, hippocampus, amygdala septal region, corpus striatum, thalamus, HT, brainstem, and spinal cord (Cui et al., 2012; Kishi et al., 2003; Liu et al., 2003; Mountjoy, 2015; Mountjoy et al., 1994, 1992; Siljee et al., 2013). Data from human and animal studies show that in the periphery MC4R expression is barely detectable (Mountjoy & Wild, 1998). MC4R mRNA expression has been detected in heart, lung, kidney, testis, and in intestinal L cells (Caruso, Lagerstrom, Olszewski, Fredriksson, & Schioth, 2014; Panaro et al., 2014; Qu et al., 2014). Mapping the brain using in situ hybridization MC4R expression has been localized in HT regions such as the preoptic nucleus, periventricular nucleus, supraoptic nucleus, paraventricular nucleus (PVN), ventromedial nucleus (VMH), dorsomedial nucleus (DMH), lateral hypothalamic areas, and the N.Arc (Cui et al., 2012; Kishi et al., 2003; Mountjoy et al., 1994; Siljee et al., 2013). In the brainstem, expression of MC4R has been shown in the superior colliculus, dorsal motor nucleus of the vagus, raphe, substantia nigra (SN), NTS, lateral parabrachial nucleus, and reticular formation (Gautron et al., 2010; Kishi et al., 2003; Mountjoy et al., 1994; Wan et al., 2008;

Rossi et al., 2011; Tatro & Entwistle, 1994). Because of its distribution in the brain, MC4R is involved in the regulation of different neurobiological functions such as sexual behaviors (Argiolas, Melis, Murgia, & Schioth, 2000), reproduction (Schioth & Watanobe, 2002), grooming (Vergoni, Bertolini, Mutulis, Wikberg, & Schioth, 1998), cachexia (Scarlett et al., 2010), pain (Bertorelli et al., 2005; Bertolini, Poggioli, & Ferrari, 1979), neuroprotection (Adan et al., 1996; Giuliani et al., 2009), memory (Giuliani et al., 2011; Zhao, Sedman, Gibbs, & Ng, 1995), inflammation, and fever (Catania & Lipton, 1993). However, most research has focused on the role of MC4R in the regulation of food intake and energy homeostasis. The first gene from the MC system that was deleted in mice was the MC4R. Deletion of MC4R results in an obese phenotype (Chen, Metzger, et al., 2000; Huszar et al., 1997). From pharmacological studies, it has been shown that administration of MC4R agonists decreases and antagonists increases food intake and body weight (Fan, Boston, Kesterson, Hruby, & Cone, 1997; Grill, Ginsberg, Seeley, & Kaplan, 1998; Lu et al., 1994; Marsh et al., 1999; for review, see Anderson et al., 2016; Krashes, Lowell, & Garfield, 2016). The role of MC4R in the regulation of food intake and energy homeostasis has been well documented, and as a result, this receptor has been proposed as a drug target for obesity (Fani, Bak, Delhanty, van Rossum, & van den Akker, 2014).

The MC5R is a 325-amino acid protein whose gene is localized in the 18q11.2 chromosome (Barret, MacDonald, Helliwell, Davidson, & Morgan, 1994; Griffon et al., 1994). In order of ligand potency, α-MSH is more potent than ACTH, which is equally as potent as β-MSH, and γ-MSH is the least potent MC peptide at the MC5R (Abdel-Malek, 2001; Schioth, Muceniece, Wikberg, & Chhajlani, 1995) (Table 1). The expression of MC5R in the central nervous system is limited, but it is widely express in the periphery (Griffon et al., 1994; Labbé, Desarnaud, Eggerickx, Vassart, & Parmentier, 1994). In the brain, the expression of MC5R has been detected in the HT, cortex, cerebellum hippocampus, and pituitary gland (Chhajlani et al., 1993; Griffon et al., 1994). In the peripheral tissues, MC5R is expressed in high levels in the exocrine glands (lacrimal and Harderian glands), sebaceous glands, and skeletal muscles (Chen et al., 1997; Griffon et al., 1994). In mice, expression of MC5R has been found in adrenal glands and adipocytes (Chen et al., 1997). The functions of the MC5R are not well understood, but data coming from genetic studies have shown the role of this receptor in the production of sebaceous lipids, water repulsion, and thermal regulation. MC5R knockout mice show impaired water repulsion which generates problems in regulating core body temperature (Chen et al., 1997). The presence of MC5R in lacrimal glands suggests a role for this receptor in the stress response of stimulating tear secretion (Entwistle, Hann, Sullivan, & Tatro, 1990).

3. MELANOCORTINS AND DRUGS OF ABUSE

The distribution of the MCRs in the brain suggests a role of the MC system in neurobiological responses of drugs of abuse (Hao, Tian, Liu, Xiang, & Zhand, 2015; Lippert, Ellacot, & Cone, 2014; Oude Ophuis, Boender, van Rozen, & Adan, 2014; for review, see Roseberry, Stuhrman, & Dunugan, 2015). Anatomical data confirm the expression of MC3R and MC4R in brain regions that modulate reward such as the VTA and the N.Ac (Alvaro et al., 1996; Griffon et al., 1994; Lindblom et al., 1998; Mountjoy et al., 1994). Pharmacological data have shown that chronic intracerebral ventricular infusion (i.c.v.) of Melanotan-II (MTII), a nonspecific MC agonist for MC3R and MC4R, alters dopamine (DA) receptor binding in the N.Ac (Lindblom, Kask, et al., 2002). Additionally, direct administration of α-MSH into the VTA increases dopamine levels in the N.Ac (Lindblom et al., 2001; Sánchez, Barontini, Armando, & Celis, 2001), and this effect can be blocked by HS131, an selective antagonist for MC4R (Lindblom et al., 2001). Further supporting the interaction between the MC and DA systems, chronic administration of MTII alters dopamine 1 (D1) and dopamine 2 (D2) receptor expression in the striatum, VTA, and brainstem (Lindblom, Kask, et al., 2002). Recently, electrophysiological studies have shown that application of α-MSH into the N.Ac modifies excitatory synapses in D1 dopamine receptor-expressing medium spiny neurons (Lim, Huang, Grueter, Rothwell, & Malenka, 2012).

The opioid system also shares anatomical distribution with the MC system in the central nervous system, and their interaction has been well documented (Alvaro, Tatro, & Duman, 1997; Contreras & Takemori, 1984; Lindblom et al., 1998; Mountjoy et al., 1994; Navarro et al., 2015; Starowicz, Sieja, Bilecki, Obara, & Przewlocka, 2003). Blockade of MC4R prevents the development of morphine tolerance and dependence (Kalange et al., 2007; Niu et al., 2012; Starowicz, Obara, Przewlocki, & Przewlocka, 2005) as well as enhances the antinociceptive effects of opioids (Ercil, Galici, & Kesterson, 2005; Vrinten, Gispen, Kalkman, & Adan, 2003). Supporting this interaction, it has been demonstrated that chronic administration of morphine decreases MC4R mRNA expression in the striatum and PAG area (Alvaro et al., 1996).

The MC system has also been reported to play a role in behavioral and neurochemical responses to psychostimulant drugs such as cocaine and amphetamine. Acute i.c.v. administration of MTII augments amphetamine's threshold lowering effect on lateral hypothalamic self-administration which demonstrates that activation of MCRs potentiates amphetamine reward (Cabeza de Vaca, Kim, & Carr, 2002). Consistent with the role of MCs in processing the rewarding effect of psychostimulants, it has been shown that N.Ac infusion of the MC3 and MC4R antagonist SHU9119 blocks cocaine self-administration and cocaine-induced locomotor sensitization (Hsu et al., 2005). However, these data are opposite to a recent study that suggests a role for the MC4R in the modulation of the locomotor sensitization induced by cocaine, but not in the N.Ac (Alserda, Adan, & Ramakers, 2013). Also, the role of MC4R in mediating the behavioral responses to cocaine has been demonstrated recently using MC4R-null mice (Cui & Lutter, 2013). In addition, chronic cocaine administration increases the expression of MC4R mRNA in the striatum and hippocampus (Alvaro, Taylor, & Duman, 2003; Hsu et al., 2005).

In the last 15 years, a growing body of evidence has been presented suggesting a role for the MC system in the neurobiological responses of ethanol (Olney et al., 2014). In the rest of the chapter we discuss the latest evidence demonstrating a role for MC signaling in behavioral and neurobiological responses to ethanol.

4. MELANOCORTINS AND ETHANOL

4.1 The Effects of Ethanol on the MC System

Numerous studies have shown that ethanol exposure produces direct effects on the MC system (Gangisetty, Bekdash, Maglakelidze, & Sarkar, 2014; Scanlon, Lazar-Wesley, Grant, & Kunos, 1992; Seizinger, Bovermann, Maysinger, Hollt, & Herz, 1983; Wilkinson, Crabbe, Keith, Kendall, & Dorsa, 1986; Winkler, Roske, Furkert, Fickel, & Melzig, 1995). However, the results are mixtures and ethanol exposure has been shown to increase (Angelogianni & Gianoulakis, 1993), decrease (Rasmussen et al., 1998), or not change (Popp & Erickson, 1998) the expression of POMC and its derived peptides (Table 2). For example, animal studies have shown that acute intraperitoneally (i.p) ethanol injection does not alter POMC mRNA expression or α-MSH immunoreactivity (IR) in the N.Arc (Cubero,

Table 2 Effects of Ethanol Exposure on the MC System

Method of Ethanol Delivery	**Brain Area**	**Alterations in Protein Expression**
Acute ethanol diet	Amygdala	= α-MSH[a]
	Arc	↓ α-MSH[a], = AgRP[b], ↓ POMC[c], ↓ PC1/3[c], = PC2[c]
	BNST	↓ α-MSH[b]
	CeA	= α-MSH[b], ↓ α-MSH[d]
	DMH	= α-MSH[b], ↓ α-MSH[d]
	Frontal cortex	= α-MSH[a]
	HC	= α-MSH[a]
	LH	= α-MSH[a,b]
	NAc	= α-MSH[a]
	PAG	= α-MSH[b]
	Pituitary	↓ α-MSH[a]
	PVN	= α-MSH[b]
	PVT	↓ α-MSH[b]
	Striatum	= α-MSH[a]
	SN	↓ α-MSH[a]
Chronic ethanol diet	Arc	↓ α-MSH[b,d] ↓ POMC[c]
	BNST	↓ α-MSH[b]
	CeA	↑ α-MSH[d], ↓ α-MSH[b]
	DMH	↑ α-MSH[d], = α-MSH[b]
	LH	↑ α-MSH[d], ↓ α-MSH[b]
	PAG	= α-MSH[b]
	PVN	↑ α-MSH[d], = α-MSH[b]
	PVT	↓ α-MSH[b], = α-MSH[d]
Withdrawal from ethanol diet	Arc	↓ α-MSH[d]
	CeA	↓ α-MSH[d]
	DMH	↓ α-MSH[d]
	LH	= α-MSH[d]
	PVN	↑ α-MSH[b]
	PVT	= α-MSH[b]

Table 2 Effects of Ethanol Exposure on the MC System—cont'd

Method of Ethanol Delivery	Brain Area	Alterations in Protein Expression
Vapor chamber	HT	↓ POMC[e]
	Pituitary	↓ α-MSH[f]
Voluntary consumption	HT	↑ POMC[g]

[a]Rainero et al. (1990).
[b]Navarro, Cubero, Knapp, Breese, and Thiele (2008).
[c]Navarro, Cubero, and Thiele (2013).
[d]Kokare, Singru, Dandekar, Chopde, and Subhedar (2008).
[e]Scanlon et al. (1992).
[f]Wilkinson et al. (1986).
[g]De Waele and Gianoulakis (1994).
(↑) increase, (↓) decrease, or (=) no change on peptide/protein levels in response to ethanol exposure.

Navarro, Carvajal, Lerma-Cabrera, & Thiele, 2010; Kinoshita et al., 2000; Popp & Erickson, 1998). Opposite to these data, and using similar doses, acute i.p ethanol administration significantly reduces α-MSH IR in the N.Arc (Kokare et al., 2008). Intragastric administration of ethanol that produces between 120 and 150 mg/dL of plasma ethanol levels increases mediobasal hypothalamus cytoplasmic POMC mRNA concentrations at 180 and 360 min after initiation of the infusion. Lower ethanol levels do not alter mediobasal hypothalamus POMC mRNA (Rasmussen et al., 1998). Acute "binge" intragastric ethanol administration (1.5 g/kg of intragastric infusion of ethanol three times over 1 day) significantly reduces POMC mRNA levels in the hypothalamus. However, the same treatment for 14 days does not produce any alteration of POMC mRNA in the same region (Zhou et al., 2000). Chronic ethanol administration with 2 weeks of ethanol liquid diet (ED) significantly decreases POMC mRNA expression in the N.Arc (Chen, Boyadjieva, Advis, & Sarkar, 2006). Consistent with these data, 21 days of exposure to an ED significantly decreases α-MSH immunoreactivity in the N.Arc and SN of rats (Rainero et al., 1990). Similarly, 18 days of ED reduces POMC and PC1/3 IR in the N.Arc. Rats exposure to either 4 days or 18 days of ethanol diet show a decrease of α-MSH IR in the N.Arc, bed nucleus of the stria terminals (BNST), and the paraventricular nucleus of the thalamus (PVT); however, the lateral hypothalamus (LH) and the central amygdala (CeA) only show a decrease on α-MSH IR after 18 days of ethanol diet (Navarro et al., 2008). Contradictory with these data, chronic ethanol treatment with a liquid diet for 15 days increases POMC mRNA (Angelogianni & Gianoulakis, 1993) and α-MSH IR

(Kokare et al., 2008) in the N.Arc, and further augmentation is seen after 24 h from ethanol withdrawal (Kokare et al., 2008). Daily voluntary ethanol consumption for 21 days induces an increase in the hypothalamic expression of POMC mRNA (De Waele & Gianoulakis, 1994). Mediobasal hypothalamic POMC mRNA is decreased after 7 weeks of daily ethanol consumption but increases after 3 weeks of ethanol withdrawal (Rasmussen, Boldt, Wilkinson, & Milton, 2002). Recently, it has been shown that in rats exposure to binge-like ethanol drinking during adolescence significantly reduces basal α-MSH in the CeA, N.Arc, and PVN during adulthood (Lerma-Cabrera, Carvajal, Alcaraz-Iborra, et al., 2013). In mice, binge-like drinking significantly reduces α-MSH IR in the LH, PVN, DMH, and the N.Arc. In addition, binge-like ethanol drinking mice demonstrate an increase in the AgRP IR in the PVN (Sprow et al., 2015). These data suggest that after a period of ethanol drinking, blunted α-MSH signaling and increased AgRP activity compromise the protective effect of the MC system against excessive ethanol consumption.

More research needs to be conducted to reconcile all these results; however, it is clear that the direction of differences in MC levels is brain site dependent, and differences between studies could be related to numerous factors including voluntary vs involuntary ethanol exposure, the use of different species (rats vs mice), routes of ethanol administration and doses, duration of treatments, levels of ethanol concentration achieved, and even the methods used to detect POMC and its peptides (IHC vs mRNA). Importantly, in studies that entail voluntary ethanol consumption, the emerging phenomenon is that repeated bouts of ethanol drinking compromise the MC system, which may be part of a mechanism contributing to increasing ethanol drinking during the transition to dependence.

4.2 The Effects of the MC System on Ethanol Consumption

The first evidence to suggest a role of the MC system in controlling alcohol consumption came from studies conducted on Alko Alcohol (AA) and Alko-non Alcohol (ANA) rats. These rats were selectively bred for high ethanol drinking (AA) or low ethanol drinking (ANA) (Sinclair, Le, & Kiianmaa, 1989). Using these rats, Lindblom, Wikberg, and Bergstrom (2002) showed that compared with the ANA rats, AA rats have low levels of MC3R in the N.Ac shell and high levels of MC3R in the PVN, VNH, and N.Arc. Regarding MC4R, AA rats show higher levels in the VMH compared with the ANA rats. Also, in situ hybridization studies indicated that AA rats have higher expression of POMC mRNA and less expression of AgRP mRNA in the N.Arc. These results suggest that the higher ethanol consumption by the AA rats may be, in part, modulated by inherent alterations of the MC system. In line with these data, other alcohol preferring rodent lines exhibit alterations of the central MC system. For example, High alcohol preferring rats show a higher POMC mRNA expression in the N.Arc compared with the low alcohol preferring rats (Kinoshita et al., 2004). Similarly, Sardinians alcohol-preferring (sP) and no-preferring (sNP) rats also display differences in POMC mRNA expression in the N.Arc with higher expression in the sP compared to the sNP rats (Zhou et al., 2013).

The first direct assessment of the effects of manipulating the MC system on ethanol consumption came from Ploj et al. (2002). Using a two-bottle choice paradigm and AA rats, they demonstrated that i.c.v. infusion of MTII significantly reduced voluntary ethanol intake and preference for alcohol without altering water consumption. Another alcohol preferring inbred strain commonly used for alcohol studies is the C57BL/6 strain of mice. These mice show high ethanol preference with daily consumption between 3 and 10 g/kg depending on the alcohol concentration (Belknap, Crabbe, & Young, 1993). These mice also show a high basal level of α-MSH IR in the N.Arc (Cubero et al., 2010). Using these mice, we examined the effect of MTII on voluntary ethanol drinking. We found that both central (i.c.v.) and peripheral (i.p) administration of MTII significantly reduces ethanol consumption. Further, this effect can be blocked by pretreatment with AgRP-(83–132), a potent and nonselective MCR antagonist (Quillan et al., 1998). No differences in blood ethanol concentration (BEC) were found between saline and MTII groups that were injected with ethanol, which suggest that the effects of MTII on ethanol drinking are not related to alterations of ethanol metabolism (Navarro, Cubero, Knapp, & Thiele, 2003). In general, these data suggest that the MC signaling exerts a protective effect against ethanol consumption.

5. THE ROLE OF MC3R AND MC4R IN THE MODULATION OF ETHANOL CONSUMPTION

5.1 MC3R and Ethanol Consumption

The data above confirm the role of the MC system in the modulation of voluntary ethanol consumption; however, because both the agonist or antagonist used above are not selective for MC3R or MC4R, thus the next step was to investigate the contribution of each receptor in the control of ethanol consumption. The role of MC3R on voluntary ethanol consumption was evaluate using MC3R-knockout ($Mc3r^{-/-}$) vs wild-type ($Mc3r^{+/+}$) mice. After peripheral and central administration of MTII, $Mc3r^{-/-}$ mice show similar reductions of ethanol intake as $Mc3r^{+/+}$ mice. These results suggest that MC3R is not a likely receptor for mediating agonist-induced reductions of voluntary ethanol drinking (Navarro et al., 2005). To further characterize the role of MC3R in ethanol consumption, the influence of MC3R on excessive ethanol drinking has been study using a model of binge-like ethanol drinking, the "drinking in the dark" (DID) protocol (Rhodes, Best, Belknap, Finn, & Crabbe, 2005). Binge drinking has been defined by the National Institute of Alcohol Abuse and Alcoholism (NIAAA) as a pattern of drinking that brings a person's blood alcohol concentration to 0.08 grams percent (80 mg/dL) or above. This typically happens when men consume five or more drinks, and when women consume four or more drinks, in about 2 h (National Institute on Alcohol Abuse and Alcoholism, 2004). Briefly, the DID protocol involves a 4-day procedure, and during the first 3 days, 3 h into the dark cycle water bottles are removed and replaced with a single bottle of ethanol (20% v/v) for 2 h. The fourth day is typically the test day on which experimental manipulations are performed just prior to ethanol access. Water bottles are removed and alcohol bottles are applied, and ethanol consumption is assessed for either 2 or 4 h depending of the experimental manipulations. Immediately after ethanol exposure, tails blood samples are collected to assess BECs. After the 4-h test ethanol is removed and water bottles return to the mice. It has been shown that mice C57BL/6J achieve BECs that reach or exceed the 80 mg/dL criterion for binge drinking using DID procedures (Rhodes et al., 2005). Using the DID procedure it has been shown that central administration of MTII (0.25, 0.50, or 1.0 μg/kg) blunts binge-like ethanol drinking in both to $Mc3r^{+/+}$ and $Mc3r^{-/-}$ mice. This is consistent with previous data using a 24 h two-bottle choice protocol and suggests that the MC3R is neither critical for the modulation of voluntary ethanol intake nor binge-like ethanol drinking. However, only the highest dose of MTII (1 μg) reduces binge-like ethanol drinking in the $Mc3r^{+/+}$ mice, while lower doses of MTII tested reduce binge-like ethanol drinking in the $Mc3r^{-/-}$ mice. These findings suggested that deletion of the MC3R leaves mice more sensitive to the protecting effect of MTII against excessive ethanol drinking in these mice (Olney et al., 2014), indicating that the MC3R may act as an inhibitory autoreceptor on POMC neurons acting as a negative feedback to inhibit the transmission of α-MSH from POMC neurons (Marks et al., 2006). Viewed this way, in $Mc3r^{-/-}$ mice both endogenous α-MSH and MTII act together to reduce binge-like ethanol as these mice lack the negative feedback provided by the MC3R, thus increasing their sensitivity to the effects of MTII on ethanol intake (Olney et al., 2014). One alternative explanation for these opposite results respect to the role of MC3R in the modulation of alcohol consumption could be the fact that different procedures have been used (24-h exposure voluntary consumption against limited asses binge–ethanol drinking.

5.2 MC4R and Ethanol Consumption

To assess the contribution of the MC4R we used a genetic approach in combination with pharmacological manipulations. We first assessed voluntary ethanol consumption in MC4R-knockout ($Mc4r^{-/-}$) mice vs wild-type ($Mc4r^{+/+}$) mice. Briefly, with voluntary consumption procedures animals have access to two bottles on their home cage, one containing water and the other an ethanol solution. The concentration of ethanol is typically increased over testing days, and the bottles are alternated every 2 days to avoid position preferences. Once the desires final ethanol concentration is reached and a stable base line is established, animals are ready for any experimental manipulations. Using this protocol, it has been demonstrated that in both knockout and wild-type MC4R mice, exposure to different ethanol concentrations (3%, 5%, 8%, 15%, and 20%) yields similar amount of ethanol consumption between the genotypes. However, when these animals received an i.c.v. administration of MTII, only the $Mc4r^{+/+}$ mice show a significant reduction on ethanol intake (Navarro et al., 2011). Because of the lack of differences on the amount of voluntary alcohol consumed at different ethanol concentrations, one initial conclusion could be that endogenous MC4R signaling does not play a critical role in modulation voluntary ethanol consumption. Nonetheless, one explanation could be developmental compensation that may be present when using constitutive knockout mice, masking the true contributions of the deleted gene (Gerlai, 1996, 2001). Importantly though, the lack of an effect of MTII on ethanol drinking in Mc4r −/− mice suggests that the MC4R is critical in modulating MC agonist-induced reduction of ethanol intake. These data are consistent with the observation that i.c.v. administration of a specific agonist for MC4R cyclo-(NH-CH_2-CH_2-CO-His-D-Phe-Arg-Trp-Glu)-NH_2 significantly reduces ethanol intake, and this effect was blocked by the nonspecific antagonist AgRP (83–132) (Navarro et al., 2005). Consistent with the role of the MC system in the modulation of ethanol consumption, deletion of AgRP has been shown to blunt ethanol self-administration and binge-like ethanol drinking (Navarro, Cubero, Ko, & Thiele, 2009).

These data make it clear that central manipulation of MC4R signaling modulates ethanol consumption; however, the next question was where specifically in the brain this effect is taking place. One of the brain region in which MCR has been implicated as having a role in excessive drinking is the LH. It has been shown that after 1, 3, or 6 cycles ofbinge-like drinking, α-MSH levels are significantly decreased in the LH. Further, this same exposure to ethanol causes an increase in AgRP IR in the same region (Sprow et al., 2015). Furthermore, in C57Bl/6J mice, intra-LH infusion of MTII blunts, whereas AgRP increases, binge–ethanol consumption compared to vehicle-treated animals (Sprow et al., 2015). These data suggest that endogenous α-MSH signaling in the LH modulates binge-like ethanol drinking. Two other areas in which MCR signaling modulates ethanol intake are the N.Ac and the VTA. In rats, and using the two-bottle choice protocol, administration of MC4R selective agonist, *cyclo*-(NH-CH_2-CH_2-CO-His-D-Phe-Arg-Trp-Glu)-NH_2 into the N.Ac and the VTA significantly reduces ethanol consumption (Lerma-Cabrera et al., 2012). However, unlike observations with mice this group show that administration on this selective MC4R agonist into the LH did not reduced ethanol consumption (Lerma-Cabrera, Carvajal, Chotro, et al., 2013). Again, the fact that different species and protocols have been used could explain the differences between studies. One interest result found in the Lerma-Cabrera 2013 study is that N.Ac administration of the selective MC4R agonist, but not LH administration, decreased the duration of hedonic responses and increased the aversive responses to intraorally infused ethanol. These results raise the possibility that the MC system may influence ethanol consumption, in part, by modulating the palatability of ethanol (Brasser, Silbaugh, Ketchum, Olney, & Lemon, 2012).

Another region in which administration of MCR compounds modulates ethanol intake is the CeA. In P rats exposed to the two-bottle choice paradigm, direct administration of MTII into the CeA significantly reduces ethanol consumption (York, Boghossian, & Park-York, 2011). However, these data are difficult to interpret. The CeA is essential for integrating emotions and its role in anxiety and stress has been well documented (for review, see Gilpin, Herman, & Roberto, 2015). Previous studies have shown that MCR agonists elicit anxiogenic-like effects (Chaki & Okuyama, 2005; Rao et al., 2003). It has been shown that CeA infusion of the selective MC4R antagonist, HS014, protects against anxiety-like behavior stemming from ethanol withdrawal, even potentiated the anxiolytic-like responses of ethanol (Kokare, Chopde, & Subhedar, 2006). These data suggest that MC4R signaling in the CeA is anxiogenic, thus MCR agonist-induced reductions of ethanol drinking, when infused into the CeA, may be secondary to increases of anxiety. In light of the different areas where the MC system can modulate ethanol consumption, one aspect to consider is that MCR signaling may modulate different neurobiological responses to ethanol, including ethanol ingestion, the rewarding properties of ethanol and the anxiety responses to ethanol; hence, more research is necessary to determine the specific brain regions and systems by which the MC system modulates ethanol consumption and other neurobiological responses to ethanol.

6. TARGETING THE MCRs TO TREAT ALCOHOL ABUSE DISORDERS

As it was mentioned earlier, MC3R and MC4R are the most highly expressed MCRs in the brain. Further, evidence shows that MC4Rs play the main role in modulating neurobiological responses to ethanol. Because of the wide distribution of MC4Rs in the brain, MC4R signaling is poised interact to numerous neurobiological systems (Roseberry et al., 2015), and these interactions should be considered when considering possible mechanisms by which the MC system modulates ethanol consumption.

The first possible mechanism by which MC4R signaling modulates ethanol consumption is by its actions involved in the regulation of food intake and energy homeostasis. As with food, ethanol is a caloric compound, and an extensive research has demonstrated that both ethanol and food intake entail both appetitive and consummatory behaviors. As it has been suggested previously (Thiele et al., 2003), it is possible that overlapping or common pathways are involved in uncontrolled eating and excessive ethanol drinking. In fact, α-MSH, the endogenous MCR agonist, modulated both alcohol and food intake. α-MSH binds to MC3R and MC4R and as with MTII, central infusion of α-MSH induces anorexia (Pétervári, Garami, Soós, Székely, & Balaskó, 2010; Poggioli, Vergoni, & Bertolini, 1986; Tsujii & Bray, 1989). In animal models, central and peripheral administration of MTII reduce food intake (Giraudo, Billington, & Levine, 1998; Grill et al., 1998; Wirth, Olszewski, Yu, Levine, & Giraudo, 2001) and ethanol consumption (Navarro et al., 2005, 2003; Sprow et al., 2015). I.c.v. administration of MTII in Mc4r$^{-/-}$ mice fails to alter food (Marsh et al., 1999) and ethanol intake (Navarro et al., 2011). All these data suggest that MC4R signaling may regulate feeding and ethanol consumption via similar pathways. However, ethanol not only has caloric properties, but it is also a drug of abuse that entails psychoactive properties. It has been shown that alcohol alters MC expression in brain regions that modulate reward (Navarro et al., 2008; Sprow et al., 2015). Thus, one question that has still not been completely answered is if MC4R signaling controls ethanol drinking through a mechanism involving the regulation of calories or by a mechanism that modulates the pharmacodynamics effects of ethanol, or both.

In this regard, the second explanation of how the MC system modulates ethanol consumption is by acting on pathways involved with modulating the rewarding properties of drugs of abuse. As it was mentioned in Section 1, the distribution of the MCRs in areas, such as the VTA and N.Ac, suggests a role of the MC system in modulating the rewarding properties of drugs of abuse. MCR signaling has been implicated in opioid tolerance and dependence (Contreras & Takemori, 1984; Szekely et al., 1979) as well as in neurochemical responses of psychostimulant drugs such as cocaine and amphetamine (Cabeza de Vaca et al., 2002). Ploj et al. (2002) proposed a mechanism that involves an interaction between endogenous opioids and MCs in the modulation of ethanol intake. They suggested that MCR signaling may regulate ethanol consumption by modulating endogenous opioid activity within mesolimbic dopamine pathways (Ploj et al., 2002). A large amount of data support the role of opioid signaling in ethanol consumption (De Waele & Gianoulakis, 1994; De Waele, Papachristou, & Gianoulakis, 1992; Gianoulakis, 1990; Rasmussen et al., 1998). Naltrexone, an opioid receptor antagonist, is an Food and Drug Administration (FDA) approved drug used to treat opioid and ethanol dependence (Bouza, Angeles, Magro, Muñoz, & Amate, 2004). As mentioned earlier, stimulation of MCRs blocks the antinociceptive effects of opioids, while MC4R antagonist enhances opioid antinociception (Ercil et al., 2005; Kalange et al., 2007). Central administration of MTII not only decreases ethanol drinking but also prevents ethanol-induced changes in opioid levels in the VTA and the SN of AA rats (Ploj et al., 2002). Putting these data together it is tempting to speculate that blockade of the MCR signaling may enhance endogenous opioid signaling in response to ethanol, promoting an increase of ethanol consumption, whereas stimulation of MCR signaling may blunt ethanol-induced opioid signaling, resulting in a decrease of ethanol intake. This possibility has been supported by a recent study that shows that MTII synergistically augments the ability of Naltrexone to blunt binge-like ethanol drinking. Using the DID model and C57BL/6J mice, this study found that both MTII and Naltrexone individually blunt binge-like ethanol drinking in a dose-dependent manner. However, giving in combination, MTII increases the effectiveness of Naltrexone in reducing excessive ethanol intake (Navarro et al., 2015).

While opioid receptor antagonists, such as naltrexone, are approved for treating alcohol use disorders (AUDs), naltrexone-based therapies are not successful with all individuals (Garbutt, 2009). Thus, approaches that increase the effectiveness of naltrexone in the treatment of AUDs would have high clinical value. Combined targeting of MCR and opioid pathways has been used as a pharmacological therapy for obesity by combining naltrexone with Bupropion, a drug that has been reported to increase MC signaling (Billes & Cowley, 2007; Hasegawa et al., 2005). Interestingly, Bupropion is also an FDA approved medication for depression (Davidson & Connor, 1998), smoking cessation (Roddy, 2004), and weight loss (Anderson et al., 2002). This drug combination, called Contrave, has been demonstrated to be an effective treatment regimen for obesity. It has been reported to have more efficacy that the use of either drug alone (Apovian et al., 2013; Greenway et al., 2009, 2010; Hollander et al., 2013; McElroy et al., 2013; Smith et al., 2013) has been recently approved by the FDA for treating eating disorders (Gohil, 2014). We have proposed that a combined naltrexone/Bupropion approach may also be effective in treating AUDs in addition to eating disorders (Navarro et al., 2015). Interestingly, MK-0493, an oral bioavailable MC4R agonist, has been tested in human phase I/II clinical trial studies for obesity treatment (Krishna et al., 2009) and may also be an effective treatment approach for AUDs. It is important to note that approximately 50% of individuals with an eating disorder abuse are dependent on alcohol (Holderness, Brooks-Gunn, & Warren, 1994). Thus, drugs such as Contrave could be especially of benefit to those suffering from comorbid AUDs and eating disorders.

REFERENCES

Abdel-Malek, Z. A. (2001). Melanocortin receptors: Their functions and regulation by physiological agonist and antagonist. *Cellular and Molecular Life Sciences*, *58*, 434–441.

Adan, R. A., & Gispen, W. H. (2000). Melanocortins and the brain: From effects via receptors to drug targets. *European Journal of Pharmacology*, *405*, 13–24.

Adan, R. A., & Kas, M. J. (2003). Inverse agonism gains weight. *Trends in Pharmacological Sciences*, *24*, 315–321.

Adan, R. A., van der Krann, M., Doornbos, R. P., Bar, P. R., Burbach, J. P., & Gispen, H. W. (1996). Melanocortin receptors mediate α-MSH-induced stimulation of neurite outgrowth in neuro 2A cells. *Brain Research. Molecular Brain Research*, *36*, 37–44.

Akil, H., Shiomi, H., & Matthews, J. (1985). Induction of the intermediate pituitary by stress: Synthesis and release of a nonopiod form beta-endorphin. *Science*, *227*, 424–426.

Alserda, E., Adan, R. A., & Ramakers, G. M. (2013). Repeated agouti related peptide (83–132) injections inhibit cocaine-induced locomotor sensitization, but not via the nucleus accumbens. *The European Journal of Neuroscience*, *719*, 187–191.

Alvaro, J. D., Tatro, J. B., & Duman, R. S. (1997). Melanocortins and opiate addiction. *Life Sciences*, *61*, 1–9.

Alvaro, J. D., Tatro, J. B., Quillan, J. M., Fogliano, M., Eisenhard, M., Lenner, M. R., et al. (1996). Morphine down-regulates melanocortin-4 receptor expression in brain regions that mediate opiate addiction. *Molecular Pharmacology*, *50*, 583–591.

Alvaro, J. D., Taylor, J. R., & Duman, R. S. (2003). Molecular and behavioral interactions between central melanocortins and cocaine. *The Journal of Pharmacology and Experimental Therapeutics*, *304*, 391–399.

Anderson, E. J. P., Çakir, I., Carrington, S. J., Cone, R. D., Ghamari-Langroudi, M., Gillyard, T., et al. (2016). Regulation of feeding and energy homeostasis by α-MSH. *Journal of Molecular Endocrinology*, *56*, T157–T174.

Anderson, J. W., Greenway, F. L., Fujioka, K., Gadde, K. M., McKenney, J., & O'Neil, P. M. (2002). Bupropion SR enhances weight loss: A 48-week double-blind, placebo-controlled trial. *Obesity Research*, *10*, 633–641.

Angelogianni, P., & Gianoulakis, C. (1993). Chronic ethanol increases proopiomelanocortin gene expression in the rat hypothalamus. *Neuroendocrinology*, *57*, 106–114.

Apovian, M. C., Aronne, L., Rubino, D., Still, C., Wyatt, H., Burns, C., et al. (2013). A randomized, phase 3 trial of naltrexone SR/bupropion SR on weight and obesity-related risk factors (COR-II). *Obesity (Silver Spring)*, *21*, 935–943.

Argiolas, A., Melis, M. R., Murgia, S., & Schioth, H. B. (2000). ACTH and alpha-MSH-induced grooming, stretching, yawning and penile erection in male rats: Site of action in the brain and role of melanocortin receptors. *Brain Research Bulletin*, *51*, 425–431.

Bagnol, D., Lu, X. Y., Kaelin, C. B., Day, H. E., Ollmann, M., Gantz, I., et al. (1999). Anatomy of an endogenous antagonist: Relationship between agouti-related protein and proopiomelanocortin in brain. *The Journal of Neuroscience*, *19*, 1–7.

Barret, P., MacDonald, A., Helliwell, R., Davidson, G., & Morgan, P. (1994). Cloning and expression of a new member of the melanocyte-stimulating hormone receptor family. *Journal of Molecular Endocrinology*, *12*, 203–213.

Begriche, K., Levasseur, P. R., Zhang, J., Rossi, J., Skorupa, J. D., Solt, L. A., et al. (2011). Genetic dissection of the functions of the melanocortin-3 receptor, a seven-transmembrane G-protein-coupled receptor, suggests roles for central and peripheral receptors in energy homeostasis. *The Journal of Biological Chemistry*, *286*(47), 40771–40781.

Belknap, J. K., Crabbe, J. C., & Young, E. R. (1993). Voluntary consumption of ethanol in 15 inbred mouse strains. *Psychopharmacology*, *112*, 503–510.

Benjannet, S., Rondeau, N., Day, R., Chretien, M., & Seidah, N. G. (1991). PC1 and PC2 are proprotein convertases capable of cleaving proopiomelanocortin at distinct pairs of basis residues. *Proceedings of the National Academy of Sciences of the United States of America*, *88*, 3564–3568.

Bertolini, A., Poggioli, R., & Ferrari, W. (1979). ACTH-induced hyperalgesia in rats. *Experientia*, *35*, 1216–1217.

Bertorelli, R., Fredduzzi, S., Tarozzo, G., Campanella, M., Grundy, R., Beltramo, M., et al. (2005). Endogenous and exogenous melanocortin antagonists induce anti-allodynic effects in a model of rat neuropathic pain. *Behavioural Brain Research*, *157*, 55–62.

Billes, S. K., & Cowley, M. A. (2007). Inhibition of dopamine and norepinephrine reuptake produces additive effects on energy balance in lean and obese mice. *Neuropsychopharmacology, 32*, 822–834.

Böhm, M., Brzoska, T., Schulte, U., Schiller, M., Kubitscheck, U., & Luger, T. A. (1999). Characterization of a polyclonal antibody raised against the human melanocortin-1 receptor. *Annals of the New York Academy of Sciences, 885*, 372–382.

Boston, B. A., & Cone, R. D. (1996). Characterization of melanocortin receptor subtype expression on murine adipose tissues and in the 3T3-L1 cell line. *Endocrinology, 137*, 2043–2050.

Bousquet-Melou, A., Galizky, J., La Fontan, M., & Berlan, M. (1995). Control of lipolysis in intra-abdominal fat cells of non-human primates: Comparison with humans. *Journal of Lipid Research, 36*, 541–561.

Bouza, C., Angeles, M., Magro, A., Muñoz, A., & Amate, J. M. (2004). Efficacy and safety of naltrexone and acamprosate in the treatment of alcohol dependence: A systematic review. *Addiction, 99*, 811–828.

Brasser, S. M., Silbaugh, B. C., Ketchum, M. J., Olney, J. J., & Lemon, C. H. (2012). Chemosensory responsiveness to ethanol and its individual sensory components in alcohol-preferring, alcohol-nonpreferring and genetically heterogeneous rats. *Addiction Biology, 17*, 423–436.

Browne, C. A., Bennett, H. P., & Solomon, S. (1981). Isolation and characterization of corticotropin- and melanotropin-related peptides from intermediary lobe of the rat pituitary by reversed-phase liquid chromatography. *Biochemistry, 20*, 4538–4546.

Butler, A. A., Kesterson, R. A., Khing, K., Cullen, M. J., Pelleymounter, M. A., Dekoning, J., et al. (2000). A unique metabolic syndrome causes obesity in the melanocortin-3 receptor-deficient mouse. *Endocrinology, 141*, 3518–3521.

Cabeza de Vaca, S., Kim, G. Y., & Carr, K. D. (2002). The melanocortin receptor agonist MTII augments the rewarding effect of amphetamine in ad-libitum-fed and food-restricted rats. *Psychopharmacology, 161*, 77–85.

Caruso, V., Lagerstrom, M. C., Olszewski, P. K., Fredriksson, R., & Schioth, H. B. (2014). Synaptic changes induced by melanocortin signaling. *Nature Reviews. Neuroscience, 15*, 98–110.

Castro, M. G., & Morrison, E. (1997). Post-translational processing of proopiomelanocortin in the pituitary and in the brain. *Critical Reviews in Neurobiology, 11*, 35–57.

Catania, A., & Lipton, J. M. (1993). Alpha-melanocyte stimulating hormone in the modulation of host reactions. *Endocrine Reviews, 14*, 564–576.

Catania, A., Rajora, N., Capsoni, F., Minonzio, F., Star, R. A., & Lipton, J. M. (1996). The neuropeptide alpha-MSH has specific receptors on neutrophils and reduces chemotaxis in vivo. *Peptides, 17*(4), 675–679.

Chaki, S., & Okuyama, S. (2005). Involvement of melanocortin-4 receptor in anxiety and depression. *Peptides, 26*, 1952–1964.

Chen, C. P., Boyadjieva, N. I., Advis, J. P., & Sarkar, D. K. (2006). Ethanol suppression of the hypothalamic proopiomelanocortin level and the splenic NK cell cytolytic activity is associated with a reduction in the expression of proinflammatory cytokines but not anti-inflammatory cytokines in neuroendocrine and immune cells. *Alcoholism, Clinical and Experimental Research, 30*, 1925–1932.

Chen, W., Kelly, M. A., Opitz-Araya, X., Thomas, R. E., Low, M. J., & Cone, R. D. (1997). Exocrine gland dysfunction in MC5-R-deficient mice: Evidence for coordinated regulation of exocrine gland function by melanocortin peptides. *Cell, 91*, 789–798.

Chen, A. S., Marsh, D. J., Trumbauer, M. E., Frazier, E. G., Guan, X. M., Yu, H., et al. (2000). Inactivation of the mouse melanocortin-3 receptor results in increased fat mass and reduced lean body mass. *Nature Genetics, 26*, 97–102.

Chen, A. S., Metzger, J. M., Trumbauer, M. E., Guan, X. M., Frazier, E. G., Marsh, D. J., et al. (2000). Role of the melanocortin-4 receptor in metabolic rate and food intake in mice. *Transgenic Research*, *9*, 145–154.

Chhajlani, V. (1996). Distribution of cDNA for melanocortin receptor subtypes in human tissues. *Biochemistry and Molecular Biology International*, *38*, 73–80.

Chhajlani, V., Muceniece, R., & Wikberg, J. E. (1993). Molecular cloning of a novel human melanocortin receptor. *Biochemical and Biophysical Research Communications*, *195*, 866–873.

Chhajlani, V., & Wikberg, J. E. (1992). Molecular cloning and expression of the human melanocyte stimulating hormone receptor cDNA. *FEBS Letters*, *309*, 417–420.

Cone, R. D. (2006). Studies on the physiological functions of the melanocortin system. *Endocrine Reviews*, *27*, 736–749.

Contreras, P. C., & Takemori, A. E. (1984). Antagonism of morphine-induced analgesia, tolerance and dependence by alpha-melanocyte-stimulating hormone. *The Journal of Pharmacology and Experimental Therapeutics*, *229*, 21–26.

Cubero, I., Navarro, M., Carvajal, F., Lerma-Cabrera, J. M., & Thiele, T. E. (2010). Ethanol-induced increase of agouti-related protein (AgRP) immunoreactivity in the arcuate nucleus of the hypothalamus of C57BL/6J, but not 129/SvJ, inbred mice. *Alcoholism, Clinical and Experimental Research*, *34*, 693–701.

Cui, H., & Lutter, M. (2013). The expression of MC4Rs in D1R neurons regulates food intake and locomotor sensitization to cocaine. *Genes, Brain, and Behavior*, *12*, 658–665.

Cui, H., Sohn, J. W., Gautron, L., Funahashi, H., Williams, K. W., Elmquist, J. K., et al. (2012). Neuroanatomy of melanocortin-4 receptor pathway in the lateral hypothalamic area. *The Journal of Comparative Neurology*, *520*, 4168–4183.

Davidson, J. R., & Connor, K. M. (1998). Bupropion sustained release: A therapeutic overview. *The Journal of Clinical Psychiatry*, *59*(Suppl. 4), 25–31.

De Waele, J. P., & Gianoulakis, C. (1994). Enhanced activity of the brain beta-endorphin system by free-choice ethanol drinking in C57BL/6 but not DBA/2 mice. *European Journal of Pharmacology*, *258*, 119–129.

De Waele, J. P., Papachristou, D. N., & Gianoulakis, C. (1992). The alcohol-preferring C57BL/6 mice present an enhanced sensitivity of the hypothalamic beta-endorphin system to ethanol than the alcohol-avoiding DBA/2 mice. *The Journal of Pharmacology and Experimental Therapeutics*, *261*, 788–794.

Dores, R. M., Liang, L., Davis, P., Thomas, A. L., & Petko, B. (2016). Melanocortin receptors: Evolution of ligand selectivity for melanocortin peptides. *Journal of Molecular Endocrinology*, *56*, T119–T133.

Dores, R. M., Londraville, R. L., Prokop, J., Davis, P., Dewey, N., & Lesinski, N. (2014). Melanocortin/melanocortin receptors. *Journal of Molecular Endocrinology*, *52*, T29–T42.

Eipper, B. A., & Mains, R. E. (1978). Analysis of the common precursor to corticotropin and endorphin. *The Journal of Biological Chemistry*, *253*, 5732–5744.

Eipper, B. A., & Mains, R. E. (1981). Further analysis of post-translational processing of beta-endorphin in rat intermediate pituitary. *The Journal of Biological Chemistry*, *256*, 5689–5695.

Emerson, R. B., & Eipper, B. A. (1986). Characterization of pro-ACTH/endorphin-derived peptides in rat hypothalamus. *The Journal of Neuroscience*, *6*, 837–849.

Entwistle, M. L., Hann, L. E., Sullivan, D. A., & Tatro, J. B. (1990). Characterization of functional melanotropin receptors in lacrimal glands of the rats. *Peptides*, *11*, 477–483.

Ercil, N. E., Galici, R., & Kesterson, R. A. (2005). HS014, a selective melanocortin-4 (MC4) receptor antagonist, modulates the behavioral effects of morphine in mice. *Psychopharmacology*, *180*, 279–285.

Fan, W., Boston, B. A., Kesterson, R. A., Hruby, V. J., & Cone, R. D. (1997). Role of melanocortinergic neurons in feeding and the agouti obesity syndrome. *Nature*, *385*, 165–168.

Fani, L., Bak, S., Delhanty, P., van Rossum, E. F. C., & van den Akker, E. L. T. (2014). The melanocortin-4 receptor as target for obesity treatment: A systematic review of emerging pharmacological therapeutic options. *International Journal of Obesity, 38*, 163–169.

Fong, T. M., Mao, C., MacNeil, T., Kalyani, R., Smith, T., Weinberg, D., et al. (1997). ART (protein product of agouti-related transcript) as an antagonist of MC-3 and MC-4 receptors. *Biochemical and Biophysical Research Communications, 237*, 629–631.

Gangisetty, O., Bekdash, R., Maglakelidze, G., & Sarkar, D. K. (2014). Fetal alcohol exposure alters proopiomelanocortin gene expression and hypothalamic-pituitary-adrenal axis function via increasing MeCP2 expression in the hypothalamus. *PLoS One, 9*, e113228.

Gantz, I., Konda, Y., Tashiro, T., Shimoto, Y., Miwa, H., Munzert, G., et al. (1993). Molecular cloning of a novel melanocortin receptor. *The Journal of Biological Chemistry, 268*, 8246–8250.

Gantz, I., Miwa, H., Konda, Y., Shimoto, T., Watson, S. J., DelValle, J., et al. (1993). Molecular cloning expression and gene localization of a fourth melanocortin receptor. *The Journal of Biological Chemistry, 268*, 15174–15179.

Gantz, I., Yamada, T., Tashiro, T., Konda, Y., Shimoto, Y., Miwa, H., et al. (1994). Mapping of the gene encoding the melanocortin-1 (α-melanocyte stimulating hormone) receptor (MC1R) to human chromosome 16q24.3 by fluorescence in situ hybridization. *Genomics, 19*(2), 394–395.

Garbutt, J. C. (2009). The state of pharmacotherapy for the treatment of alcohol dependence. *Journal of Substance Abuse Treatment, 36*, S15–S23. quiz 24-S25.

García-Borrón, J. C., Sánchez-Laorden, B. L., & Jiménez-Cervantes, C. (2005). Melanocortin-1 receptor structure and functional regulation. *Pigment Cell Research, 18*, 393–410.

Gautron, L., Lee, C., Funahashi, H., Friedman, J., Lee, S., & Elmquist, J. (2010). Melanocortin-4 receptor expression in a vago-vagal circuitry involved in postprandial functions. *The Journal of Comparative Neurology, 518*, 6–24.

Gerlai, R. (1996). Gene-targeting studies of mammalian behavior: Is it the mutation or the background genotype? *Trends in Neurosciences, 19*, 177–181.

Gerlai, R. (2001). Gene targeting: Technical confounds and potential solutions in behavioral brain research. *Behavioural Brain Research, 125*, 13–21.

Gianoulakis, C. (1990). Characterization of the effects of acute ethanol administration on the release of beta-endorphin peptides by the rat hypothalamus. *European Journal of Pharmacology, 180*, 21–29.

Gilpin, N. W., Herman, M., & Roberto, M. (2015). The central amygdala as an integrative hub for anxiety and alcohol use disorders. *Biological Psychiatry*, 77, 859–869.

Girardet, C., Begriche, K., Ptitsyn, R. A., & Butler, A. A. (2014). Unravelling the mysterious roles of melanocortin-3 receptors in metabolic homeostasis and obesity using mouse genetics. *International Journal of Obesity, 4*, 37–44.

Giraudo, S. Q., Billington, C. J., & Levine, A. S. (1998). Feeding effects of hypothalamic injection of melanocortin 4 receptor ligands. *Brain Research, 809*, 302–306.

Giuliani, D., Ottani, A., Minutoli, L., Di Stefano, V., Galantucci, M., Bitto, A., et al. (2009). Functional recovery after delayed treatment of ischemic stroke with melanocortins is associated with overexpression of the activity-dependent gene Zif268. *Brain, Behavior, and Immunity, 23*, 844–850.

Giuliani, D., Zaffe, D., Ottani, A., Spaccapelo, L., Galantucci, M., Minutoli, L., et al. (2011). Treatment of cerebral ischemia with melanocortins acting at MC4 receptors induces marked neurogenesis and long-lasting functional recovery. *Acta Neuropathologica, 122*, 443–453.

Gohil, K. (2014). Pharmaceutical approval update. *P & T: A Peer-Reviewed Journal for Formulary Management, 39*, 746–772.

Greenway, E. L., Dunayevich, E., Tollefson, G., Erickson, J., Guttadauria, M., Fujioka, K., et al. (2009). Comparison of combined bupropion and naltrexone therapy for obesity with monotherapy and placebo. *The Journal of Clinical Endocrinology and Metabolism, 94*, 4898–4906.

Greenway, F. L., Fujioka, K., Plodkowski, R. A., Mudaliar, S., Guttadauria, M., Erickson, J., et al. (2010). Effect of naltrexone and plus bupropion on weight loss in overweight and obese adults (COR-I): A multicenter, randomized, double-blind, placebo-controlled, phase 3 trial. *Lancet, 376*, 595–605.

Griffon, N., Mignon, V., Facchinetti, P., Diaz, J., Schwartz, J. C., & Sokoloff, P. (1994). Molecular cloning and characterization of the rat fifth melanocortin receptor. *Biochemical and Biophysical Research Communications, 200*, 1007–1014.

Grill, H. J., Ginsberg, A. B., Seeley, R. J., & Kaplan, J. M. (1998). Brainstem application of melanocortin receptor ligands produce long-lasting effects on feeding and body weight. *The Journal of Neuroscience, 18*, 10128–10135.

Hadley, M. E., & Haskell-Luevano, C. (1999). The proopiomelanocortin system. *Annals of the New York Academy of Sciences, 885*, 1–21.

Hao, Y., Tian, X. B., Liu, C., Xiang, H. B., & Zhand, J. G. (2015). MC4R expression in pedunculopontine nucleus involved in the modulation of midbrain dopamine system. *International Journal of Clinical and Experimental Pathology, 8*, 2039–2043.

Hasegawa, H., Meeusen, R., Sarre, S., Diltoer, M., Piacentini, M. F., & Michotte, Y. (2005). Acute dopamine/norepinephrine reuptake inhibition increases brain and core temperature in rats. *Journal of Applied Physiology, 99*, 1397–1401.

Hoch, M., Eberle, A. M., Wagner, U., Bussmann, C., Peters, T., & Peterli, R. (2007). Expression and localization of melanocortin-1 receptor in human adipose tissues of severely obese patients. *Obesity, 15*(1), 40–49.

Holderness, C. C., Brooks-Gunn, J., & Warren, M. P. (1994). Co-morbidity of eating disorders and substance abuse review of the literature. *The International Journal of Eating Disorders, 16*, 1–34.

Hollander, P., Gupta, A. K., Plodkowski, R., Greenway, F., Bays, H., Burns, C., et al. (2013). Effects of naltrexone sustained-release/bupropion sustained-release combination therapy on body weight and glycemic parameters in overweight and obese patients with type 2 diabetes. *Diabetes Care, 36*, 4022–4029.

Hsu, R., Taylor, J. R., Newton, S. S., Alvaro, J. D., Haile, C., Han, G., et al. (2005). Blockade of melanocortin transmission inhibits cocaine reward. *The European Journal of Neuroscience, 21*, 2233–2242.

Huszar, D., Lynch, C. A., Fairchild-Huntress, V., Dunmore, J. H., Fang, Q., Berkemeier, L. R., et al. (1997). Targeted disruption of the melanocortin-4 receptor results in obesity in mice. *Cell, 88*, 131–141.

Kalange, A. S., Kokare, D. M., Singru, P. S., Upadhya, M. A., Chopde, C. T., & Subhedar, N. K. (2007). Central administration of selective melanocortin 4 receptor antagonist HS014 prevents morphine tolerance and withdrawal hyperalgesia. *Brain Research, 1181*, 10–20.

Kinoshita, H., Harbuz, M. S., Nishiguchi, M., Ouchi, H., Minami, T., Utsumi, T., et al. (2004). High alcohol preferring (HAP) and low alcohol preferring (LAP) rats show altered proopiomelanocortin (POMC) messenger RNA expression in the arcuate nucleus. *Alcohol and Alcoholism, 39*, 406–409.

Kinoshita, H., Jessop, D. S., Finn, D. P., Coventry, T. L., Roberts, D., Ameno, K., et al. (2000). Acute ethanol decreases NPY mRNA but not POMC mRNA in the arcuate nucleus. *Neuroreport, 11*, 3517–3519.

Kishi, T., Aschkenase, C. J., Lee, C. E., Mountjouy, K. G., Saper, C. B., & Elmquist, J. K. (2003). Expression of melanocortin 4 receptor mRNA in the central nervous system of the rat. *The Journal of Comparative Neurology, 457*, 213–235.

Kokare, D. M., Chopde, C. T., & Subhedar, N. K. (2006). Participation of α-melanocyte stimulating hormone in ethanol-induced anxiolysis and withdrawal anxiety in rats. *Neuropharmacology*, *51*, 536–545.
Kokare, D. M., Singru, P. S., Dandekar, M. P., Chopde, C. T., & Subhedar, N. K. (2008). Involvement of alpha-melanocyte stimulating hormone (α-MSH) in differential ethanol exposure and withdrawal related depression in rat: Neuroanatomical-behavioral correlates. *Brain Research*, *1216*, 53–67.
Krashes, M., Lowell, B. D., & Garfield, A. S. (2016). Melanocortin-4 receptor-regulated energy homeostasis. *Nature Neuroscience*, *19*(2), 206–219.
Krishna, R., Gumbiner, B., Stevens, C., Musser, B., Mallick, M., Suryawanshi, S., et al. (2009). Potent and selective agonism of the melanocortin receptor 4 with MK-0493 does not induce weight loss in obese human subjects: Energy intake predicts: Lack of weight loss efficacy. *Clinical Pharmacology and Therapeutics*, *86*, 659–666.
Labbé, O., Desarnaud, F., Eggerickx, D., Vassart, G., & Parmentier, M. (1994). Molecular cloning of a mouse melanocortin 5 receptor gene widely expressed in peripheral tissues. *Biochemistry*, *33*, 4543–4549.
Lee, M., Kim, A., Conwell, I. M., Hruby, V., Mayorok, A., Cai, M., et al. (2008). Effects of selective modulation of the central melanocortin-3-receptor on food intake and hypothalamic POMC expression. *Peptides*, *29*, 440–447.
Lerma-Cabrera, J. M., Carvajal, F., Alcaraz-Iborra, M., de la Fuente, L., Navarro, M., Thiele, T. E., et al. (2013). Adolescent binge-like ethanol exposure reduces basal α-MSH expression in the hypothalamus and the amygdala of adult rats. *Pharmacology, Biochemistry, and Behavior*, *110*, 66–74.
Lerma-Cabrera, J. M., Carvajal, F., Chotro, G., Gaztañaga, M., Navarro, M., Thiele, T. E., et al. (2013). MC4R signaling within the nucleus accumbens shell, but not the lateral hypothalamus, modulates ethanol palatability in rats. *Behavioural Brain Research*, *239*, 51–54.
Lerma-Cabrera, J. M., Carvajal, F., de la Torre, L., de la Fuente, L., Navarro, M., Thiele, T. E., et al. (2012). Control of food intake by MC4-R signaling in the lateral hypothalamus, nucleus accumbens shell and ventral tegmental area: Interactions with ethanol. *Behavioural Brain Research*, *234*, 51–60.
Lim, B. K., Huang, K. W., Grueter, B. A., Rothwell, P. E., & Malenka, R. C. (2012). Anhedonia requires MC4R-mediated synaptic adaptations in nucleus accumbens. *Nature*, *487*, 183–189.
Lindblom, J., Kask, A., Hagg, E., Harmark, L., Bergstrom, L., & Wikberg, J. (2002). Chronic infusion of a melanocortin receptor agonist modulates dopamine receptor binding in the rat brain. *Pharmacological Research*, *45*, 119–124.
Lindblom, J., Opmane, B., Mutulis, F., Mutule, I., Petrovska, R., Klusa, V., et al. (2001). The MC4 receptor mediates alpha-MSH induced release of nucleus accumbens dopamine. *Neuroreport*, *12*, 2155–2158.
Lindblom, J., Schiot, H. B., Larsson, A., Wikberg, J. E., & Bergstrom, L. (1998). Autoradiographic discrimination of melanocortin receptors indicates that the MC3 subtype dominates in the medial brain. *Brain Research*, *810*, 161–171.
Lindblom, J., Wikberg, J. E. S., & Bergstrom, L. (2002). Alcohol-preferring AA rats show a derangement in their central melanocortin signaling system. *Pharmacology, Biochemistry, and Behavior*, *72*, 491–496.
Lippert, R. N., Ellacot, K. L. J., & Cone, R. D. (2014). Gender-specific roles for the melanocortin-3 receptor in the regulation of the mesolimbic dopamine system in mice. *Endocrinology*, *155*, 1718–1727.
Liu, H., Kishi, T., Roseberry, A. G., Cai, X., Lee, C. E., Montez, J. M., et al. (2003). Transgenic mice expressing green fluorescent protein under the control of the melanocortin-4-receptor promoter. *The Journal of Neuroscience*, *23*, 7143–7154.

Loir, B., Pérez Sánchez, C., Ghanem, G., Lozano, J. A., García-Borrón, J. C., & Jiménez-Cervantes, C. (1999). Expression of the MC1 receptor gene in normal and malignant human melanocytes. A semiquantitative RT-PCR study. *Cellular and Molecular Biology (Noisy-le-Grand, France)*, *45*, 1083–1092.

Lu, D., Willard, D., Patel, I. R., Kadwell, S., Overton, L., Kost, T., et al. (1994). Agouti protein is an antagonist of the melanocyte-stimulating-hormone receptor. *Nature*, *371*, 799–802.

Mains, E. E., & Eipper, B. A. (1976). Biosynthesis of adrenocorticotropic hormone in mouse pituitary tumor cells. *The Journal of Biological Chemistry*, *251*, 4115–4120.

Mains, E. E., & Eipper, B. A. (1981). Differences in the post-translational processing of beta-endorphin in rat anterior and intermediate pituitary. *The Journal of Biological Chemistry*, *256*, 5683–8688.

Marks, D. L., Hruby, V., Brookhart, G., & Cone, R. D. (2006). The regulation of food intake by selective stimulation of the type 3 melanocortin receptor (MC3R). *Peptides*, *27*, 259–264.

Marsh, D. J., Hollopeter, G., Huszar, D., Laufer, R., Yagaloff, K. A., Fisher, S. L., et al. (1999). Response of melanocortin-4 receptor-deficient mice to anorectic and orexigenic peptides. *Nature Genetics*, *21*, 119–122.

McElroy, S., Guerdjikova, A. I., Kim, D. D., Burns, C., Harris-Collazo, R., Landbloom, R., et al. (2013). Naltrexone/Bupropion combination therapy in overweight or obese patients with major depressive disorder: Results of a pilot study. *The Primary Care Companion for CNS Disorders*, *15*, e1–e8.

Mogil, J. S., Ritchie, J., Smith, S. B., Strasburg, K., Kaplan, L., Wallace, M. R., et al. (2005). Malanocortin-1 receptor gene variants affect pain and mu-opioid analgesia in mice and humans. *Journal of Medical Genetics*, *42*(7), 583–587.

Mogil, J. S., Wilson, S. G., Chesler, E. J., Rankin, A. L., Nemmani, K. V. S., William, R., et al. (2003). The melanocortin-1 receptor gene mediates female-specific mechanisms of analgesia in mice and humans. *Proceedings of the National Academy of Sciences of the United States of America*, *100*(8), 4867–4872.

Møller, C. L., Raun, K., Jacobsen, M. L., Pedersen, T. A., Holst, B., Conde-Frieboes, K. W., et al. (2011). Characterization of murine melanocortin receptors mediating adipocyte lipolysis and examination of signaling pathways involved. *Molecular and Cellular Endocrinology*, *341*(1–2), 9–17.

Mounien, L., Bizet, P., Boutelet, I., Vaudry, H., & Jegou, S. (2005). Expression of melanocortin MC3 and MC4 receptor mRNAs by neuropeptide Y neurons in the rat arcuate nucleus. *Neuroendocrinology*, *82*, 164–170.

Mountjoy, K. G. (2015). Pro-opiomelanocortin (POMC) neurones, POMC-derived peptides, melanocortin receptors and obesity: How understanding of this system has changed over the last decade. *Journal of Neuroendocrinology*, *27*, 406–418.

Mountjoy, K. G., Mortrud, M. T., Low, M. J., Simerly, R. B., & Cone, R. D. (1994). Localization of the melanocortin-4 receptor (MC4-R) in neuroendocrine and autonomic control circuits in the brain. *Molecular Endocrinology*, *8*, 1298–1308.

Mountjoy, K. G., Robbins, L. S., Mortrud, M. T., & Cone, R. D. (1992). The cloning of a family of genes that encode the melanocortin receptors. *Science*, *257*, 1248–1251.

Mountjoy, K. G., & Wild, J. M. (1998). Melanocortin-4 receptor mRNA expression in the developing autonomic and central nervous systems. *Brain Research. Developmental Brain Research*, *15*, 309–314.

Navarro, M., Carvajal, F., Lerma-Cabrera, J. M., Cubero, I., Picker, M. J., & Thiele, T. E. (2015). Evidence that melanocortin receptor agonist Melanotan-II synergistically augments the ability of naltrexone to blunt binge-like ethanol intake in male C57BL/6J mice. *Alcoholism, Clinical and Experimental Research*, *39*, 1425–1433.

Navarro, M., Cubero, I., Chen, A. S., Chen, H. Y., Knapp, D. J., Breese, G. R., et al. (2005). Effects of the melanocortin receptor activation and blockade of ethanol intake: A possible role for the melanocortin-4 receptor. *Alcoholism, Clinical and Experimental Research, 29*, 949–957.

Navarro, M., Cubero, I., Knapp, D. J., Breese, G. R., & Thiele, T. E. (2008). Decreased immunoreactivity of the melanocortin neuropeptides α-melanocyte-stimulating hormone (α-MSH) after chronic ethanol exposure in Sprague-Dawley rats. *Alcoholism, Clinical and Experimental Research, 32*, 266–276.

Navarro, M., Cubero, I., Knapp, D. J., & Thiele, T. E. (2003). MTII-induced reduction of voluntary ethanol drinking is blocked by pretreatment with AgRP-(83-132). *Neuropeptides, 37*, 338–344.

Navarro, M., Cubero, I., Ko, L., & Thiele, T. E. (2009). Deletion of agouti-related protein blunts ethanol self-administration and binge-like drinking in mice. *Genes, Brain, and Behavior, 8*, 450–458.

Navarro, M., Cubero, I., & Thiele, T. E. (2013). Decreased immunoreactivity of the polypeptide precursor pro-opiomelanocortin (POMC) and the prohormonee convertase PC1/3 after chronic ethanol exposure in Sprague-Dawley rats. *Alcoholism, Clinical and Experimental Research, 37*, 399–406.

Navarro, M., Lerma-Cabrera, J. M., Carvajal, F., Lowery, E. G., Cubero, I., & Thiele, T. E. (2011). Assessment of voluntary ethanol consumption and the effects of a melanocortin (mc) receptor agonist on ethanol intake in mutant C57BL/6J mice lacking the MC-4 receptor. *Alcoholism, Clinical and Experimental Research, 35*, 1058–1066.

NIAAA. (2004). National Institute on Alcohol Abuse and Alcoholism Council approves definition of binge drinking. *NIAAA Newsletter, 3*(Winter).

Niu, Z., Ma, J., Chu, H., Zhao, Y., Feng, W., & Cheng, Y. (2012). Melanocortin 4 receptor antagonist attenuates morphine antinociceptive tolerance, astroglial activation and cytokine expression in the spinal cord rat. *Neuroscience Letters, 529*, 112–117.

Ollmann, M. M., Wilson, B. D., Yang, Y. K., Kerns, J. A., Chen, Y., Gantz, I., et al. (1997). Antagonism of central melanocortin receptors in vitro and vivo by agouti-related protein. *Science, 278*, 135–138.

Olney, J. J., Navarro, M., & Thiele, T. E. (2014). Targeting central melanocortin receptors: A promising novel approach for treating alcohol abuse disorders. *Frontiers in Neuroscience, 3*(8), 1–9.

Oude Ophuis, R. J. A., Boender, A. J., van Rozen, A. J., & Adan, R. A. H. (2014). Cannabinoid, melanocortin and opioid receptor expression on DRD1 and DRD2 subpopulations in rat striatum. *Frontiers in Neuroanatomy, 26*, 8–14.

Panaro, B. L., Tough, I. R., Engelstoft, M. S., Matthews, R. T., Digby, G. J., Moller, C. L., et al. (2014). The melanocortin-4-receptor is expressed in enteroendocrine L cells and regulates the release of peptide YY and glucagon-like peptides 1 in vivo. *Cell Metabolism, 20*, 1018–1029.

Pétervári, E., Garami, A., Soós, S., Székely, M., & Balaskó, M. (2010). Age-dependence of alpha-MSH-induced anorexia. *Neuropeptides, 44*, 315–322.

Ploj, K., Roman, E., Kask, A., Hyytia, P., Schioth, H. B., Wikberg, J., et al. (2002). Effects of melanocortin receptor ligands on ethanol intake and opioid levels in alcohol preferring AA rats. *Brain Research Bulletin, 59*, 97–104.

Poggioli, R., Vergoni, A. V., & Bertolini, A. (1986). ACTH-(1-24) and alpha-MSH antagonize feeding behavior stimulated by kappa opiate agonist. *Peptides, 7*, 843–848.

Popp, R. L., & Erickson, C. K. (1998). The effect of an acute ethanol exposure on the rat brain POMC opiopeptide system. *Alcohol, 16*, 139–148.

Pritchard, L. E., Turnbull, A. V., & White, A. (2002). Pro-opiomelanocortin processing in the hypothalamus: Impact on melanocortin signalling and obesity. *The Journal of Endocrinology, 172*, 411–421.

Pritchard, L. E., & White, A. (2007). Minireview: Neuropeptide processing and its impact on melanocortin pathways. *Endocrinology, 148*, 4201–4207.

Qu, H., Li, J., Chen, W., Li, Y., Jiang, Q., Jiang, H., et al. (2014). Differential expression of the melanocortin-4 receptor in male and female C57BL/6J mice. *Molecular Biology Reports, 41*, 3245–3256.

Quillan, J. M., Sadee, W., Wei, E. T., Jiménez, C., Ji, L., & Chang, J. K. (1998). A synthetic human agouti-related protein (83-132)-NH2 fragments is a potent inhibitor of melanocortin receptor function. *FEBS Letters, 428*, 59–62.

Rainero, I., De Gennaro, T., Visenti, G., Brunetti, E., Cerrato, P., Torre, E., et al. (1990). Effects of chronic ethanol treatment on alpha-MSH concentrations in rat brain and pituitary. *Neuropeptides, 15*, 139–141.

Rao, T. L., Kokare, D. M., Sarkar, S., Khisti, R. T., Chopde, C. T., & Subhedar, N. (2003). GABAergic agents prevent alpha-melanocyte stimulating hormone induced anxiety and anorexia in rats. *Pharmacology, Biochemistry, and Behavior, 76*, 417–423.

Rasmussen, D. D., Boldt, B. M., Wilkinson, C. W., & Milton, D. R. (2002). Chronic daily ethanol and withdrawal: 3. Forebrain pro-opiomelanocortin gene expression and implications for dependence, relapse, and deprivation effect. *Alcoholism, Clinical and Experimental Research, 26*, 535–546.

Rasmussen, D. D., Bryant, C. A., Boldt, B. M., Colasurdo, E. A., Levin, N., & Wilkinson, C. W. (1998). Acute ethanol effects on opiomelanocortinergic regulation. *Alcoholism, Clinical and Experimental Research, 22*, 789–801.

Rees, J. L. (2003). Genetics of hair and skin color. *Annual Review of Genetics, 37*, 67–90.

Renquist, B. J., Lippert, R., Sebag, J. A., Ellacott, K. L. J., & Cone, R. D. (2011). Physiological roles of the melanocortin MC3 receptor. *European Journal of Pharmacology, 660*(1), 13–20.

Rhodes, J. S., Best, K., Belknap, J. K., Finn, D. A., & Crabbe, J. C. (2005). Evaluation of a simple model of ethanol drinking to intoxication in C57BL/6J mice. *Physiology & Behavior, 84*, 53–63.

Roddy, E. (2004). Bupropion and other non-nicotine pharmacotherapies. *BMJ, 328*, 509–511.

Roseberry, A. G., Stuhrman, K., & Dunugan, A. I. (2015). Regulation of the mesocorticolimbic and mesostriatal dopamine system by α-melanocyte stimulating hormone and agouti-related protein. *Neuroscience and Biobehavioral Reviews, 56*, 15–25.

Roselli-Rehfuss, L., Mountjoy, K. G., Robbins, L. S., Mortrud, M. T., Low, M. J., Tatro, J. B., et al. (1993). Identification of a receptor for gamma melanotropin and other proopiomelanocortin peptides in the hypothalamus and limbic system. *Proceedings of the National Academy of Sciences of the United States of America, 90*, 8856–8860.

Rossi, J., Balthasar, N., Olson, D., Scott, M., Berglund, E., Lee, C. E., et al. (2011). Melanocortin-4 receptors expressed by cholinergic neurons regulate energy balance and glucose homeostasis. *Cell Metabolism, 13*, 195–204.

Sánchez, M. S., Barontini, M., Armando, I., & Celis, M. E. (2001). Correlation of increased grooming behavior and motor activity with alterations in nigrostriatal and mesolimbic catecholamines after alpha-melanotropin and neuropeptide glutamine-isoleucine injection in the rat tegmental area. *Cellular and Molecular Neurobiology, 5*, 523–533.

Scanlon, M. N., Lazar-Wesley, E., Grant, K. A., & Kunos, G. (1992). Proopiomelanocortin messenger mRNA is decreased in the mediobasal hypothalamus of rats made dependent on ethanol. *Alcoholism, Clinical and Experimental Research, 16*, 1147–1151.

Scarlett, J. M., Bowe, D. D., Zhu, X., Batra, A. K., Grant, W. F., & Marks, D. L. (2010). Genetic and pharmacologic blockade of central melanocortin signaling attenuates cardiac cachexia in rodent models of heart failure. *The Journal of Endocrinology, 206*, 121–130.

Schioth, H. B., Chhajlani, V., Muceniece, R., Klusa, V., & Wikberg, J. E. (1996). Major pharmacological distinction of the ACTH receptor from other melanocortin receptors. *Life Sciences, 59*(10), 797–801.
Schioth, H. B., Muceniece, R., Wikberg, J. E., & Chhajlani, V. (1995). Characterization of melanocortin receptor subtypes by radioligand binding analysis. *European Journal of Pharmacology, 288*, 311–317.
Schioth, H. B., & Watanobe, H. (2002). Melanocortin and reproduction. *Brain Research. Brain Research Reviews, 38*, 340–350.
Seizinger, B. R., Bovermann, K., Maysinger, D., Hollt, V., & Herz, A. (1983). Differential effects of acute and chronic ethanol treatment on particular opioid peptide systems in discrete regions of rats brain and pituitary. *Pharmacology, Biochemistry, and Behavior, 18*, 361–369.
Siljee, J. E., Unmehopa, U. A., Kalsbeek, A., Swaab, D. F., Fliers, E., & Alkemade, A. (2013). Melanocortin 4 receptor distribution in the human hypothalamus. *European Journal of Endocrinology, 168*, 361–369.
Sinclair, J. D., Le, A. D., & Kiianmaa, K. (1989). The AA and ANA rat lines, selected for differences in voluntary alcohol consumption. *Experientia, 45*, 798–805.
Smith, S. R., Fujioka, K., Gupta, A. K., Billes, S. K., Burns, C., Kim, D., et al. (2013). Combination therapy with naltrexone and bupropion for obesity reduces total and visceral adiposity. *Diabetes, Obesity & Metabolism, 15*, 863–866.
Smith, A. I., & Funder, J. W. (1988). Proopiomelanocortin processing in the pituitary, central nervous system, and peripheral tissues. *Endocrine Reviews, 9*, 159–179.
Sprow, G. M., Rinker, J. A., Lowery-Gionta, E. G., Sparrow, A. M., Navarro, M., & Thiele, T. E. (2015). Lateral hypothalamic melanocortin receptor signaling modulates binge-like ethanol drinking in C57BL/6J mice. *Addiction Biology, 21*, 835–846.
Stander, S., Böhm, M., Brzoska, T., Zimmer, K. P., Luger, T., & Metze, D. (2002). Expression of melanocortin-1 receptor in normal, malformed and neoplastic skin glands and hair follicles. *Experimental Dermatology, 11*, 42–51.
Starowicz, K., Obara, I., Przewlocki, R., & Przewlocka, B. (2005). Inhibition of morphine tolerance by spinal melanocortin receptor blockade. *Pain, 117*, 401–411.
Starowicz, K., Sieja, A., Bilecki, W., Obara, I., & Przewlocka, B. (2003). The effect of morphine on MC4 and CRF receptor mRNAs in the rat amygdala and attenuation of tolerance after their blockade. *Brain Research, 990*, 113–119.
Suzuki, I., Cone, R., Im, S., Nordlund, J., & Abdel-Malek, Z. (1996). Binding capacity and activation of the MC1 receptors by melanotropic hormones correlate directly with their mitogenic and melanogenic effects on human melanocytes. *Endocrinology, 137*, 1627–1633.
Szekely, J. I., Miglecz, E., Dunai-Kovacs, Z., Tarnawa, I., Ronai, A. Z., Graf, L., et al. (1979). Attenuation of morphine tolerance and dependence by alpha-melanocyte stimulating hormone (alpha-MSH). *Life Sciences, 24*, 1931–1938.
Tatro, J. B. (1990). Melanotropin receptors in the brain are differentially distributed and recognize both corticotropin and alpha-melanocyte stimulating hormone. *Brain Research, 536*, 124–132.
Tatro, J. B., & Entwistle, M. L. (1994). Distribution of melanocortin receptors in the lower brainstem of the rat. *Annals of the New York Academy of Sciences, 739*, 311–314.
Thiele, T. E., Navarro, M., Sparta, D. R., Fee, J. R., Knapp, D. J., & Cubero, I. (2003). Alcoholism and obesity: Overlapping neuropeptide pathways? *Neuropeptides, 37*, 321–337.
Tsujii, S., & Bray, G. A. (1989). Acetylation alters the feeding response to MSH and beta-endorphin. *Brain Research Bulletin, 23*, 165–169.
Vergoni, A. V., Bertolini, A., Mutulis, F., Wikberg, J. E., & Schioth, H. B. (1998). Differential influence of a selective melanocortin MC4 receptor antagonist (HS014) on

melanocortin-induced behavioral effects in rats. *European Journal of Pharmacology, 362*, 95–101.

Voisey, J., Carrol, L., & van Daal, A. (2003). Melanocortins and their receptors and antagonists. *Current Drug Targets, 4*, 586–897.

Vrinten, D. H., Gispen, W. H., Kalkman, C. J., & Adan, R. A. (2003). Interaction between the spinal melanocortin and opioid system in a rat model neuropathic pain. *Anesthesiology, 99*, 449–454.

Wan, S., Browning, K. N., Coleman, F. H., Sutton, G., Zheng, H., Butler, A., et al. (2008). Presynaptic melanocortin-4 receptors on vagal afferent fibers modulate the excitability of rat nucleus tractus solitarius neurons. *The Journal of Neuroscience, 28*, 4957–4966.

Wardlaw, S. L. (2011). Hypothalamic proopiomelanocortin processing and the regulation of energy balance. *European Journal of Pharmacology, 660*, 213–219.

Wikberg, J. E., Muceniece, R., Mandrika, I., Prusis, P., Lindblom, J., Post, C., et al. (2000). New aspects on the melanocortins and their receptor. *Pharmacological Research, 42*, 393–420.

Wilkinson, C. W. (2006). Roles of acetylation and other post-translational modifications in melanocortin function and interactions with endorphins. *Peptides, 27*, 453–471.

Wilkinson, C. W., Crabbe, J. C., Keith, L. D., Kendall, J. W., & Dorsa, D. M. (1986). Influence of ethanol dependence on regional brain content of β-endorphin in the mouse. *Brain Research, 378*, 107–114.

Winkler, A., Roske, I., Furkert, J., Fickel, J., & Melzig, M. F. (1995). Effects of voluntary ethanol ingestion on the POMC gene expression in the rat pituitary and on the plasma β-endorphin content. *Alcohol and Alcoholism, 30*, 231–238.

Wirth, M. M., Olszewski, P. K., Yu, C., Levine, A. S., & Giraudo, S. Q. (2001). Paraventricular hypothalamic alpha-melanocyte-stimulating hormone and MTII reduce feeding without causing aversive effects. *Peptides, 22*, 129–134.

Wolf Horrell, E. M. W., Boulanger, M. C., & D'Orazio, J. A. (2016). Melanocortin 1 receptor: Structure, function and regulation. *Frontiers in Genetics*, 7(95), 1–16.

Xia, Y., & Wikberg, J. E. (1997). Postnatal expression of melanocortin-3 receptor in rat diencephalon and mesencephalon. *Neuropharmacology, 36*, 217–224.

Xia, Y., Wikberg, J. E., & Chhajlani, V. (1995). Expression of melanocortin 1 receptor in periaqueductal gray matter. *Neuroreport, 6*, 2193–2196.

Yeo, G. S., Farroqi, I. S., Challis, B. G., Jackson, R. S., & O'Rahilly, S. (2000). The role of melanocortin signalling in the control of body weight: Evidence from human and murine genetic models. *The Quarterly Journal of Medicine, 93*, 7–14.

York, D. A., Boghossian, S., & Park-York, M. (2011). Melanocortin activity in the amygdala influences alcohol intake. *Pharmacology, Biochemistry, and Behavior, 98*, 112–119.

Zhao, W., Sedman, G., Gibbs, M., & Ng, K. T. (1995). Phosphorylation changes following weakly reinforced learning and ACTH-induced memory consolidation for a weak learning experience. *Brain Research Bulletin, 36*, 161–168.

Zhou, Y., Colombo, G., Niikura, K., Carai, M. A. M., Femenis, T., García-Gutiérrez, M. S., et al. (2013). Voluntary alcohol drinking enhances proopiomelanocortin gene expression in nucleus accumbens shell and hypothalamus of Sardinian alcohol-preferring rats. *Alcoholism, Clinical and Experimental Research, 37*, E131–E140.

Zhou, Y., Franck, J., Spangler, R., Maggos, C. E., Ho, A., & Kreek, M. J. (2000). Reduced hypothalamic POMC and anterior pituitary CRF1 receptor mRNA levels after acute, but not chronic, daily "binge" intragastric alcohol administration. *Alcoholism, Clinical and Experimental Research, 24*, 1575–1582.

CHAPTER SIX

Substance P and the Neurokinin-1 Receptor: The New CRF

Jesse R. Schank*[,1], Markus Heilig[†]
*University of Georgia, Athens, GA, United States
[†]Center for Social and Affective Neuroscience, Linkoping University, Linkoping, Sweden
[1]Corresponding author: e-mail address: jschank@uga.edu

Contents

Abstract

Substance P (SP) is an 11-amino acid neuropeptide of the tachykinin family that preferentially activates the neurokinin-1 receptor (NK1R). First isolated 85 years ago and sequenced 40 years later, SP has been extensively studied. Early studies identified a role for SP and the NK1R in contraction of intestinal smooth muscle, central pain processing, and neurogenic inflammation. An FDA-approved NK1R antagonist, aprepitant, is used clinically for the treatment of chemotherapy-induced nausea, as the NK1R influences the activity of the brain stem emesis centers. More recently, SP and the NK1R have gained attention for their role in complex psychiatric processes including stress, anxiety, and depression. However, clinical development of NK1R antagonists for these indications has so far been unsuccessful. Several preclinical studies have also demonstrated a role of the NK1R in drug taking and drug seeking, especially as it relates to escalated consumption and stress-elicited seeking. This line of research developed in parallel with

International Review of Neurobiology, Volume 136
ISSN 0074-7742
http://dx.doi.org/10.1016/bs.irn.2017.06.008

findings supporting a role of corticotropin-releasing factor (CRF) in stress-induced drug seeking. Over this time, CRF arguably gained more attention as a target for development of addiction pharmacotherapies. However, this effort has not resulted in a viable drug for use in human populations. Given promising clinical findings for the efficacy of NK1R antagonists on craving in alcoholics, along with recent data suggesting that a number of negative results from NK1R trials were likely due to insufficient receptor occupancy, the NK1R merits being revisited as a target for the development of novel pharmacotherapeutics for addiction.

1. INTRODUCTION

The isolation of substance P (SP) in 1931 and its sequencing in 1971 both precede that of corticotropin-releasing factor (CRF), so it may seem counterintuitive to refer to SP as "the new CRF." However, CRF became widely studied in the context of affective and addictive disorders shortly following its isolation in 1981. In contrast, SP and its neurokinin-1 receptor (NK1R) began garnering attention as a signaling system involved in stress-related conditions such as anxiety and depression only with the arrival of the knockout mouse era in the 1990s. The increasing realization of the influence that stress-related processes play in addiction-related behaviors then spurred research that has examined the role of the NK1R in drug seeking and taking. In the sections that follow, we will introduce this system, discuss its role in affective disorders and addiction, outline neural mechanisms and system interactions that mediate its effects, and survey the clinical research that has been performed using NK1R antagonists.

2. BASIC CHARACTERISTICS OF THE SP/NK1R SYSTEM

2.1 Neurokinin Receptors and Peptides

SP is an 11-amino acid peptide that belongs to a class of signaling molecules known as neurokinins. Other neurokinin peptides include neurokinin A (NKA) and neurokinin B (NKB), and collectively, the neurokinins belong to the larger family of peptides called tachykinins (Pennefather et al., 2004). SP is a cleavage product of the precursor peptide preprotachykinin A (PPTA), which is processed to the mature molecule during axonal transport to the terminal and also generates NKA. The gene for PPTA is *TAC1* in humans and *Tac1* mice. NKB is produced from the *TAC3* gene (*Tac2* in mice). There are a number of other, less studied peptides that fall into the tachykinin family and are produced by the *TAC1* and *TAC2* genes, but these will not be discussed in this chapter.

Three receptors exist for the neurokinin peptides: NK1R (*TACR1* gene), NK2R (*TACR2* gene), and NK3R (*TACR3* gene). Each neurokinin receptor has at least moderate affinity for each neurokinin ligand, but is preferentially activated by one of these. NK1R is the primary receptor for SP, NK2R is the primary receptor for NKA, and NK3R is the primary receptor for NKB (see Fig. 1 for a summary of neurokinin genes, transmitters, and receptors). While other neurokinin receptors have shown some role in affective disorders and drug-related behaviors, most notably NKB and the NK3R, this chapter will focus specifically on SP and the NK1R.

SP and the NK1R are highly expressed throughout the central nervous system. For example, the terminals of primary afferent pain fibers coexpress SP and glutamate, and make synaptic contact with NK1R-expressing, ascending neurons of the dorsal horn of the spinal cord. Expression in other brain areas suggests a possible role of NK1R in stress, anxiety, and motivated behaviors. For example, the NK1R is expressed in many forebrain and brainstem regions including the dorsal and ventral striatum, hippocampus, amygdala, periaqueductal gray, locus coeruleus, habenula, septal nuclei, nucleus of the solitary tract, and dorsal raphe (DR) nucleus (Mantyh, 2002; Mantyh, Hunt, & Maggio, 1984; Quirion et al., 1983; Yip & Chahl, 2000). SP peptide is expressed in many of the same areas including the dorsal and ventral striatum, amygdala, DR nucleus, bed nucleus of the

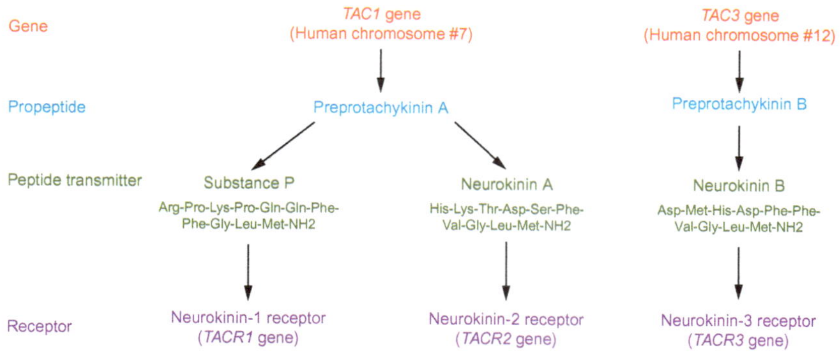

Fig. 1 Components of the neurokinin signaling systems. The neurokinin family of neuropeptides includes substance P, neurokinin A, and neurokinin B. Each of these peptides has a preferred endogenous receptor, although they do have limited affinity across all neurokinin receptors. While neurokinin B is produced by a unique propeptide, substance P and neurokinin A are both end products of cleavage of the common propeptide preprotachykinin-A (*TAC1* gene).

stria terminalis, septal nuclei, substantia nigra, habenula, nucleus of the solitary tract, and various hypothalamic nuclei (Hurd, Keller, Sotonyi, & Sedvall, 1999; Shults, Quirion, Chronwall, Chase, & O'Donohue, 1984).

The NK1R couples primarily to the $G\alpha_q$ signaling pathway which activates phospholipase C, protein kinase C, and mobilizes intracellular calcium. However, under some conditions, this receptor can also active adenylyl cyclase- and arachidonic acid-related signaling systems. Endogenously, the NK1R can be found either as the full-length version or as a truncated form which lacks much of the C-terminus. While the truncated form is not believed to differ from the full length in terms of receptor affinity, the lack of critical C-terminal residues leads to altered desensitization and internalization processes.

2.2 Physiological Functions of SP and NK1R

Although the main body of this chapter will discuss the role of SP and the NK1R in affective disorders and addiction, it is beneficial to review some of the peripheral effects of this signaling system. Major roles of SP and the NK1R in peripheral and reflexive neural processes include an effect on pain processing, inflammation, intestinal contraction, and emesis. For a comprehensive review of the role of neurokinins in physiological processes, the reader is referred to Steinhoff, von Mentzer, Geppetti, Pothoulakis, and Bunnett (2014).

The role of SP and the NK1R in pain sensation is well established. Primary afferents that express peripheral nociceptors and transmit pain signals into the central nervous system express SP along with the classical neurotransmitter glutamate (see earlier). These primary afferents have cell bodies in the dorsal root ganglion, just lateral to the spinal column, and possess bifurcated axons with one terminal in the dorsal horn of the spinal cord and one in the periphery. The peripheral terminals target the skin and contain nociceptors that sense pain and temperature stimuli. SP can be released both centrally in the dorsal horn synapse and in the peripheral terminal. At the dorsal horn spinal synapse, SP action at NK1Rs located on ascending neurons of the anterolateral pain pathways potentiates the effect of postsynaptic glutamatergic stimulation. SP release in the periphery from nociceptive neurons can induce neurogenic inflammation via stimulation of histamine release from mast cells, among other mechanisms.

In addition to these pain processing and inflammation promoting effects, the NK1R also plays a role in gastrointestinal function and the vomiting reflex. Increased intestinal contraction following SP application was one of the earliest identified physiological effects of this peptide. NK1R receptors are also located centrally in the brainstem vomiting center of the medulla, and have their influence on nausea and emesis via their activity in these regions. Accordingly, chemotherapy-induced emesis is the indication for FDA-approved NK1R antagonists (for example, aprepitant).

While the role of the NK1R in pain processing, emesis, stress, anxiety, depression, and addiction has seemed to be consistent between species, there are important differences in expression pattern and ligand affinity. For example, interspecies variation in a small number of amino acid residues of the NK1R influences a divergence between humans and rats in affinity for specific antagonists (Jensen, Gerard, Schwartz, & Gether, 1994). In general, rat and mouse NK1Rs seem to differ in affinity for specific NK1R antagonists from a second cluster of species including human, gerbil, guinea pig, and dog (Leffler et al., 2009). Interspecies differences in receptor structure are not believed to impact the endogenous role of the NK1R in behaviors or receptor distribution, but are predicted to affect the affinity of small molecule antagonists. For this reason, while mice and rats can be used for mechanistic studies to examine the role of the NK1R in specific behavioral responses, different NK1R antagonists must be used from those that might be clinical candidates in humans.

3. PRECLINICAL FINDINGS ON SP AND NK1R

3.1 Preclinical Findings in Affective Disorders

Beginning in the mid-1990s, several groups began examining a potential role of SP and the NK1R in more complex behaviors related to stress responses, anxiety, and depression. To date, a rich literature has accumulated supporting a critical role of the SP/NK1R system in stress sensitivity and anxiety-like behavior in rodents and other species (reviewed in Ebner & Singewald, 2006). For example, infusion of exogenous SP or other NK1R agonists into the brain induces anxiogenic-like responses in rodents (Duarte, Testolin, & De Lima, 2004; Mello et al., 2007; Teixeira et al., 1996). Immobilization stress and elevated platform exposure induce SP release in the medial amygdala (MeA), an important hub in the brain stress circuitry, and NK1R antagonists reverse the anxiogenic effects of prior stress exposure and intra-MeA SP infusion (Ebner, Rupniak, Saria, & Singewald, 2004; Singewald et al., 2008). Fear potentiated startle is also mediated by the actions of SP in the amygdala and other stress-sensitive brain regions (Zhao, Yang, Walker, & Davis, 2009).

NK1R inhibition, either by genetic deletion or by pharmacological antagonism, attenuates stress reactivity, anxiety-related behavior, and stress-induced glucocorticoid release (Santarelli et al., 2001). For example, NK1R antagonism attenuates neuronal activation by exposure to a mild stressor in regions of the brain stress circuitry (Ebner, Muigg, Singewald, & Singewald, 2008). Interestingly, NK1R antagonists also have anxiolytic-like properties in the absence of prior stress, an effect which is not always observed with other stress-related transmitter systems (Teixeira et al., 1996). NK1R antagonists also suppress fear responses in gerbils (Rupniak, Webb, Fisher, Smith, & Boyce, 2003), a species often used to study the behavioral effects of NK1R antagonists because their receptors are more homologous to the human receptor than those of mice or rats (see earlier). Another study observed a strong NK1R activation by maternal separation in guinea pig pups, and also found that infusion of a specific NK1R antagonist into the amygdala suppressed stress-induced vocalizations in these maternally separated offspring (Boyce et al., 2001).

Forced swim stress is a behavioral exposure used both as a stressor and as a preclinical screen for antidepressant properties of pharmacological agents. Forced swim has been shown to induce SP release in the lateral septum (LS), which influences coping behavior and suppresses serotonin (5HT) release via its actions at the NK1R (Ebner, Singewald, Whittle, Ferraguti, & Singewald, 2008). In this particular study, NK1R antagonism induced active coping behavior during the forced swim stressor, which is typically considered an antidepressant response, and prevented the stress-induced suppression of cortical 5HT output. The effects of NK1R antagonists in this model are consistent with what is observed with treatment by clinically approved antidepressant drugs including selective serotonin reuptake inhibitors and tricyclic antidepressants. Microinjection of an NK1R antagonist specifically in the lateral habenula (LHb) has also been shown to produce antidepressant-like behavioral effects in rodent models and induces increased 5HT levels in the DR (Yang, Yu, Jin, & Zhao, 2014). Given this and other preclinical findings in the 1990s and early 2000s, there was much enthusiasm about the antidepressant potential of NK1R-acting drugs, and several clinical development programs were initiated, most notably by Merck. These efforts are discussed in detail in later sections of this chapter.

3.2 Preclinical Findings in Addiction

As the behavioral findings detailed earlier began to accumulate, focus on the behavioral effects of the NK1R was then expanded to assess drug-related behaviors. In addition to the clinical comorbidity of anxiety and depression with drug abuse, the neuroanatomical localization of this receptor further suggested a potential role in addiction. Also, some of the earliest phenotypes identified in NK1R knockout mice were an attenuation of the analgesic and rewarding properties of morphine (Murtra, Sheasby, Hunt, & De Felipe, 2000). Anatomical and behavioral findings concerning addictive behaviors will be discussed in detail later, but the reader is also referred to excellent recent reviews on the topic (Commons & Serock, 2009; Schank, 2014; Schank, Ryabinin, Giardino, Ciccocioppo, & Heilig, 2012).

3.2.1 Opiates

As mentioned earlier, some of the earliest phenotypes tested following the development of mice with a genetic deletion of the NK1R (NK1R$^{-/-}$ mice) were the rewarding and reinforcing properties of opiate drugs. It was found that morphine reward, as measured by the conditioned place preference method, was absent in these mice, and morphine-induced locomotion was suppressed (Murtra et al., 2000). However, reward for cocaine was not affected and was expressed normally in these mice. NK1R$^{-/-}$ mice are also less sensitive to some analgesic properties of morphine (De Felipe et al., 1998). However, these findings were obtained with constitutive knockouts, and might therefore be confounded by developmental changes. It is therefore noteworthy that the findings regarding the role for NK1Rs in morphine reward were replicated using delivery of SP-saporin, a neurotoxin that specifically targets NK1R-expressing cells (Gadd, Murtra, De Felipe, & Hunt, 2003). This study additionally suggested that the effect of the NK1R on morphine reward is localized to the amygdala and not the nucleus accumbens. Generally consistent with the finding described earlier, NK1R antagonism has also been shown to suppress morphine-induced locomotion in rats and morphine-induced potentiation of intracranial self-stimulation in mice (Placenza, Fletcher, Vaccarino, & Erb, 2006; Robinson et al., 2012). Intracranial self-stimulation is an additional model used to assess the state of the endogenous reward systems in rodents.

Operant models that have been used to examine the reinforcing effects of morphine and cocaine are largely consistent with these findings. Specifically,

$NK1R^{-/-}$ mice exhibit lower rates of morphine self-administration compared to wild types, but cocaine self-administration is unaffected (Ripley, Gadd, De Felipe, Hunt, & Stephens, 2002). This study also demonstrated a role of the NK1R in locomotor sensitization to morphine, but not cocaine. In the locomotor sensitization model, repeated exposure to the drug induces a long-lasting potentiation of locomotor activity following subsequent injection. This is believed to be the result of alterations in spine morphology in brain regions including the striatum (Robinson & Kolb, 1997). Taken together with the results earlier, the NK1R appears to play a role in opiate-induced reward and reinforcement, as well as neuroadaptations that result from repeated opiate administration. More recent work has found that pharmacological NK1R antagonism in rats attenuates heroin self-administration (Barbier et al., 2013; but see Placenza et al., 2006). This effect was generalized to both primary reinforcement in short-access sessions and escalated self-administration induced by long-access sessions (for description and interpretation of short vs long access models, see Ahmed, Walker, & Koob, 2000). Briefly, long-access models, which consist of self-administration sessions of 6 h or more in duration as opposed to 1 or 2 h typically used in short-access sessions, lead to escalated drug intake and negative affective states upon drug withdrawal. The fact that NK1R antagonists impact both types of self-administration could either be evidence of a collective effect on primary reinforcement, or a dual effect on both baseline and escalated components of drug intake. While it seems likely that NK1R antagonists will impact relapse-like behavior for opiate seeking as measured by the operant-based preclinical model of reinstatement, this has to date not been examined directly.

The cellular mechanisms of these effects are largely unknown, but a few potential targets have emerged. For example, it has been shown that NK1R activation by SP prevents mu-opioid receptor inactivation and internalization induced by concurrent morphine exposure, an effect that results from NK1R sequestration of β-arrestin during its own internalization process (Yu, Arttamangkul, Evans, Williams, & von Zastrow, 2009). Also, NK1R antagonism blocks morphine-induced activation of nuclear factor kappa light chain enhancer of activated B cells (NFκB) (Wang et al., 2004), a transcription factor that is commonly associated with inflammation, but has also been shown to mediate morphine reward (Zhang et al., 2011). Thus, NK1R and opioid receptor intracellular interactions and cross talk may explain the effect of the NK1R in opiate drug responses.

3.2.2 Cocaine

For the most part, based on early negative findings in NK1R$^{-/-}$ mice, relatively little attention has been paid to the role of the NK1R in cocaine-related behavior, as compared to the number of studies that have examined opiates or alcohol. As stated earlier, NK1R deletion has not been found to affect cocaine reward or self-administration and NK1R-specific lesions do not impact the development of cocaine reward. Other, more recent studies have demonstrated that NK1R antagonism does not affect the maintenance of operant cocaine self-administration in rats (Placenza et al., 2006; Schank et al., 2014). It should be noted that these studies used short-access sessions exclusively, and as such did not assess escalated drug intake. It will be critical, therefore, to assess the role of this receptor in escalation induced by long-access self-administration sessions, as this process is thought to recruit the brain stress systems, and the NK1R has been closely associated with stress responses (see earlier). Conflicting results have been published in regards to the role of the NK1R in cocaine-induced locomotion, with some studies showing a decrease and some showing no effect (Kraft, Ahluwahlia, & Angulo, 2001; Placenza et al., 2006). Similarly, there are conflicting reports regarding the effect of NK1R inhibition on locomotor sensitization to repeated cocaine injection (Davidson, Lee, & Ellinwood, 2004; Ripley et al., 2002).

Taken together, these findings suggest that the NK1R does not mediate baseline reward and reinforcement for cocaine. However, the exogenous application of NK1R agonists either intracerebroventricularly or into the ventral tegmental area can serve as a stimulus that can induce cocaine-seeking behavior after extinction in the reinstatement procedure, which is a preclinical model of relapse-like behavior (Placenza, Fletcher, Rotzinger, & Vaccarino, 2004; Placenza, Vaccarino, Fletcher, & Erb, 2005). Reinstatement of drug seeking can be triggered by a variety of stimuli, including noncontingent injection of the drug that was previously self-administered (drug priming), exposure to discrete or contextual cues that were paired with drug availability, or exposure to stress (Bossert, Marchant, Calu, & Shaham, 2013; Kalivas & McFarland, 2003; Kalivas & Volkow, 2005; Shaham, Shalev, Lu, De Wit, & Stewart, 2003). While NK1R agonism triggers reinstatement of cocaine seeking, NK1R antagonists do not affect drug-primed reinstatement of cocaine seeking (Placenza et al., 2005). NK1R blockade does, however, suppress reinstatement of drug seeking that is induced by yohimbine, a pharmacological stressor. It is unknown what effect the NK1R has on footshock-induced (another form of stress-induced reinstatement) or cue-primed reinstatement of cocaine seeking. Collectively, the limited available findings suggest that the NK1R specifically mediates cocaine seeking that is induced by stress. Consistent with this, it may be hypothesized that NK1R antagonism could be effective at attenuating escalated cocaine self-administration induced by extended access sessions because there is usually strong mechanistic overlap between stress-induced reinstatement and escalated consumption. Data to evaluate this hypothesis are presently not available.

3.2.3 Alcohol

Compared to other abused substances, alcohol has been most studied in regards to NK1R effects in recent years. In general, the NK1R seems to have a specific role in stress-elicited alcohol seeking and escalated alcohol consumption. For example, genetic deletion of the NK1R attenuates alcohol intake and alcohol reward (George et al., 2008; Thorsell, Schank, Singley, Hunt, & Heilig, 2010). The genetically modified mice used in these studies were maintained on a C57BL6/J background, which is among the highest alcohol consuming mouse strains. This finding was replicated with pharmacological antagonism and micro-RNA-based NK1R inhibition in this strain (Baek et al., 2010; Thorsell et al., 2010). When systemic NK1R antagonism was used in two outbred rat strains (Wistar and Long-Evans), this treatment did not decrease alcohol self-administration or voluntary intake until doses were administered that also attenuated sucrose intake (Augier et al., 2014; Schank et al., 2011, 2013; Steensland et al., 2010). However, NK1R antagonism is effective at attenuating escalated alcohol self-administration in two *alcohol-preferring* rat strains, the alcohol-preferring (P) rat and the Marchigian-Sardinian alcohol-preferring (msP) rat (Ayanwuyi et al., 2015; Schank et al., 2013). These and other alcohol-preferring strains have been selectively bred over many generations with the goal of generating lines of animals with high alcohol intake/preference. These lines have been used to study escalated alcohol intake and genetic factors that influence this behavior. In both P and msP rats, the effect of the NK1R was localized to specific amygdala subregions. For P rats specifically, increased NK1R expression and receptor levels were found in the CeA, and delivery of an NK1R antagonist specifically to this region attenuated escalated alcohol intake. This study also supported the possibility of a single-nucleotide polymorphism in the promoter of the

Tacr1 gene that influences transcription factor binding, providing a genetic mechanism by which NK1Rs are upregulated.

Another way to induce escalated intake is to use intermittent access schedules (Monday, Wednesday, Friday access, 20% ethanol solution) as a variation of the basic two-bottle choice procedure (Simms et al., 2008; Wise, 1973). Under these conditions, most animals increase their voluntary intake of alcohol by 50%–100% (Simms et al., 2008). Using these methods, one group has identified that SP in the paraventricular thalamus (PVT) mediates alcohol consumption on an intermittent-access schedule (Barson et al., 2015). This activation was found to lie downstream of orexin-containing projections from the hypothalamus to the PVT. This finding is intriguing, as it suggests a potential source of SP innervation that activates NK1R during escalated alcohol intake. Future studies should examine this and other sources of innervation, such as the LHb and DR, to determine the specific inputs that mediate NK1R-induced effects on drug and alcohol seeking. These are questions that lend themselves well to being addressed using recently developed circuit-specific chemogenetic and optogenetic strategies.

NK1R antagonism has been consistently found to attenuate stress-induced reinstatement, either by intermittent footshock exposure or by yohimbine injection, an effect that is mediated at least in part by NK1Rs in the nucleus accumbens shell subregion (Ayanwuyi et al., 2015; Schank et al., 2014, 2015, 2011). This effect has been observed in outbred Wistar rats, Long-Evans rats, as well as in alcohol-preferring msP rats. Thus, the NK1R appears to play a role in stress-elicited alcohol seeking in both preferring rats and in outbred strains. In contrast, NK1R antagonism has no effect on reinstatement triggered by alcohol-paired cues, neither in outbred rats nor in selected alcohol-preferring strains. In addition to relapse-like behavior in response to acute stress, escalated alcohol intake as a result of chronic stress may be influenced by the NK1R. For example, when mice are exposed to the chronic stressor of repeated social defeat stress, animals that were susceptible to this exposure and consumed more alcohol also showed increased expression levels of the NK1R in the ventral striatum (Nelson, Sequeira, & Schank, 2017). This study also demonstrated that in stress-naïve animals, increased NK1R expression in the amygdala correlated with increased alcohol consumption, and decreased social interaction. Future studies should use interventional methods to support these correlational findings in stress-exposed mice. These results, taken together with findings that alcohol-preferring rats, have increased NK1R expression in

Drug class	Effects of NK1R
Psychostimulants	Stress-induced reinstatement
Opiates	Reward Reinforcement Escalated self-administration Locomotor activation
Alcohol	Escalated intake Stress-induced reinstatement Stress-induced escalation

Fig. 2 Neurokinin-1 receptor effects on drug-related behaviors. Neurokinin-1 receptors have diverse effects on drug-related behaviors. In regards to those behaviors that have been examined, it seems that the neurokinin-1 receptor has a general effect on most opiate-induced behaviors including reward, reinforcement, and locomotor activation. For alcohol and psychostimulants, the effect of the neurokinin-1 receptor is more restricted to behaviors that are associated with stress-elicited drug seeking and escalated intake.

the amygdala, suggest that high NK1R levels in this region may be a predisposing factor to elevated alcohol consumption.

3.2.4 Summary

The SP/NK1R system has been consistently found to most prominently affect stress-induced drug seeking. This is very similar to what is observed for other stress-related signaling molecules such as CRF, norepinephrine (NE), and dynorphin. This pattern holds especially for alcohol and cocaine, while for opiate drugs, the NK1R seems to have a more general role in reward and reinforcement. For summary of drug-related effects of NK1R signaling, see Fig. 2.

4. MECHANISMS OF ACTION: MONOAMINE SIGNALING AND INTERACTIONS WITH OTHER TRANSMITTERS

Due to localization of the NK1R in specific brain regions and the fact that the NK1R was found to mediate depression-like, anxiety-like, and drug-related behaviors, a number of mechanistic studies have examined the monoamine neurotransmitters as potential systems with which NK1R interacts. It has been found that SP and the NK1R influence monoamine function in general, as well as in response to stress exposure or drug administration. For the most part, these findings are parallel to what is observed with the effects of CRF on monoamine activity.

4.1 Serotonin

A number of groups that focus on the role of SP in stress and anxiety have examined the effect of SP/NK1R on serotonergic systems. It has been found that SP infusion into the DR nucleus, a major serotonergic nucleus of the brainstem, attenuates DR neuronal activity and 5HT output from DR terminals (Guiard et al., 2007; Valentino, Bey, Pernar, & Commons, 2003). Conversely, NK1R inhibition increases DR neuron firing (Conley et al., 2002; Gobbi et al., 2007). The facilitatory effect of NK1R antagonism is likely either to be the result of local autoreceptor-driven inhibition or to be secondary to NK1R effects in the locus coeruleus (Gobbi et al., 2007; Guiard et al., 2007; Santarelli et al., 2001). In regards to stress exposure, forced swim stress induces SP release in the LS, which in turn suppresses extracellular 5HT levels. This effect is likely the result of NK1R stimulation of GABAergic interneurons, which negatively impact transmitter release from 5HT terminals (Ebner, Singewald, et al., 2008). Interestingly, SP and 5HT are also coexpressed in some neurons of the brainstem raphe nuclei in both the rat, cat, and human (Arvidsson et al., 1994; Chan-Palay, Jonsson, & Palay, 1978; Hokfelt et al., 1978; Sergeyev, Hokfelt, & Hurd, 1999). As such, SP and 5HT may have direct interactions at their receptors in common target regions.

4.2 Catecholamines

There is also evidence to suggest that the NK1R can interact with the catecholamine neurotransmitters, which include dopamine (DA) and NE. For example, stressors such as restraint induce the release of catecholamines into cortical regions and NK1R antagonism attenuates this stress-induced DA and NE release in the frontal cortex (Hutson, Patel, Jay, & Barton, 2004; Renoldi & Invernizzi, 2006). This interaction also extends to drug-induced catecholamine activity. Specifically, the induction of cocaine seeking that is triggered by intra-VTA NK1R agonist infusion (see earlier) is attenuated by dopamine type 1 receptor (D1R) antagonism (Placenza et al., 2004). Also, NK1R antagonists prevent the increased striatal DA that is induced by cocaine administration (Kraft, Noailles, & Angulo, 2001; Noailles & Angulo, 2002). Intracranial SP infusion induces increased locomotion, and this is reversed by DA antagonism or 6-OH-DA lesions of dopamine cells (Kelley, Stinus, & Iversen, 1979).

For NE function specifically, SP infusion into the locus coeruleus increases neuronal firing in this noradrenergic brainstem nucleus

(Cheeseman, Pinnock, & Henderson, 1983; Guyenet & Aghajanian, 1977). In this nucleus, NK1Rs are expressed directly on NE neurons to influence their activity and output (Chen, Wei, Liu, & Rao, 2000). This contrasts with findings concerning the control of 5HT function in the DR, which is indirect and acts via intrinsic circuits (Ma & Bleasdale, 2002). However, some studies have suggested some level of direct expression of NK1Rs on 5HT neurons of the DR (Lacoste, Riad, & Descarries, 2006).

4.3 Other Neurotransmitters and Peptides

The identity of major subtypes of striatal neurons highlights another way in which SP and the NK1R could link to both monoamine neurotransmitters and other neuropeptides. Specifically, medium spiny neurons of the striatum have been found to cluster into two subtypes, those that express the D1R along with SP and dynorphin, and those that express the dopamine type 2 receptor along with encephalin (Anderson & Reiner, 1990; Le Moine & Bloch, 1995; Lu, Ghasemzadeh, & Kalivas, 1998). As such, cells of the striatum that receive dopaminergic innervation from the midbrain and are activated by D1R binding, could potentially release SP and dynorphin along with GABA (the classical neurotransmitter of medium spiny neurons) in their terminal fields. It is interesting to note that dynorphin, like SP, is involved in stress responses, drug seeking, and prodepressive phenotypes. Additionally, the NK1R is highly expressed in the striatum and may respond to either local SP collaterals or extrinsic innervation. In the nucleus accumbens, for example, SP innervation and NK1R expression is found in GABAergic and cholinergic nonspiny interneurons, as well as dynorphin/SP-expressing medium spiny neurons (Commons & Serock, 2009; Pickel, Douglas, Chan, Gamp, & Bunnett, 2000; Van Bockstaele, Gracy, & Pickel, 1995). In the CeA, SP action at NK1Rs has been shown to directly activate the GABAergic cells of this region (Sosulina et al., 2015).

There is also some published evidence to suggest that SP has direct intracellular interactions with CRF itself. For example, SP activation of NK1Rs can drive an upregulation of CRF type 1 receptors (Hamke, Herpfer, Lieb, Wandelt, & Fiebich, 2006). SP projections have also been shown to stimulate neuronal activity in the paraventricular hypothalamus, the nucleus where CRF-containing cell bodies are primarily located (Womack & Barrett-Jolley, 2007; Womack, Morris, Gent, & Barrett-Jolley, 2007). Further evidence of an action on hypothalamic–pituitary–adrenal axis function is provided by findings that intracerebroventricular SP infusion increases circulating glucocorticoid levels (Mello et al., 2007), while administration of an NK1R antagonist in humans markedly attenuates stress-induced cortisol responses (George et al., 2008).

5. CLINICAL RESEARCH

Despite the promise of preclinical findings for a role of NK1R antagonism as an anxiolytic or antidepressant treatment, clinical results have been inconsistent, and development for this mechanism has by now largely been abandoned both in depression and other psychiatric indications, including addictive disorders. In the following, we will review these data, attempt to reconcile them, and argue that a considerable potential of this mechanism in fact remains to be exploited.

5.1 Affective Disorders

The seminal study in this area was the report from Merck that the specific NK1R antagonist MK-869 (aprepitant) was efficacious for the treatment of moderate to severe depression (Kramer et al., 1998). However, several subsequent clinical trials for the treatment of depression, generalized anxiety disorder and posttraumatic stress disorder showed no significant effect of NK1R antagonism when compared to placebo (Keller et al., 2006; Kwako et al., 2015; Mathew et al., 2011; Michelson et al., 2013).

An important insight has since emerged that may help account for these seemingly inconsistent clinical results with NK1R antagonists. It is commonly assumed that a central receptor occupancy (RO) of >90% is required for GPCR antagonists to achieve their clinical effects. Aprepitant, at doses administered in most of the published trials, occupies 60%–90% of NK1Rs (Bergstrom et al., 2004). This level of occupancy may be sufficient for achieving antiemesis effects. Because emesis is mediated by brainstem regions with increased blood brain barrier permeability (the medullary vomiting center includes the chemosensitive trigger zone of the area postrema, one of the circumventricular organs of the brain), it may be easier to achieve high levels of RO in these regions, compared to monoaminergic nuclei or forebrain structures which are presumably the site of NK1R mediation of affective and addictive disorders.

Most importantly, however, it appears that the level of central RO required for efficacy may vary between different therapeutic targets. Studies by the GSK group utilized NK1R PET-ligand displacement methodology that allowed an analysis of the relationship between systemic exposure, central RO, and clinical outcomes in depression. These studies indicated that in the case of the NK1R system, near complete central RO is required for therapeutic efficacy, probably as a result of the high receptor reserve within this system (Ratti et al., 2011, 2013; Ridler et al., 2014; Trist, Ratti, & Bye, 2013). Illustrating the point, the initial positive results with Merck's aprepitant was achieved using a daily dose of 300 mg (Kramer et al., 1998); while, in contrast, the subsequent Phase III program studies all used 160 mg/day as their highest dose (Keller et al., 2006). The dose selection in this program was driven by a combination of assumptions about the requirement for central RO, challenges with drug synthesis, and safety concerns with the higher aprepitant doses in women. Given the high expectations this program held, and the level of disappointment when it failed, it is unlikely that this mechanism will be able to attract the investment required for an adequate reevaluation of its efficacy in depression. Nevertheless, based on the GSK studies, it remains a distinct possibility that an NK1R antagonist administered to achieve a near-complete central NK1R RO could be expected to have efficacy in depression.

5.2 Addiction

Genetic polymorphisms of the *TACR1* gene have been found to associate with increased risk of alcohol dependence as well as increased responsiveness to alcohol-paired cues, suggesting a possible role of this receptor and its variants in alcohol dependence in humans, and its potential as a therapeutic target (Blaine, Claus, Harlaar, & Hutchison, 2013; Seneviratne et al., 2009; Sharp et al., 2014). A promising initial finding with NK1R antagonism in clinical alcoholism was obtained using an experimental medicine approach at the NIH Clinical Center, where we found that NK1R antagonism with Eli Lilly's selective, brain penetrant NK1R antagonist LY686017 was capable of decreasing spontaneous cravings, stress/cue-induced cravings, and brain activation in response to negative images in detoxified alcoholics with comorbid anxiety (George et al., 2008). It is, however, important to note that voluntary alcohol consumption was not assessed in this study, as the patient population was treatment seeking, inpatient alcoholics. A subsequent experimental medicine study examined aprepitant in comorbid alcoholism and PTSD, and was negative for a range of surrogate markers of efficacy (Kwako et al., 2015). This study was designed prior to the analysis of central RO requirements by the GSK group, and used an aprepitant dose of 125 mg daily, making it subject to the same limitations as the Merck Phase III program in depression. It is, therefore, unclear if NK1R antagonism would be effective at promoting abstinence and preventing relapse in an outpatient population, but a strong rationale remains for testing these endpoints in a properly designed study, based on the preclinical data and the promising results with LY686017.

The two key design features that are prompted by the prior work in this field are adequate selection of the patient population, since the original efficacy signal was obtained in a highly anxious population of patients; and a dose regiment that achieves the near-complete central RO that is prompted as requirement by the prior depression studies. A third, more speculative design feature is pharmacogenetic. As stated earlier, specific polymorphism of the *TACR1* gene associate with increased risk of alcohol dependence and other psychiatric conditions. It is, therefore, possible that the response to NK1R antagonists is influenced by pharmacogenetic interactions driven by polymorphisms of the *TACR1* gene. It is possible that this factor is functionally related to the efficacy signal in anxious alcoholism and that selecting for anxious alcoholics would generate the same subpopulation that would be selected using genetic prescreening.

The therapeutic potential of NK1R antagonism in other addictions is less clear. A recent study found that NK1R antagonism with aprepitant *increased* rather than decreased the subjective properties of an opiate analgesic, oxycodone (Walsh, Heilig, Nuzzo, Henderson, & Lofwall, 2013). This result is contrary to the suppression of heroin self-administration results in rats reviewed earlier, but it should be noted that the human oxycodone study examined a single (as opposed to steady state) dose of aprepitant, and also did not examine the consequences on drug self-administration.

6. SUMMARY AND CONCLUSIONS

SP and the NK1R play a significant role in diverse physiological, neurochemical, and behavioral effects in the central and peripheral nervous system (for summary, see Fig. 3). This peptide was one of the first to be isolated from biological tissue and its sequence has been known for nearly 50 years. The SP/NK1R system has been found to play a role in the peripheral nervous system through its mediation of pain sensitivity, neurogenic inflammation, and intestinal contraction. At the level of the brainstem, this system mediates the function of the monoamine transmitter systems, and plays a role in emesis. Indeed, the only FDA-approved

Nervous system region	Behavioral/physiological effects
Dorsal horn spinal	Pain sensitivity
Peripheral terminals of pain fibers	Neurogenic inflammation
Medulla	Emesis
Locus coeruleus	Regulation of norepinephrine transmission
Dorsal raphe	Regulation of serotonin transmission
Midbrain	Regulation of dopamine transmission
Amygdala	Stress reactivity Anxiety Escalated alcohol intake Morphine reward
Nucleus accumbens	Stress-induced alcohol seeking
Lateral septum	Stress coping Depressive-like behavior Serotonin transmission
Paraventricular hypothalamus	Hypothalamic stress response
Lateral habenula	Depressive-like behavior

Fig. 3 Neurokinin-1 receptor effects in the central and peripheral nervous system. Neurokinin-1 receptors are found throughout the central and peripheral nervous system, where they have diverse effects on behavior and physiology. While this table is not exhaustive, it provides some major examples from the literature concerning the effects mediated by neurokinin-1 receptors in specific regions of the nervous system.

indication for a NK1R antagonist is for the treatment of chemotherapy-induced nausea. More recent work has identified a role for the NK1R in complex psychiatric processes including stress responsivity, depression-like behavior, and addiction.

The flurry of preclinical behavioral research that took place in the 1990s lead to enthusiasm for translation but this excitement was dampened by a series of negative clinical trials. However, specific factors not directly related to clinical efficacy may have influenced these results. New generation, high potency NK1R antagonists have been developed, and when sufficient RO is achieved, clinical antidepressant effects are observed. Thus, it seems important to follow-up on the plethora of promising preclinical data for addiction, and especially alcohol addiction. The limited clinical trials that have been performed in the realm of addiction therapy have thus far produced mixed results, and this may be the result of the specific antagonist used, psychiatric comorbidities of specific subpopulations, or pharmacogenetic interactions. As such, it is critical that we reexamine the therapeutic potential of NK1R antagonists for anxiety, depression, and addiction in well-designed clinical trials that take into account these important factors.

REFERENCES

Ahmed, S. H., Walker, J. R., & Koob, G. F. (2000). Persistent increase in the motivation to take heroin in rats with a history of drug escalation. *Neuropsychopharmacology*, *22*(4), 413–421. http://dx.doi.org/10.1016/S0893-133X(99)00133-5.

Anderson, K. D., & Reiner, A. (1990). Extensive co-occurrence of substance P and dynorphin in striatal projection neurons: An evolutionarily conserved feature of basal ganglia organization. *The Journal of Comparative Neurology*, *295*(3), 339–369. http://dx.doi.org/10.1002/cne.902950302.

Arvidsson, U., Cullheim, S., Ulfhake, B., Luppi, P. H., Kitahama, K., Jouvet, M., et al. (1994). Quantitative and qualitative aspects on the distribution of 5-HT and its coexistence with substance P and TRH in cat ventral medullary neurons. *Journal of Chemical Neuroanatomy*, 7(1–2), 3–12.

Augier, E., Flanigan, M., Dulman, R. S., Pincus, A., Schank, J. R., Rice, K. C., et al. (2014). Wistar rats acquire and maintain self-administration of 20% ethanol without water deprivation, saccharin/sucrose fading, or extended access training. *Psychopharmacology*, *231*(23), 4561–4568. http://dx.doi.org/10.1007/s00213-014-3605-3.

Ayanwuyi, L. O., Stopponi, S., Ubaldi, M., Cippitelli, A., Nasuti, C., Damadzic, R., et al. (2015). Neurokinin 1 receptor blockade in the medial amygdala attenuates alcohol drinking in rats with innate anxiety but not in Wistar rats. *British Journal of Pharmacology*, *172*(21), 5136–5146. http://dx.doi.org/10.1111/bph.13280.

Baek, M. N., Jung, K. H., Halder, D., Choi, M. R., Lee, B. H., Lee, B. C., et al. (2010). Artificial microRNA-based neurokinin-1 receptor gene silencing reduces alcohol consumption in mice. *Neuroscience Letters*, *475*(3), 124–128. http://dx.doi.org/10.1016/j.neulet.2010.03.051.

Barbier, E., Vendruscolo, L. F., Schlosburg, J. E., Edwards, S., Juergens, N., Park, P. E., et al. (2013). The NK1 receptor antagonist L822429 reduces heroin reinforcement. *Neuropsychopharmacology*, *38*(6), 976–984. http://dx.doi.org/10.1038/npp.2012.261.

Barson, J. R., Poon, K., Ho, H. T., Alam, M. I., Sanzalone, L., & Leibowitz, S. F. (2015). Substance P in the anterior thalamic paraventricular nucleus: Promotion of ethanol drinking in response to orexin from the hypothalamus. *Addiction Biology*. http://dx.doi.org/10.1111/adb.12288.

Bergstrom, M., Hargreaves, R. J., Burns, H. D., Goldberg, M. R., Sciberras, D., Reines, S. A., et al. (2004). Human positron emission tomography studies of brain neurokinin 1 receptor occupancy by aprepitant. *Biological Psychiatry*, *55*(10), 1007–1012. http://dx.doi.org/10.1016/j.biopsych.2004.02.007.

Blaine, S., Claus, E., Harlaar, N., & Hutchison, K. (2013). TACR1 genotypes predict fMRI response to alcohol cues and level of alcohol dependence. *Alcoholism, Clinical and Experimental Research*, *37*(Suppl. 1), E125–E130. http://dx.doi.org/10.1111/j.1530-0277.2012.01923.x.

Bossert, J. M., Marchant, N. J., Calu, D. J., & Shaham, Y. (2013). The reinstatement model of drug relapse: Recent neurobiological findings, emerging research topics, and translational research. *Psychopharmacology*, *229*(3), 453–476. http://dx.doi.org/10.1007/s00213-013-3120-y.

Boyce, S., Smith, D., Carlson, E., Hewson, L., Rigby, M., O'Donnell, R., et al. (2001). Intra-amygdala injection of the substance P [NK(1) receptor] antagonist L-760735 inhibits neonatal vocalisations in guinea-pigs. *Neuropharmacology*, *41*(1), 130–137.

Chan-Palay, V., Jonsson, G., & Palay, S. L. (1978). Serotonin and substance P coexist i, neurons of the rat's central nervous system. *Proceedings of the National Academy of Sciences of the United States of America*, *75*(3), 1582–1586.

Cheeseman, H. J., Pinnock, R. D., & Henderson, G. (1983). Substance P excitation of rat locus coeruleus neurones. *European Journal of Pharmacology*, *94*(1–2), 93–99.

Chen, L. W., Wei, L. C., Liu, H. L., & Rao, Z. R. (2000). Noradrenergic neurons expressing substance P receptor (NK1) in the locus coeruleus complex: A double immunofluorescence study in the rat. *Brain Research, 873*(1), 155–159.

Commons, K. G., & Serock, M. R. (2009). Coincidence of neurokinin 1 receptor with the vesicular glutamate transporter 3 (VGLUT3) in the rat forebrain. *Neuroscience Letters, 464*(3), 188–192. http://dx.doi.org/10.1016/j.neulet.2009.08.042.

Conley, R. K., Cumberbatch, M. J., Mason, G. S., Williamson, D. J., Harrison, T., Locker, K., et al. (2002). Substance P (neurokinin 1) receptor antagonists enhance dorsal raphe neuronal activity. *The Journal of Neuroscience, 22*(17), 7730–7736.

Davidson, C., Lee, T. H., & Ellinwood, E. H. (2004). The NK(1) receptor antagonist WIN51708 reduces sensitization after chronic cocaine. *European Journal of Pharmacology, 499*(3), 355–356. http://dx.doi.org/10.1016/j.ejphar.2004.08.006.

De Felipe, C., Herrero, J. F., O'Brien, J. A., Palmer, J. A., Doyle, C. A., Smith, A. J., et al. (1998). Altered nociception, analgesia and aggression in mice lacking the receptor for substance P. *Nature, 392*(6674), 394–397. http://dx.doi.org/10.1038/32904.

Duarte, F. S., Testolin, R., & De Lima, T. C. (2004). Further evidence on the anxiogenic-like effect of substance P evaluated in the elevated plus-maze in rats. *Behavioural Brain Research, 154*(2), 501–510. http://dx.doi.org/10.1016/j.bbr.2004.03.020.

Ebner, K., Muigg, P., Singewald, G., & Singewald, N. (2008a). Substance P in stress and anxiety: NK-1 receptor antagonism interacts with key brain areas of the stress circuitry. *Annals of the New York Academy of Sciences, 1144*, 61–73. http://dx.doi.org/10.1196/annals.1418.018.

Ebner, K., Rupniak, N. M., Saria, A., & Singewald, N. (2004). Substance P in the medial amygdala: Emotional stress-sensitive release and modulation of anxiety-related behavior in rats. *Proceedings of the National Academy of Sciences of the United States of America, 101*(12), 4280–4285. http://dx.doi.org/10.1073/pnas.0400794101.

Ebner, K., & Singewald, N. (2006). The role of substance P in stress and anxiety responses. *Amino Acids, 31*(3), 251–272. http://dx.doi.org/10.1007/s00726-006-0335-9.

Ebner, K., Singewald, G. M., Whittle, N., Ferraguti, F., & Singewald, N. (2008b). Neurokinin 1 receptor antagonism promotes active stress coping via enhanced septal 5-HT transmission. *Neuropsychopharmacology, 33*(8), 1929–1941. http://dx.doi.org/10.1038/sj.npp.1301594.

Gadd, C. A., Murtra, P., De Felipe, C., & Hunt, S. P. (2003). Neurokinin-1 receptor-expressing neurons in the amygdala modulate morphine reward and anxiety behaviors in the mouse. *The Journal of Neuroscience, 23*(23), 8271–8280.

George, D. T., Gilman, J., Hersh, J., Thorsell, A., Herion, D., Geyer, C., et al. (2008). Neurokinin 1 receptor antagonism as a possible therapy for alcoholism. *Science, 319*(5869), 1536–1539. http://dx.doi.org/10.1126/science.1153813.

Gobbi, G., Cassano, T., Radja, F., Morgese, M. G., Cuomo, V., Santarelli, L., et al. (2007). Neurokinin 1 receptor antagonism requires norepinephrine to increase serotonin function. *European Neuropsychopharmacology, 17*(5), 328–338. http://dx.doi.org/10.1016/j.euroneuro.2006.07.004.

Guiard, B. P., Guilloux, J. P., Reperant, C., Hunt, S. P., Toth, M., & Gardier, A. M. (2007). Substance P neurokinin 1 receptor activation within the dorsal raphe nucleus controls serotonin release in the mouse frontal cortex. *Molecular Pharmacology, 72*(6), 1411–1418. http://dx.doi.org/10.1124/mol.107.040113.

Guyenet, P. G., & Aghajanian, G. K. (1977). Excitation of neurons in the nucleus locus coeruleus by substance P and related peptides. *Brain Research, 136*(1), 178–184.

Hamke, M., Herpfer, I., Lieb, K., Wandelt, C., & Fiebich, B. L. (2006). Substance P induces expression of the corticotropin-releasing factor receptor 1 by activation of the neurokinin-1 receptor. *Brain Research, 1102*(1), 135–144. http://dx.doi.org/10.1016/j.brainres.2006.03.026.

Hokfelt, T., Ljungdahl, A., Steinbusch, H., Verhofstad, A., Nilsson, G., Brodin, E., et al. (1978). Immunohistochemical evidence of substance P-like immunoreactivity in some 5-hydroxytryptamine-containing neurons in the rat central nervous system. *Neuroscience*, *3*(6), 517–538.

Hurd, Y. L., Keller, E., Sotonyi, P., & Sedvall, G. (1999). Preprotachykinin-A mRNA expression in the human and monkey brain: An in situ hybridization study. *The Journal of Comparative Neurology*, *411*(1), 56–72.

Hutson, P. H., Patel, S., Jay, M. T., & Barton, C. L. (2004). Stress-induced increase of cortical dopamine metabolism: Attenuation by a tachykinin NK1 receptor antagonist. *European Journal of Pharmacology*, *484*(1), 57–64.

Jensen, C. J., Gerard, N. P., Schwartz, T. W., & Gether, U. (1994). The species selectivity of chemically distinct tachykinin nonpeptide antagonists is dependent on common divergent residues of the rat and human neurokinin-1 receptors. *Molecular Pharmacology*, *45*(2), 294–299.

Kalivas, P. W., & McFarland, K. (2003). Brain circuitry and the reinstatement of cocaine-seeking behavior. *Psychopharmacology*, *168*(1–2), 44–56. http://dx.doi.org/10.1007/s00213-003-1393-2.

Kalivas, P. W., & Volkow, N. D. (2005). The neural basis of addiction: A pathology of motivation and choice. *The American Journal of Psychiatry*, *162*(8), 1403–1413. http://dx.doi.org/10.1176/appi.ajp.162.8.1403.

Keller, M., Montgomery, S., Ball, W., Morrison, M., Snavely, D., Liu, G., et al. (2006). Lack of efficacy of the substance p (neurokinin1 receptor) antagonist aprepitant in the treatment of major depressive disorder. *Biological Psychiatry*, *59*(3), 216–223. http://dx.doi.org/10.1016/j.biopsych.2005.07.013.

Kelley, A. E., Stinus, L., & Iversen, S. D. (1979). Behavioural activation induced in the rat by substance P infusion into ventral tegmental area: Implication of dopaminergic A10 neurones. *Neuroscience Letters*, *11*(3), 335–339.

Kraft, M., Ahluwahlia, S., & Angulo, J. A. (2001a). Neurokinin-1 receptor antagonists block acute cocaine-induced horizontal locomotion. *Annals of the New York Academy of Sciences*, *937*, 132–139.

Kraft, M., Noailles, P., & Angulo, J. A. (2001b). Substance P modulates cocaine-evoked dopamine overflow in the striatum of the rat brain. *Annals of the New York Academy of Sciences*, *937*, 121–131.

Kramer, M. S., Cutler, N., Feighner, J., Shrivastava, R., Carman, J., Sramek, J. J., et al. (1998). Distinct mechanism for antidepressant activity by blockade of central substance P receptors. *Science*, *281*(5383), 1640–1645.

Kwako, L. E., George, D. T., Schwandt, M. L., Spagnolo, P. A., Momenan, R., Hommer, D. W., et al. (2015). The neurokinin-1 receptor antagonist aprepitant in co-morbid alcohol dependence and posttraumatic stress disorder: A human experimental study. *Psychopharmacology*, *232*(1), 295–304. http://dx.doi.org/10.1007/s00213-014-3665-4.

Lacoste, B., Riad, M., & Descarries, L. (2006). Immunocytochemical evidence for the existence of substance P receptor (NK1) in serotonin neurons of rat and mouse dorsal raphe nucleus. *The European Journal of Neuroscience*, *23*(11), 2947–2958. http://dx.doi.org/10.1111/j.1460-9568.2006.04833.x.

Le Moine, C., & Bloch, B. (1995). D1 and D2 dopamine receptor gene expression in the rat striatum: Sensitive cRNA probes demonstrate prominent segregation of D1 and D2 mRNAs in distinct neuronal populations of the dorsal and ventral striatum. *The Journal of Comparative Neurology*, *355*(3), 418–426. http://dx.doi.org/10.1002/cne.903550308.

Leffler, A., Ahlstedt, I., Engberg, S., Svensson, A., Billger, M., Oberg, L., et al. (2009). Characterization of species-related differences in the pharmacology of tachykinin NK

receptors 1, 2 and 3. *Biochemical Pharmacology*, 77(9), 1522–1530. http://dx.doi.org/10.1016/j.bcp.2009.01.020.
Lu, X. Y., Ghasemzadeh, M. B., & Kalivas, P. W. (1998). Expression of D1 receptor, D2 receptor, substance P and enkephalin messenger RNAs in the neurons projecting from the nucleus accumbens. *Neuroscience*, *82*(3), 767–780.
Ma, Q. P., & Bleasdale, C. (2002). Modulation of brain stem monoamines and gamma-aminobutyric acid by NK1 receptors in rats. *Neuroreport*, *13*(14), 1809–1812.
Mantyh, P. W. (2002). Neurobiology of substance P and the NK1 receptor. *The Journal of Clinical Psychiatry*, *63*(Suppl. 11), 6–10.
Mantyh, P. W., Hunt, S. P., & Maggio, J. E. (1984). Substance P receptors: Localization by light microscopic autoradiography in rat brain using [3H]SP as the radioligand. *Brain Research*, *307*(1–2), 147–165.
Mathew, S. J., Vythilingam, M., Murrough, J. W., Zarate, C. A., Jr., Feder, A., Luckenbaugh, D. A., et al. (2011). A selective neurokinin-1 receptor antagonist in chronic PTSD: A randomized, double-blind, placebo-controlled, proof-of-concept trial. *European Neuropsychopharmacology*, *21*(3), 221–229. http://dx.doi.org/10.1016/j.euroneuro.2010.11.012.
Mello, D. M., Marcinichen, D. R., Madruga, D., Branco, R., Paschoalini, M. A., & De Lima, T. C. (2007). Involvement of NK1 receptors in metabolic stress markers after the central administration of substance P. *Behavioural Brain Research*, *181*(2), 232–238. http://dx.doi.org/10.1016/j.bbr.2007.04.010.
Michelson, D., Hargreaves, R., Alexander, R., Ceesay, P., Hietala, J., Lines, C., et al. (2013). Lack of efficacy of L-759274, a novel neurokinin 1 (substance P) receptor antagonist, for the treatment of generalized anxiety disorder. *The International Journal of Neuropsychopharmacology*, *16*(1), 1–11. http://dx.doi.org/10.1017/S1461145712000065.
Murtra, P., Sheasby, A. M., Hunt, S. P., & De Felipe, C. (2000). Rewarding effects of opiates are absent in mice lacking the receptor for substance P. *Nature*, *405*(6783), 180–183. http://dx.doi.org/10.1038/35012069.
Nelson, B. S., Sequeira, M. K., & Schank, J. R. (2017). Bidirectional relationship between alcohol intake and sensitivity to social defeat: Association with Tacr1 and Avp expression. *Addiction Biology*. http://dx.doi.org/10.1111/adb.12494.
Noailles, P. A., & Angulo, J. A. (2002). Neurokinin receptors modulate the neurochemical actions of cocaine. *Annals of the New York Academy of Sciences*, *965*, 267–273.
Pennefather, J. N., Lecci, A., Candenas, M. L., Patak, E., Pinto, F. M., & Maggi, C. A. (2004). Tachykinins and tachykinin receptors: A growing family. *Life Sciences*, *74*(12), 1445–1463.
Pickel, V. M., Douglas, J., Chan, J., Gamp, P. D., & Bunnett, N. W. (2000). Neurokinin 1 receptor distribution in cholinergic neurons and targets of substance P terminals in the rat nucleus accumbens. *The Journal of Comparative Neurology*, *423*(3), 500–511.
Placenza, F. M., Fletcher, P. J., Rotzinger, S., & Vaccarino, F. J. (2004). Infusion of the substance P analogue, DiMe-C7, into the ventral tegmental area induces reinstatement of cocaine-seeking behaviour in rats. *Psychopharmacology*, *177*(1–2), 111–120. http://dx.doi.org/10.1007/s00213-004-1912-9.
Placenza, F. M., Fletcher, P. J., Vaccarino, F. J., & Erb, S. (2006). Effects of central neurokinin-1 receptor antagonism on cocaine- and opiate-induced locomotor activity and self-administration behaviour in rats. *Pharmacology, Biochemistry, and Behavior*, *84*(1), 94–101. http://dx.doi.org/10.1016/j.pbb.2006.04.011.
Placenza, F. M., Vaccarino, F. J., Fletcher, P. J., & Erb, S. (2005). Activation of central neurokinin-1 receptors induces reinstatement of cocaine-seeking behavior. *Neuroscience Letters*, *390*(1), 42–47. http://dx.doi.org/10.1016/j.neulet.2005.07.050.
Quirion, R., Shults, C. W., Moody, T. W., Pert, C. B., Chase, T. N., & O'Donohue, T. L. (1983). Autoradiographic distribution of substance P receptors in rat central nervous system. *Nature*, *303*(5919), 714–716.

Ratti, E., Bellew, K., Bettica, P., Bryson, H., Zamuner, S., Archer, G., et al. (2011). Results from 2 randomized, double-blind, placebo-controlled studies of the novel NK1 receptor antagonist casopitant in patients with major depressive disorder. *Journal of Clinical Psychopharmacology*, *31*(6), 727–733. http://dx.doi.org/10.1097/JCP.0b013e31823608ca.

Ratti, E., Bettica, P., Alexander, R., Archer, G., Carpenter, D., Evoniuk, G., et al. (2013). Full central neurokinin-1 receptor blockade is required for efficacy in depression: Evidence from orvepitant clinical studies. *Journal of Psychopharmacology*, *27*(5), 424–434. http://dx.doi.org/10.1177/0269881113480990.

Renoldi, G., & Invernizzi, R. W. (2006). Blockade of tachykinin NK1 receptors attenuates stress-induced rise of extracellular noradrenaline and dopamine in the rat and gerbil medial prefrontal cortex. *Journal of Neuroscience Research*, *84*(5), 961–968. http://dx.doi.org/10.1002/jnr.20997.

Ridler, K., Gunn, R. N., Searle, G. E., Barletta, J., Passchier, J., Dixson, L., et al. (2014). Characterising the plasma-target occupancy relationship of the neurokinin antagonist GSK1144814 with PET. *Journal of Psychopharmacology*, *28*(3), 244–253. http://dx.doi.org/10.1177/0269881113517953.

Ripley, T. L., Gadd, C. A., De Felipe, C., Hunt, S. P., & Stephens, D. N. (2002). Lack of self-administration and behavioural sensitisation to morphine, but not cocaine, in mice lacking NK1 receptors. *Neuropharmacology*, *43*(8), 1258–1268.

Robinson, J. E., Fish, E. W., Krouse, M. C., Thorsell, A., Heilig, M., & Malanga, C. J. (2012). Potentiation of brain stimulation reward by morphine: Effects of neurokinin-1 receptor antagonism. *Psychopharmacology*, *220*(1), 215–224. http://dx.doi.org/10.1007/s00213-011-2469-z.

Robinson, T. E., & Kolb, B. (1997). Persistent structural modifications in nucleus accumbens and prefrontal cortex neurons produced by previous experience with amphetamine. *The Journal of Neuroscience*, *17*(21), 8491–8497.

Rupniak, N. M., Webb, J. K., Fisher, A., Smith, D., & Boyce, S. (2003). The substance P (NK1) receptor antagonist L-760735 inhibits fear conditioning in gerbils. *Neuropharmacology*, *44*(4), 516–523.

Santarelli, L., Gobbi, G., Debs, P. C., Sibille, E. T., Blier, P., Hen, R., et al. (2001). Genetic and pharmacological disruption of neurokinin 1 receptor function decreases anxiety-related behaviors and increases serotonergic function. *Proceedings of the National Academy of Sciences of the United States of America*, *98*(4), 1912–1917. http://dx.doi.org/10.1073/pnas.041596398.

Schank, J. R. (2014). The neurokinin-1 receptor in addictive processes. *The Journal of Pharmacology and Experimental Therapeutics*, *351*(1), 2–8. http://dx.doi.org/10.1124/jpet.113.210799.

Schank, J. R., King, C. E., Sun, H., Cheng, K., Rice, K. C., Heilig, M., et al. (2014). The role of the neurokinin-1 receptor in stress-induced reinstatement of alcohol and cocaine seeking. *Neuropsychopharmacology*, *39*(5), 1093–1101. http://dx.doi.org/10.1038/npp.2013.309.

Schank, J. R., Nelson, B. S., Damadzic, R., Tapocik, J. D., Yao, M., King, C. E., et al. (2015). Neurokinin-1 receptor antagonism attenuates neuronal activity triggered by stress-induced reinstatement of alcohol seeking. *Neuropharmacology*, *99*, 106–114. http://dx.doi.org/10.1016/j.neuropharm.2015.07.009.

Schank, J. R., Pickens, C. L., Rowe, K. E., Cheng, K., Thorsell, A., Rice, K. C., et al. (2011). Stress-induced reinstatement of alcohol-seeking in rats is selectively suppressed by the neurokinin 1 (NK1) antagonist L822429. *Psychopharmacology*, *218*(1), 111–119. http://dx.doi.org/10.1007/s00213-011-2201-z.

Schank, J. R., Ryabinin, A. E., Giardino, W. J., Ciccocioppo, R., & Heilig, M. (2012). Stress-related neuropeptides and addictive behaviors: Beyond the usual suspects. *Neuron*, *76*(1), 192–208. http://dx.doi.org/10.1016/j.neuron.2012.09.026.

Schank, J. R., Tapocik, J. D., Barbier, E., Damadzic, R., Eskay, R. L., Sun, H., et al. (2013). Tacr1 gene variation and neurokinin 1 receptor expression is associated with antagonist efficacy in genetically selected alcohol-preferring rats. *Biological Psychiatry, 73*(8), 774–781. http://dx.doi.org/10.1016/j.biopsych.2012.12.027.

Seneviratne, C., Ait-Daoud, N., Ma, J. Z., Chen, G., Johnson, B. A., & Li, M. D. (2009). Susceptibility locus in neurokinin-1 receptor gene associated with alcohol dependence. *Neuropsychopharmacology, 34*(11), 2442–2449. http://dx.doi.org/10.1038/npp.2009.65.

Sergeyev, V., Hokfelt, T., & Hurd, Y. (1999). Serotonin and substance P co-exist in dorsal raphe neurons of the human brain. *Neuroreport, 10*(18), 3967–3970.

Shaham, Y., Shalev, U., Lu, L., De Wit, H., & Stewart, J. (2003). The reinstatement model of drug relapse: History, methodology and major findings. *Psychopharmacology, 168*(1–2), 3–20. http://dx.doi.org/10.1007/s00213-002-1224-x.

Sharp, S. I., McQuillin, A., Marks, M., Hunt, S. P., Stanford, S. C., Lydall, G. J., et al. (2014). Genetic association of the tachykinin receptor 1 TACR1 gene in bipolar disorder, attention deficit hyperactivity disorder, and the alcohol dependence syndrome. *American Journal of Medical Genetics. Part B, Neuropsychiatric Genetics, 165B*(4), 373–380. http://dx.doi.org/10.1002/ajmg.b.32241.

Shults, C. W., Quirion, R., Chronwall, B., Chase, T. N., & O'Donohue, T. L. (1984). A comparison of the anatomical distribution of substance P and substance P receptors in the rat central nervous system. *Peptides, 5*(6), 1097–1128.

Simms, J. A., Steensland, P., Medina, B., Abernathy, K. E., Chandler, L. J., Wise, R., et al. (2008). Intermittent access to 20% ethanol induces high ethanol consumption in long-Evans and Wistar rats. *Alcoholism, Clinical and Experimental Research, 32*(10), 1816–1823. http://dx.doi.org/10.1111/j.1530-0277.2008.00753.x.

Singewald, N., Chicchi, G. G., Thurner, C. C., Tsao, K. L., Spetea, M., Schmidhammer, H., et al. (2008). Modulation of basal and stress-induced amygdaloid substance P release by the potent and selective NK1 receptor antagonist L-822429. *Journal of Neurochemistry, 106*(6), 2476–2488. http://dx.doi.org/10.1111/j.1471-4159.2008.05596.x.

Sosulina, L., Strippel, C., Romo-Parra, H., Walter, A. L., Kanyshkova, T., Sartori, S. B., et al. (2015). Substance P excites GABAergic neurons in the mouse central amygdala through neurokinin 1 receptor activation. *Journal of Neurophysiology, 114*(4), 2500–2508. http://dx.doi.org/10.1152/jn.00883.2014.

Steensland, P., Simms, J. A., Nielsen, C. K., Holgate, J., Bito-Onon, J. J., & Bartlett, S. E. (2010). The neurokinin 1 receptor antagonist, ezlopitant, reduces appetitive responding for sucrose and ethanol. *PloS One, 5*(9). http://dx.doi.org/10.1371/journal.pone.0012527.

Steinhoff, M. S., von Mentzer, B., Geppetti, P., Pothoulakis, C., & Bunnett, N. W. (2014). Tachykinins and their receptors: Contributions to physiological control and the mechanisms of disease. *Physiological Reviews, 94*(1), 265–301. http://dx.doi.org/10.1152/physrev.00031.2013.

Teixeira, R. M., Santos, A. R., Ribeiro, S. J., Calixto, J. B., Rae, G. A., & De Lima, T. C. (1996). Effects of central administration of tachykinin receptor agonists and antagonists on plus-maze behavior in mice. *European Journal of Pharmacology, 311*(1), 7–14.

Thorsell, A., Schank, J. R., Singley, E., Hunt, S. P., & Heilig, M. (2010). Neurokinin-1 receptors (NK1R:s), alcohol consumption, and alcohol reward in mice. *Psychopharmacology, 209*(1), 103–111. http://dx.doi.org/10.1007/s00213-010-1775-1.

Trist, D. G., Ratti, E., & Bye, A. (2013). Why receptor reserve matters for neurokinin1 (NK1) receptor antagonists. *Journal of Receptor and Signal Transduction Research, 33*(6), 333–337. http://dx.doi.org/10.3109/10799893.2013.843194.

Valentino, R. J., Bey, V., Pernar, L., & Commons, K. G. (2003). Substance P acts through local circuits within the rat dorsal raphe nucleus to alter serotonergic neuronal activity. *The Journal of Neuroscience, 23*(18), 7155–7159.

Van Bockstaele, E. J., Gracy, K. N., & Pickel, V. M. (1995). Dynorphin-immunoreactive neurons in the rat nucleus accumbens: Ultrastructure and synaptic input from terminals containing substance P and/or dynorphin. *The Journal of Comparative Neurology*, *351*(1), 117–133. http://dx.doi.org/10.1002/cne.903510111.

Walsh, S. L., Heilig, M., Nuzzo, P. A., Henderson, P., & Lofwall, M. R. (2013). Effects of the NK1 antagonist, aprepitant, on response to oral and intranasal oxycodone in prescription opioid abusers. *Addiction Biology*, *18*(2), 332–343. http://dx.doi.org/10.1111/j.1369-1600.2011.00419.x.

Wang, X., Douglas, S. D., Commons, K. G., Pleasure, D. E., Lai, J., Ho, C., et al. (2004). A non-peptide substance P antagonist (CP-96,345) inhibits morphine-induced NF-kappa B promoter activation in human NT2-N neurons. *Journal of Neuroscience Research*, *75*(4), 544–553. http://dx.doi.org/10.1002/jnr.10873.

Wise, R. A. (1973). Voluntary ethanol intake in rats following exposure to ethanol on various schedules. *Psychopharmacologia*, *29*(3), 203–210.

Womack, M. D., & Barrett-Jolley, R. (2007). Activation of paraventricular nucleus neurones by the dorsomedial hypothalamus via a tachykinin pathway in rats. *Experimental Physiology*, *92*(4), 671–676. http://dx.doi.org/10.1113/expphysiol.2007.037457.

Womack, M. D., Morris, R., Gent, T. C., & Barrett-Jolley, R. (2007). Substance P targets sympathetic control neurons in the paraventricular nucleus. *Circulation Research*, *100*(11), 1650–1658. http://dx.doi.org/10.1161/CIRCRESAHA.107.153494.

Yang, L. M., Yu, L., Jin, H. J., & Zhao, H. (2014). Substance P receptor antagonist in lateral habenula improves rat depression-like behavior. *Brain Research Bulletin*, *100*, 22–28. http://dx.doi.org/10.1016/j.brainresbull.2013.10.007.

Yip, J., & Chahl, L. A. (2000). Localization of tachykinin receptors and Fos-like immunoreactivity induced by substance P in guinea-pig brain. *Clinical and Experimental Pharmacology & Physiology*, *27*(11), 943–946.

Yu, Y. J., Arttamangkul, S., Evans, C. J., Williams, J. T., & von Zastrow, M. (2009). Neurokinin 1 receptors regulate morphine-induced endocytosis and desensitization of mu-opioid receptors in CNS neurons. *The Journal of Neuroscience*, *29*(1), 222–233. http://dx.doi.org/10.1523/JNEUROSCI.4315-08.2009.

Zhang, X., Cui, Y., Jing, J., Cui, Y., Xin, W., & Liu, X. (2011). Involvement of p38/NF-kappaB signaling pathway in the nucleus accumbens in the rewarding effects of morphine in rats. *Behavioural Brain Research*, *218*(1), 184–189. http://dx.doi.org/10.1016/j.bbr.2010.11.049.

Zhao, Z., Yang, Y., Walker, D. L., & Davis, M. (2009). Effects of substance P in the amygdala, ventromedial hypothalamus, and periaqueductal gray on fear-potentiated startle. *Neuropsychopharmacology*, *34*(2), 331–340. http://dx.doi.org/10.1038/npp.2008.55.

CHAPTER SEVEN

The Role of Neuropeptide Y (NPY) in Alcohol and Drug Abuse Disorders

Stacey L. Robinson, Todd E. Thiele[1]
University of North Carolina at Chapel Hill, Chapel Hill, NC, United States
[1]Corresponding author: e-mail address: thiele@unc.edu

Contents

Abstract

Neuropeptide Y (NPY) is a neuromodulator that is widely expressed throughout the central nervous system (CNS) and which is cosecreted with classic neurotransmitters including GABA and glutamate. There is a long history of research implicating a role for NPY in modulating neurobiological responses to alcohol (ethanol) as well as other drugs of abuse. Both ethanol exposure and withdrawal from chronic ethanol have been shown to produce changes in NPY and NPY receptor protein levels and mRNA expression in the CNS. Importantly, manipulations of NPY Y1 and Y2 receptor signaling have been shown to alter ethanol consumption and self-administration in a brain region-specific manner, with Y1 receptor activation and Y2 receptor blockade in regions of the extended amygdala promoting robust reductions of ethanol intake. Similar observations have been made in studies examining neurobiological responses to nicotine, psychostimulants, and opioids. When taken together with observations of potential genetic linkage between the NPY system and the human alcohol abuse disorders, NPY represents a promising target for treating problematic alcohol and drug use, and in protecting individuals from relapse during abstinence.

International Review of Neurobiology, Volume 136
ISSN 0074-7742
http://dx.doi.org/10.1016/bs.irn.2017.06.005

1. INTRODUCTION

Neuropeptide Y (NPY) is a 36-amino-acid polypeptide belonging to the neuropeptide tyrosine family that is phylogenetically conserved across species (Allen, Bloom, & Polak, 1986). NPY is generally colocalized with GABA in inhibitory interneurons. NPY is currently known to interact with four G protein-coupled receptor subtypes: Y1, Y2, Y4, and Y5. Upon activation, these NPY receptors inhibit production of cyclicadenosine monophosphate, decrease Ca^{2+} channel activity, and enhance G protein coupled inwardly rectifying potassium channel currents (Acuna-Goycolea, Tamamaki, Yanagawa, Obata, & van den Pol, 2005; Palmiter, Erickson, Hollopeter, Baraban, & Schwartz, 1998). Though the Y6 receptor has been found, it has not been well characterized and is not expressed in the rat brain. Y1 and Y2 receptors have been best characterized within the central nervous system (CNS) and are expressed on multiple different cell types, including GABAergic, glutamatergic, and monoaminergic neurons. Y1 receptors are generally thought to act postsynaptically to modulate cell activity, and Y2 receptors function in a predominantly presynaptic fashion, serving as both an autoreceptor regulating NPY release and a heteroreceptor regulating release of coexpressed neurotransmitters and neuromodulators (Chen, DiMaggio, Han, & Westfall, 1997; Dumont, Fournier, St-Pierre, & Quirion, 1993; Gilpin et al., 2011; Greber, Schwarzer, & Sperk, 1994; Gustafson et al., 1997; Kash & Winder, 2006; Kopp et al., 2002; Parker & Herzog, 1999).

The NPY system is expressed throughout the CNS and peripheral nervous system, including in the extended amygdala, hippocampus, basal ganglia, brainstem, and hypothalamus (Allen et al., 1986; Dumont et al., 1993; Gøtzsche & Woldbye, 2016; Kopp et al., 2002; Palmiter et al., 1998; Parker & Herzog, 1999; Roberto, Gilpin, & Siggins, 2012). Given this wide range of expression, NPY is unsurprisingly known to be involved in numerous biological functions within the CNS. NPY signaling plays a role in neuronal development, blood pressure, circadian rhythms, learning, memory, thermogenesis, pain, integration of emotional behavior, as well as feeding, reproduction, reward, and stress-related behaviors (Allen et al., 1986; Bouali, Fournier, St-Pierre, & Jolicoeur, 1995; Caberlotto & Hurd, 1999; Palmiter et al., 1998; Stanley, Chin, & Leibowitz, 1985). In recent years, NPY has further garnered substantial attention for its role in the neurobiological processes underlying the use of ethanol and other drugs of abuse.

This role in substances of abuse is thought to be predominantly regulated by the NPY system expressed in regions involved in reward- and emotion-related behavior, such as the nucleus accumbens (NAcc) and extended amygdala. Indeed, infusion of NPY directly into the NAcc shell has been shown to increase extracellular dopamine as measured by in vivo microdialysis (Sørensen et al., 2009) and is sufficient to induce conditioned place preference (Brown, Coscina, & Fletcher, 2000). Notably, this role of NPY in substances of abuse is distinct from the well-known role of the system in consumption and regulation of caloric intake. The role of NPY in feeding behavior is known to be regulated by NPY actions in the hypothalamus, particularly the paraventricular (PVN) and arcuate (ARC) nuclei. The findings that injection of NPY into the ARC or PVN, but not the NAcc or amygdala, is orexigenic support the hypothesis that the role of the NPY system in regulating reward and general feeding behavior is region specific (Stanley et al., 1985).

Here, we review first the impact of ethanol on NPY expression and signaling within the brain, followed by an examination of manipulating the NPY system as a potential treatment during acute, chronic, and withdrawal from ethanol exposure. We then summarize the role of NPY in other major drugs of abuse. Finally, we discuss the translational relevance of the NPY system.

2. NPY SYSTEM AND GENETIC PREDISPOSITION TO ETHANOL USE

In knockout and transgenic mice, NPY expression level was first demonstrated to negatively correlate with innate ethanol preference 20 years ago (Thiele, Marsh, Ste Marie, Bernstein, & Palmiter, 1998). Specifically, mice overexpressing NPY were found to have decreased and NPY knockout (NPY −/−) mice to have increased ethanol consumption and resistance to the sedative effects of ethanol (Thiele, Koh, & Pedrazzini, 2002; Thiele et al., 1998; Thiele, Miura, Marsh, Bernstein, & Palmiter, 2000; Thiele, Naveilhan, & Ernfors, 2004). This pattern further occurs between inbred strains of mice. The ethanol-preferring C57BL/6J inbred strain has significantly lower baseline levels of NPY in numerous brain regions compared to the ethanol-nonpreferring DBA/2J inbred mouse strain (Hayes, Knapp, Breese, & Thiele, 2005). This correlation between baseline NPY levels and eventual ethanol consummatory behavior has been validated in multiple different rodent genetic models. Selective breeding of high or low ethanol-consuming rats from the outbred Wistar line produced the alcohol-preferring (P) or -nonpreferring (NP) lines of rats. P-rats express notably lower brain levels of NPY compared to NP-rats (Caberlotto et al., 2001; Pandey, Zhang, Roy, & Xu, 2005; Suzuki, Lumeng, McBride, Li, & Hwang, 2004). High-alcohol-drinking and low-alcohol-drinking rats, likewise selectively bred for their ethanol-drinking behavior, also display an inverse relationship between NPY expression and ethanol preference (Ehlers et al., 1998; Hwang, Zhang, Ehlers, Lumeng, & Li, 1999). Interestingly, within ethanol-preferring rodents, differences in NPY are region dependent. NPY in ethanol-preferring animals is decreased relative to ethanol-nonpreferring animals in regions associated with drug dependence, such as the amygdala and shell of the NAcc, but not in the hypothalamus, a region more associated with general caloric intake and feeding behavior (Ehlers et al., 1998; Hayes et al., 2005; Misra & Pandey, 2006). This convergence between genetic models suggest baseline NPY expression within specific brain regions is a critical contributor to innate ethanol-directed behaviors rather than a nonspecific side effect of selective breeding. Notably, central administration of NPY is sufficient to markedly decrease nondependent ethanol intake in these ethanol-preferring rodent genetic lines, but not in nonpreferring or wild-type outbred lines (Badia-Elder, Gilpin, & Stewart, 2007; Badia-Elder et al., 2001), likely because of the inherent low levels of central NPY in preferring lines.

In addition to NPY itself, differences in NPY receptor expression impact baseline ethanol-directed behaviors. Both the Y1 and Y2 receptor have been shown to play an important role in innate ethanol-directed behaviors, while, in contrast, deletion of the Y5 receptor was not shown to impact ethanol intake (Thiele et al., 2000). Similar to the impact of deletion of NPY itself, Y1 receptor knockout (Y1 −/−) mice consume significantly more ethanol and are less responsive to the sedative effects of ethanol, while consumption of sucrose or quinine solutions, as well as metabolism of ethanol, remains unaltered (Thiele et al., 2002). In contrast, Y2 receptor knockout (Y2 −/−) mice show decreased ethanol preference relative to wild-type mice, in keeping with the role of Y2 as a NPY autoreceptor (Thiele, Naveilhan, et al., 2004). Like Y1 −/− mice, Y2 −/− mice showed no change in consumption of sucrose, quinine, or ethanol metabolism. Interestingly, P-rats display differences in Y1:Y2 receptor ratio mRNA compared to NP-rats (Caberlotto et al., 2001), suggesting the balance between these two receptors may significantly contribute to innate ethanol preference. Like changes in NPY expression, these differences in Y1:Y2 balance in ethanol-preferring animals most notably occur in regions related to reward and anxiety-like behavior, such as the hippocampus and amygdala.

Interestingly, the early interaction between the ethanol and the NPY system appears to be a highly conserved reaction as evidenced by genetic mutation studies with lower organisms. NPR-1, the *Caenorhabditis elegans* ortholog of the NPY receptor, is critically involved in the development of acute ethanol tolerance (Davies, Bettinger, Thiele, Judy, & McIntire, 2004). Further, the *Drosophila* ortholog of NPY and the Y1 receptor, NPF and NPF-1, respectively, play important roles in initial ethanol sensitivity (Chen, Zhang, & Shen, 2008). The early evolutionary roots of acute ethanol interactions with the NPY system, which have been maintained in mammalian species, underscore the critical role NPY is poised to play from the very first drug exposure in ethanol-directed behavior.

3. ETHANOL IMPACT ON THE NPY SYSTEM

A wide range of patterns and modes of ethanol exposure have been shown to alter brain NPY signaling and expression in numerous brain regions related to reward and emotionality. Acute ethanol exposure reduces NPY mRNA and peptide expression (Kinoshita et al., 2000). Binge drinking in P-rats has been found to increase NPY and decrease Y1 receptor RNA within the dorsal raphe nucleus (McClintick et al., 2015). Similarly, mice display a reduction in NPY expression within the NAcc following ethanol consumption and reduced NPY protein levels in the central amygdala (CeA) following a history of binge-like ethanol drinking (Barkley-Levenson, Ryabinin, & Crabbe, 2016; Sparrow et al., 2012).

Unsurprisingly, given the interaction between the acute ethanol and the NPY system, chronic ethanol exposure is well known to result in significant alterations in NPY peptide and receptor function and expression in a region-dependent manner. Rats chronically fed an ethanol diet show region-dependent alterations in NPY immunoreactivity (NPY-IR), with reductions most notably found in the hippocampus and cerebral cortex (Bison & Crews, 2003). Bison suggests this decrease may be related to the level of intoxication reached by the rats, greater than 200 mg/dL, as previous work utilizing chronic ethanol exposure resulting in comparatively lower blood ethanol concentrations detected no alteration in NPY-IR in numerous sampled brain regions (Ehlers et al., 1998). These results suggest a critical role of the NPY system specifically during high levels of ethanol intake, such as is expected to occur in patterns of heavy ethanol intake associated with the transition to and expression of ethanol dependence.

Withdrawal from chronic ethanol exposure has been reported to both induce elevated NPY levels in numerous brain regions, including the amygdala and hippocampus, and to decrease NPY protein and mRNA levels in areas including the amygdala and frontal cortex (Barkley-Levenson et al., 2016; Roy & Pandey, 2002; Zhang & Pandey, 2003). These opposing results are likely related to both the intensity of ethanol exposure and the length of withdrawal prior to evaluation of NPY expression. Specifically, withdrawal from chronic ethanol exposure has a potentially biphasic effect on the NPY system. For instance, the Mathé lab found an initial reduction in NPY at 24 h of ethanol withdrawal later shifts to an overexpression of NPY 2 weeks into withdrawal (Criado, Liu, Ehlers, & Mathé, 2011). A similar effect was found by the Crews lab, in which acute withdrawal from a chronic ethanol treatment decreased NPY in the hippocampus, but 72 h of withdrawal resulted in a significant increase in NPY within the same region (Bison & Crews, 2003). Interestingly, this biphasic shift in NPY expression during the transition from acute to protracted withdrawal seems to correspond with the changes in anxiety-like behavior observed during withdrawal, i.e., an initial spike of anxiogenesis followed by a slow decrease (Thiele, Sparta, Hayes, & Fee, 2004). Additionally, intensity and duration of chronic ethanol exposure is thought to be particularly important in determining ethanol-induced alterations in the NPY system during withdrawal. For example, withdrawal from 8, but not 2, weeks of ethanol vapor exposure were found to produce changes in NPY-IR in the adult rat brain (Criado et al., 2011). Overall, these studies demonstrate alterations in NPY occur throughout acute and into protracted withdrawal. Further, timing specific alterations in NPY system function may significantly impact behavior and susceptibility to relapse during multiple stages of ethanol withdrawal.

4. THE IMPACT OF NPY SYSTEM MANIPULATION ON ETHANOL-DIRECTED BEHAVIORS

The ability of pharmacological manipulations of the NPY system to alter ethanol-directed behavior is likely dependent on genetic background and history of ethanol intake. In nondependent animals, intraventricular infusion of NPY has been shown to suppress ethanol intake in ethanol-preferring P-rats, without impacting baseline intake in nonpreferring animals (Badia-Elder, Stewart, Powrozek, Murphy, & Li, 2003; Badia-Elder et al., 2001; Gilpin, Misra, & Koob, 2008; Henderson & Czachowski, 2012). A similar result has been observed in the high ethanol consuming C57/BJ line of mice, in which a central injection of either a NPY itself or an Y1 agonist reduces ethanol intake without altering water intake or the sedative properties of ethanol (Sparta et al., 2004; Thiele, Sparta, et al., 2004). The importance of ethanol history is highlighted by the finding that the ability of NPY to reduce ethanol intake is increased following successive cycles of 5 days of ethanol consumption and 2 days of withdrawal (Bertholomey, Henderson, Badia-Elder, & Stewart, 2011).

Inactivation of the Y2 receptor is likewise able to alter ethanol-directed behaviors. Intraventricular administration of an Y2 antagonist decreases ethanol intake (Sparrow et al., 2012; Thorsell, Rimondini, & Heilig, 2002). Interestingly, rather than the specificity of Y1 receptor activation in altering ethanol intake only in ethanol-experienced or genetically predisposed animals, antagonism of the Y2 receptor has been reported to suppress operant self-administration of ethanol in both dependent and nondependent outbred Wistar rats (Rimondini, Thorsell, & Heilig, 2005). Notably, the impact of the Y2 antagonist was found in this study to be strongly facilitated by ethanol dependence, suggesting ethanol-induced alterations to Y2 function. That the relative role of Y2 as an auto- and heteroreceptor may be disrupted during ethanol exposure has been suggested as a possible explanation for the complex role of Y2 in ethanol-related behaviors (Gilpin, 2012). It is of note that in contrast to Y2 antagonism, manipulations of the Y1 receptor, while decreasing ethanol intake, have in some studies been reported not to alter the appetitive aspect of ethanol-directed behavior which operant lever-pressing represents (Thorsell, Slawecki, & Ehlers, 2005). Together, these studies suggest the role of specific NPY receptors in ethanol seeking and consumption is complex and an area deserving of continued investigation.

The NPY system's ability to suppress ethanol intake is likely specifically controlled by brain regions involved in reward and emotionality. This is evidenced by the finding that activation of the NPY system through administration of NPY or a Y1 agonist into the bed nucleus of the stria terminalis (BNST), amygdala, or NAcc shell reduces ethanol intake, while in contrast hypothalamic infusion of NPY increases ethanol drinking and preference through interaction with the Y1 receptor (Gilpin et al., 2008; Kelley, Nannini, Bratt, & Hodge, 2001; Pleil et al., 2015; Sparrow et al., 2012; Thorsell et al., 2007). Investigations into the synaptic mechanisms mediating the ability of the NPY system to modulate ethanol-directed behaviors have primarily focused on the CeA. NPY or Y1 agonist administration into the

CeA is sufficient to suppress ethanol consumption without impacting general consumption or intake of other palatable reinforces, while overexpression of NPY specifically within the CeA reduces ethanol consumption in animals chronically exposed to ethanol (Gilpin et al., 2008; Sparrow et al., 2012). Animals with a history of binge or chronic ethanol exposure display alterations in the ability of NPY to modulate inhibitory signaling within the CeA. Specifically, NPY inhibition of evoked inhibitory postsynaptic currents is enhanced following binge-like ethanol consumption in mice (Sparrow et al., 2012). Further, NPY is sufficient to prevent and reverse ethanol-induced increases in CeA inhibitory signaling in both ethanol-naïve and ethanol-dependent animals through presynaptic effects on GABA release (Gilpin et al., 2011). Understanding the underlying synaptic mechanisms and brain regions mediating the specific role of the NPY system in modulating ethanol-directed behaviors following chronic ethanol exposure remains an area of intense research.

In addition to reducing ongoing ethanol intake, NPY system manipulations have been shown to alter relapse to ethanol-seeking behavior. Gilpin found repeated intraventricular administration of NPY during withdrawal from chronic ethanol exposure attenuated future increases in ethanol self-administration during subsequent withdrawals compared to nontreated cohorts (Gilpin et al., 2011). NPY administration has further been found to block stress-induced reinstatement of ethanol seeking (Cippitelli et al., 2010). The overall ability of NPY system manipulations to specifically reduce ethanol intake in animals with a genetic predisposition to high ethanol intake or history of chronic ethanol intake suggests this system is an important target in preventing the relapse to ethanol use during withdrawal.

5. NPY SYSTEM AND OTHER DRUGS OF ABUSE

5.1 Nicotine

The majority of research into the interactions between the NPY system and the nicotine has focused on the role of NPY in the well-known anorexigenic effects of smoking. Accordingly, interactions between NPY and nicotine have been conducted with a primary focus on the hypothalamus due to the critical role of this region in regulating feeding behavior and caloric intake. However, research has also found acute nicotine decreases NPY-IR in regions associated with reward and emotionality, including the NAcc and BNST, in addition to within feeding-related regions such as the PVN

(Kotagale et al., 2014). This suggests the NPY system likely also plays a role in modulating the rewarding and anxiety-altering aspects of nicotine intake. Notably, this decrease in NPY-IR was speculated to represent increased release, as opposed to decreased production, of NPY (Kotagale et al., 2014). Such an increase in release would agree with the observed dose-dependent increase in NPY mRNA levels and downregulation of Y1 receptors observed in the hypothalamus following acute nicotine administration (Li et al., 2000). Overall, these data implicate a nicotine-driven activation of the NPY system may contribute to drug-taking behavior.

As observed in numerous systems within the brain, chronic drug exposure induces alterations primarily opposite to those found following acute treatment. Chronic cigarette smoke or nicotine treatment has been found to decrease hypothalamic NPY and upregulate Y1 receptor density (Chen et al., 2007; Frankish et al., 1995; Jang et al., 2003; Kane, Parker, & Li, 2001). Interestingly, an increase in NPY following chronic nicotine exposure has been observed in some studies (Li et al., 2000). Li et al. suggests the specific impact of chronic nicotine on hypothalamic NPY levels is likely related to the dose and duration of exposure, with higher nicotine doses inducing a reduction in NPY expression. This was supported by the finding that rats chronically exposed to 6 mg/kg/day nicotine displayed an average 30% reduction in hypothalamic NPY compared to rats which received 4 mg/kg/day over the same period (Li et al., 2000). Though the specific impact of chronic nicotine on the NPY system appears dose dependent, the ability of pharmacological manipulations of the NPY system to alleviate specific symptoms of nicotine withdrawal has been repeatedly observed. Specifically, NPY administration attenuates the overall somatic signs of nicotine withdrawal (Rylkova et al., 2008). Interestingly, this study found no impact of NPY or Y1 agonist administration on deficits in reward-related behaviors associated with nicotine withdrawal. In contrast, systemic administration of Y2 receptor antagonist during abstinence was found to reduce withdrawal-induced social anxiety-like behavior, locomotor sensitization to a nicotine challenge, and alterations in NPY mRNA levels within the amygdala and hippocampus (Aydin, Oztan, & Isgor, 2011). Similar to what is observed in ethanol withdrawal, such data implicate a potentially differential role of Y1 and Y2 receptors in impacting discrete aspects of withdrawal-associated behaviors. Overall, these results suggest chronic nicotine exposure may impact aspects of NPY system signaling in a region-specific manner, highlighting the need for more research into the role of NPY in nicotine dependence and withdrawal.

5.2 Psychostimulants

The NPY system has been suggested to directly modulate activity of the striatal dopamine system. Infusion of NPY into the NAcc of freely moving rats was found to increase extracellular dopamine concentration (Kerkerian-Le Goff et al., 1992). Further, NPY increases KCl-induced and NMDA-stimulated dopamine release in striatal slices (Adewale, Macarthur, & Westfall, 2007; Ault & Werling, 1998). A considerable amount of attention has therefore been paid to the potential role of NPY in dependence on, and withdrawal from, psychostimulants, the effects of which are mediated primarily by the striatal dopaminergic system. Cocaine, methamphetamine (METH), and amphetamine have been reliably found to alter NPY system expression in numerous brain regions, including the prefrontal cortex, cingulate cortex, striatum, caudate, hippocampus, PVN, ARC, and dorsomedial nuclei of the hypothalamus (Crowley et al., 2005; Goodman & Sloviter, 1993; Thiriet et al., 2005; Wahlestedt et al., 1991; Westwood & Hanson, 1999; Yoshihara, Honma, Mitome, & Honma, 1996). These alterations in NPY were found in some studies to occur at a substantially greater degree following multiple psychostimulant administrations and to last up to 48 h after treatment, suggesting NPY system alterations persist even after the drug has been fully cleared from the brain (Westwood & Hanson, 1999). Manipulations of D1 receptor activity, a critical mediator of the rewarding effects of psychostimulants, reduced and D2 receptor antagonism failed to alter striatal NPY-IR (Midgley, Bush, Gibb, & Hanson, 1992; Midgley et al., 1994; Westwood & Hanson, 1999). Indeed, D1 receptor antagonism was found to prevent METH-induced alterations to NPY expression, suggesting NPY-D1 interaction may be involved in psychostimulant activity (Westwood & Hanson, 1999). Importantly, manipulations of the NPY system alter behavioral responses to psychostimulant use. Intracerebroventricular injection of NPY augments cocaine-induced hyperactivity and increases cocaine self-administration, while NPY knockout mice display increased sensitivity to cocaine, increased frequency of amphetamine-induced stereotypic behaviors, and increased susceptibility to METH-induced neuronal apoptosis (Cannon, Abdallah, Tecott, During, & Palmiter, 2004; Maric, Cantor, Cuccioletta, Tobin, & Shalev, 2009; Sørensen & Woldbye, 2012; Thiriet et al., 2005).

Three NPY receptors have been suggested to play a role in mediating the interaction between NPY, dopamine receptors, and psychostimulants. Antagonism of the Y1 receptor has been found to inhibit amphetamine-induced increases in locomotion (Kask & Harro, 2000). Y2 receptor antagonism abolished METH-induced memory deficits, suggesting a potentially important role of Y2 activity in the hippocampus in mediating the effects of METH use (Gonçalves et al., 2012). Interestingly, the primary focus of research in interactions between NPY receptors and psychostimulant drug use has been the Y5 receptor. Y5 mRNA and immunoreactivity have been found in cells of the ventral tegmental area and NAcc, as well as Y5 receptor binding within the latter (Durkin et al., 2000; Morin & Gehlert, 2006; Quarta, Leslie, Carletti, Valerio, & Caberlotto, 2011). Y5 knockout (Y5 −/−) mice have a blunted extracellular dopamine response following cocaine and, further, show attenuated cocaine self-administration and cocaine-induced hyperlocomotion (Sørensen et al., 2012). These effects are unlikely due to compensatory mechanisms as Y5 antagonism displays a similar behavioral profile, capable of decreasing nose poking for cocaine without impacting saline self-administration, blocking cocaine-induced increases in c-fos expression and extracellular dopamine response, and expediting extinction and attenuating reinstatement of cocaine-induced place preference (Sørensen et al., 2012; Sørensen, Wörtwein, Fink-Jensen, & Woldbye, 2013). Overall, while the NPY system is well supported as a promising target for altering psychostimulant intake and relapse, the Y5 receptor seems particularly well positioned as a potential mediator between psychostimulants, dopamine, and the NPY system.

5.3 Opioids

Numerous studies have demonstrated a potentially agonistic interaction between the NPY and opioid systems. Acute heroin injection in drug-naïve rats increases NPY-IR in the PVN and BNST (D'Este, Casini, Pontieri, & Renda, 2006), and morphine induces NPY release in the ARC, NAcc, and BNST (Saurer, Ijames, & Lysle, 2006). In addition to opioid treatment inducing NPY release, the reciprocal relationship appears likewise true in which NPY administration increases opioid-directed behavior. NPY microinjected into the NAcc was found to enhance morphine reward, while intracerebroventricular NPY administration dose dependently increased heroin self-administration and facilitated the reinstatement of heroin-seeking during extinction (Desai, Upadhya, Subhedar, & Kokare, 2013; Maric, Tobin, Quinn, & Shalev, 2008). In keeping with the role of NPY in the rewarding effects of opioids, the rewarding effects of morphine were notably reduced by an Y1 receptor antagonist (Desai et al., 2013). The NPY system also appears heavily involved in the transition to a withdrawal state during abstinence from prolonged opioid use. NPY gene expression is significantly decreased in rats at both 1 and 14 days of withdrawal from heroin self-administration, while the Y5 receptor is increased in the same animals (Kuntz-Melcavage, Brucklacher, Grigson, Freeman, & Vrana, 2009). Accordingly, exogenous administration of NPY during abstinence attenuated withdraw symptoms associated with morphine, with this effect suggested to be mediated by interaction with Y5 receptors (Clausen, Møller, & Woldbye, 2001; Woldbye, Klemp, & Madsen, 1998). All together, these studies suggest a prominent role of the NPY system in mediating not only the initial rewarding effects of opioids, but in the negative affect experienced during withdrawal from chronic use.

6. TRANSLATIONAL RELEVANCE

The substantial preclinical evidence linking the NPY system to drugs of abuse, specifically ethanol abuse and dependence, is further supported by numerous links between polymorphisms in the NPY system and ethanol-associated behaviors in clinical human population (Mottagui-Tabar et al., 2005). Notably, these links between polymorphisms in the NPY system and human ethanol-directed behavior are of particular importance given the significant genetic contribution twin and genealogical studies have demonstrated for alcohol use disorders (Heath et al., 1997; McGue, 1999; Pickens et al., 1991; Wu & Shete, 2005). The Collaborative Study on the Genetics of Alcoholism found single nucleotide polymorphisms in the genes coding for the Y2 and Y5 receptors to be strongly associated with numerous ethanol abuse-related phenotypes (Wetherill et al., 2008). Links between alterations in genes coding for NPY itself and ethanol-related behaviors have likewise been found. The 5671C/T polymorphism (rs5574) in NPY has been associated with alcohol-dependent individuals who display withdrawal-induced seizures (Okubo & Harada, 2001). Further, increased frequency of the NPY variant Leu7Pro allele has been linked to increased alcohol consumption and dependence (Kauhanen et al., 2000; Lappalainen et al., 2002), though this effect has not been replicated in all studies (Zhu et al., 2003; Zill, Preuss, Koller, Bondy, & Soyka, 2008). This lack of consistent replication has been suggested to be potentially linked to the stress history of the sampled alcoholic population (Lovinger & Roberto, 2013). Interaction between stress exposure and NPY allele variation contributing to ethanol-directed behaviors would be in keeping with the extensive literature discussed in this review linking specific alterations in NPY expression within brain areas related to emotionality in influencing ethanol intake. Further, research in nonhuman primates demonstrates a loss-of-function variant of NPY alters ethanol intake only in animals previously exposed to stress (Lindell et al., 2010). Manipulations of the NPY system may therefore be an area of particular interest in developing treatments for individuals possessing both an alcohol use disorder and a history of traumatic life experience or anxiety-related mental disorders. The recent finding that a group particularly vulnerable to the development of alcohol use disorders, veterans suffering from posttraumatic stress disorder (PTSD), have markedly lower levels of NPY in their cerebrospinal fluid compared to veterans without PTSD supports this potentially critical connection between NPY, stress, and the development of alcohol use disorders (Sah, Ekhator, Jefferson-Wilson, Horn, & Geracioti, 2014; Sah et al., 2009; Waller, McGuire, & Dobson, 2015).

7. CONCLUSION

An extensive literature supports a role of the NPY system in modulating the seeking and intake of drugs of abuse, from the baseline genetic predisposition to symptoms of withdrawal which contribute to relapse. Innate levels of NPY in regions of the brain associated with reward and emotional behavior have been shown to influence innate ethanol preference and binge-like intake. Acute and chronic drug exposure, in a dose- and pattern-specific manner, influence brain NPY peptide and receptor expression. Notably, NPY receptors do not appear to be impacted uniformly, but rather drug exposure alters Y1, Y2, and Y5 receptors in divergent patterns within distinct brain regions. Such specific effects on receptor function and expression suggest individual NPY receptor subtypes likely play distinct roles in mediating the various effects of acute and chronic drug use, as well as in withdrawal. Collectively, the findings summarized in this chapter demonstrate the immense promise of the NPY system in the pharmacological intervention and treatment of alcohol use disorders, as well as other drugs of abuse. Further, as the NPY system is well known as a critical component of anxiety- and stress-related behaviors and pathology (Thiele, Sparta, et al., 2004), this system is particularly well placed to serve as a treatment for those suffering from comorbidity of alcohol and anxiety-related disorders, such as PTSD. Further research is needed to gain a more thorough understanding of the specific components of the NPY system in discrete brain regions involved in each stage of ethanol use and dependence to develop the best possible treatments for drug use disorders within a highly diverse human population.

ACKNOWLEDGMENTS

The authors would like to thank the National Institute on Alcohol Abuse and Alcoholism for support during the writing of this chapter (Grants AA022048 & AA013573).

REFERENCES

Acuna-Goycolea, C., Tamamaki, N., Yanagawa, Y., Obata, K., & van den Pol, A. N. (2005). Mechanisms of neuropeptide Y, peptide YY, and pancreatic polypeptide inhibition of identified green fluorescent protein-expressing GABA neurons in the hypothalamic neuroendocrine arcuate nucleus. *The Journal of Neuroscience*, *25*(32), 7406–7419. http://dx.doi.org/10.1523/JNEUROSCI.1008-05.2005.

Adewale, A. S., Macarthur, H., & Westfall, T. C. (2007). Neuropeptide Y-induced enhancement of the evoked release of newly synthesized dopamine in rat striatum: Mediation by Y2 receptors. *Neuropharmacology*, *52*(6), 1396–1402. http://dx.doi.org/10.1016/j.neuropharm.2007.01.018.

Allen, Y. S., Bloom, S. R., & Polak, J. M. (1986). The neuropeptide Y-immunoreactive neuronal system: Discovery, anatomy and involvement in neurodegenerative disease. *Human Neurobiology*, *5*(4), 227–234.

Ault, D. T., & Werling, L. L. (1998). Neuropeptide Y-mediated enhancement of NMDA-stimulated [3H]dopamine release from rat prefrontal cortex is reversed by sigma1 receptor antagonists. *Schizophrenia Research*, *31*(1), 27–36.

Aydin, C., Oztan, O., & Isgor, C. (2011). Effects of a selective Y2R antagonist, JNJ-31020028, on nicotine abstinence-related social anxiety-like behavior, neuropeptide Y and corticotropin releasing factor mRNA levels in the novelty-seeking phenotype. *Behavioural Brain Research*, *222*(2), 332–341. http://dx.doi.org/10.1016/j.bbr.2011.03.067.

Badia-Elder, N. E., Gilpin, N. W., & Stewart, R. B. (2007). Neuropeptide Y modulation of ethanol intake: Effects of ethanol drinking history and genetic background. *Peptides*, *28*(2), 339–344. http://dx.doi.org/10.1016/j.peptides.2006.07.028.

Badia-Elder, N. E., Stewart, R. B., Powrozek, T. A., Murphy, J. M., & Li, T. K. (2003). Effects of neuropeptide Y on sucrose and ethanol intake and on anxiety-like behavior in high alcohol drinking (HAD) and low alcohol drinking (LAD) rats. *Alcoholism, Clinical and Experimental Research*, *27*(6), 894–899. http://dx.doi.org/10.1097/01.ALC.0000071929.17974.DA.

Badia-Elder, N. E., Stewart, R. B., Powrozek, T. A., Roy, K. F., Murphy, J. M., & Li, T. K. (2001). Effect of neuropeptide Y (NPY) on oral ethanol intake in Wistar, alcohol-preferring (P), and -nonpreferring (NP) rats. *Alcoholism, Clinical and Experimental Research*, *25*(3), 386–390.

Barkley-Levenson, A. M., Ryabinin, A. E., & Crabbe, J. C. (2016). Neuropeptide Y response to alcohol is altered in nucleus accumbens of mice selectively bred for drinking to intoxication. *Behavioural Brain Research*, *302*, 160–170. http://dx.doi.org/10.1016/j.bbr.2016.01.015.

Bertholomey, M. L., Henderson, A. N., Badia-Elder, N. E., & Stewart, R. B. (2011). Neuropeptide Y (NPY)-induced reductions in alcohol intake during continuous access and following alcohol deprivation are not altered by restraint stress in alcohol-preferring

(P) rats. *Pharmacology, Biochemistry, and Behavior, 97*(3), 453–461. http://dx.doi.org/10.1016/j.pbb.2010.10.002.

Bison, S., & Crews, F. (2003). Alcohol withdrawal increases neuropeptide Y immunoreactivity in rat brain. *Alcoholism, Clinical and Experimental Research, 27*(7), 1173–1183. http://dx.doi.org/10.1097/01.ALC.0000075827.74538.FE.

Bouali, S. M., Fournier, A., St-Pierre, S., & Jolicoeur, F. B. (1995). Effects of NPY and NPY2-36 on body temperature and food intake following administration into hypothalamic nuclei. *Brain Research Bulletin, 36*(2), 131–135.

Brown, C. M., Coscina, D. V., & Fletcher, P. J. (2000). The rewarding properties of neuropeptide Y in perifornical hypothalamus vs. nucleus accumbens. *Peptides, 21*(8), 1279–1287.

Caberlotto, L., & Hurd, Y. L. (1999). Reduced neuropeptide Y mRNA expression in the prefrontal cortex of subjects with bipolar disorder. *Neuroreport, 10*(8), 1747–1750.

Caberlotto, L., Thorsell, A., Rimondini, R., Sommer, W., Hyytiä, P., & Heilig, M. (2001). Differential expression of NPY and its receptors in alcohol-preferring AA and alcohol-avoiding ANA rats. *Alcoholism, Clinical and Experimental Research, 25*(11), 1564–1569.

Cannon, C. M., Abdallah, L., Tecott, L. H., During, M. J., & Palmiter, R. D. (2004). Dysregulation of striatal dopamine signaling by amphetamine inhibits feeding by hungry mice. *Neuron, 44*(3), 509–520. http://dx.doi.org/10.1016/j.neuron.2004.10.009.

Chen, X., DiMaggio, D. A., Han, S. P., & Westfall, T. C. (1997). Autoreceptor-induced inhibition of neuropeptide Y release from PC-12 cells is mediated by Y2 receptors. *The American Journal of Physiology, 273*(4 Pt. 2), H1737–H1744.

Chen, H., Hansen, M. J., Jones, J. E., Vlahos, R., Bozinovski, S., Anderson, G. P., et al. (2007). Regulation of hypothalamic NPY by diet and smoking. *Peptides, 28*(2), 384–389. http://dx.doi.org/10.1016/j.peptides.2006.07.034.

Chen, J., Zhang, Y., & Shen, P. (2008). A protein kinase C activity localized to neuropeptide Y-like neurons mediates ethanol intoxication in Drosophila melanogaster. *Neuroscience, 156*(1), 42–47. http://dx.doi.org/10.1016/j.neuroscience.2008.07.008.

Cippitelli, A., Damadzic, R., Hansson, A. C., Singley, E., Sommer, W. H., Eskay, R., et al. (2010). Neuropeptide Y (NPY) suppresses yohimbine-induced reinstatement of alcohol seeking. *Psychopharmacology, 208*(3), 417–426. http://dx.doi.org/10.1007/s00213-009-1741-y.

Clausen, T. R., Møller, M., & Woldbye, D. P. (2001). Inhibitory effect of neuropeptide Y on morphine withdrawal is accompanied by reduced c-fos expression in specific brain regions. *Journal of Neuroscience Research, 64*(4), 410–417. http://dx.doi.org/10.1002/jnr.1092.

Criado, J. R., Liu, T., Ehlers, C. L., & Mathé, A. A. (2011). Prolonged chronic ethanol exposure alters neuropeptide Y and corticotropin-releasing factor levels in the brain of adult Wistar rats. *Pharmacology, Biochemistry, and Behavior, 99*(1), 104–111. http://dx.doi.org/10.1016/j.pbb.2011.04.005.

Crowley, W. R., Ramoz, G., Keefe, K. A., Torto, R., Kalra, S. P., & Hanson, G. R. (2005). Differential effects of methamphetamine on expression of neuropeptide Y mRNA in hypothalamus and on serum leptin and ghrelin concentrations in ad libitum-fed and schedule-fed rats. *Neuroscience, 132*(1), 167–173. http://dx.doi.org/10.1016/j.neuroscience.2004.11.037.

Davies, A. G., Bettinger, J. C., Thiele, T. R., Judy, M. E., & McIntire, S. L. (2004). Natural variation in the npr-1 gene modifies ethanol responses of wild strains of C. elegans. *Neuron, 42*(5), 731–743. http://dx.doi.org/10.1016/j.neuron.2004.05.004.

Desai, S. J., Upadhya, M. A., Subhedar, N. K., & Kokare, D. M. (2013). NPY mediates reward activity of morphine, via NPY Y1 receptors, in the nucleus accumbens shell. *Behavioural Brain Research, 247*, 79–91. http://dx.doi.org/10.1016/j.bbr.2013.03.018.

D'Este, L., Casini, A., Pontieri, F. E., & Renda, T. G. (2006). Changes in neuropeptide FF and NPY immunohistochemical patterns in rat brain under heroin treatment. *Brain Research*, *1083*(1), 151–158. http://dx.doi.org/10.1016/j.brainres.2006.02.009.

Dumont, Y., Fournier, A., St-Pierre, S., & Quirion, R. (1993). Comparative characterization and autoradiographic distribution of neuropeptide Y receptor subtypes in the rat brain. *The Journal of Neuroscience*, *13*(1), 73–86.

Durkin, M. M., Walker, M. W., Smith, K. E., Gustafson, E. L., Gerald, C., & Branchek, T. A. (2000). Expression of a novel neuropeptide Y receptor subtype involved in food intake: An in situ hybridization study of Y5 mRNA distribution in rat brain. *Experimental Neurology*, *165*(1), 90–100. http://dx.doi.org/10.1006/exnr.2000.7446.

Ehlers, C. L., Li, T. K., Lumeng, L., Hwang, B. H., Somes, C., Jimenez, P., et al. (1998). Neuropeptide Y levels in ethanol-naive alcohol-preferring and nonpreferring rats and in Wistar rats after ethanol exposure. *Alcoholism, Clinical and Experimental Research*, *22*(8), 1778–1782.

Frankish, H. M., Dryden, S., Wang, Q., Bing, C., MacFarlane, I. A., & Williams, G. (1995). Nicotine administration reduces neuropeptide Y and neuropeptide Y mRNA concentrations in the rat hypothalamus: NPY may mediate nicotine's effects on energy balance. *Brain Research*, *694*(1–2), 139–146.

Gilpin, N. W. (2012). Neuropeptide Y (NPY) in the extended amygdala is recruited during the transition to alcohol dependence. *Neuropeptides*, *46*(6), 253–259. http://dx.doi.org/10.1016/j.npep.2012.08.001.

Gilpin, N. W., Misra, K., Herman, M. A., Cruz, M. T., Koob, G. F., & Roberto, M. (2011). Neuropeptide Y opposes alcohol effects on gamma-aminobutyric acid release in amygdala and blocks the transition to alcohol dependence. *Biological Psychiatry*, *69*(11), 1091–1099. http://dx.doi.org/10.1016/j.biopsych.2011.02.004.

Gilpin, N. W., Misra, K., & Koob, G. F. (2008). Neuropeptide Y in the central nucleus of the amygdala suppresses dependence-induced increases in alcohol drinking. *Pharmacology, Biochemistry, and Behavior*, *90*(3), 475–480. http://dx.doi.org/10.1016/j.pbb.2008.04.006.

Gonçalves, J., Baptista, S., Olesen, M. V., Fontes-Ribeiro, C., Malva, J. O., Woldbye, D. P., et al. (2012). Methamphetamine-induced changes in the mice hippocampal neuropeptide Y system: Implications for memory impairment. *Journal of Neurochemistry*, *123*(6), 1041–1053. http://dx.doi.org/10.1111/jnc.12052.

Goodman, J. H., & Sloviter, R. S. (1993). Cocaine neurotoxicity and altered neuropeptide Y immunoreactivity in the rat hippocampus; a silver degeneration and immunocytochemical study. *Brain Research*, *616*(1–2), 263–272.

Gøtzsche, C. R., & Woldbye, D. P. (2016). The role of NPY in learning and memory. *Neuropeptides*, *55*, 79–89. http://dx.doi.org/10.1016/j.npep.2015.09.010.

Greber, S., Schwarzer, C., & Sperk, G. (1994). Neuropeptide Y inhibits potassium-stimulated glutamate release through Y2 receptors in rat hippocampal slices in vitro. *British Journal of Pharmacology*, *113*(3), 737–740.

Gustafson, E. L., Smith, K. E., Durkin, M. M., Walker, M. W., Gerald, C., Weinshank, R., et al. (1997). Distribution of the neuropeptide Y Y2 receptor mRNA in rat central nervous system. *Brain Research. Molecular Brain Research*, *46*(1–2), 223–235.

Hayes, D. M., Knapp, D. J., Breese, G. R., & Thiele, T. E. (2005). Comparison of basal neuropeptide Y and corticotropin releasing factor levels between the high ethanol drinking C57BL/6J and low ethanol drinking DBA/2J inbred mouse strains. *Alcoholism, Clinical and Experimental Research*, *29*(5), 721–729.

Heath, A. C., Bucholz, K. K., Madden, P. A., Dinwiddie, S. H., Slutske, W. S., Bierut, L. J., et al. (1997). Genetic and environmental contributions to alcohol dependence risk in a national twin sample: Consistency of findings in women and men. *Psychological Medicine*, *27*(6), 1381–1396.

Henderson, A. N., & Czachowski, C. L. (2012). Neuropeptide Y (NPY) in the central nucleus of the amygdala (CeA) does not affect ethanol-reinforced responding in binge-drinking, nondependent rats. *Pharmacology, Biochemistry, and Behavior, 101*(1), 8–13. http://dx.doi.org/10.1016/j.pbb.2011.11.008.

Hwang, B. H., Zhang, J. K., Ehlers, C. L., Lumeng, L., & Li, T. K. (1999). Innate differences of neuropeptide Y (NPY) in hypothalamic nuclei and central nucleus of the amygdala between selectively bred rats with high and low alcohol preference. *Alcoholism, Clinical and Experimental Research, 23*(6), 1023–1030.

Jang, M. H., Shin, M. C., Kim, K. H., Cho, S. Y., Bahn, G. H., Kim, E. H., et al. (2003). Nicotine administration decreases neuropeptide Y expression and increases leptin receptor expression in the hypothalamus of food-deprived rats. *Brain Research, 964*(2), 311–315.

Kane, J. K., Parker, S. L., & Li, M. D. (2001). Hypothalamic orexin-A binding sites are downregulated by chronic nicotine treatment in the rat. *Neuroscience Letters, 298*(1), 1–4.

Kash, T. L., & Winder, D. G. (2006). Neuropeptide Y and corticotropin-releasing factor bi-directionally modulate inhibitory synaptic transmission in the bed nucleus of the stria terminalis. *Neuropharmacology, 51*(5), 1013–1022. http://dx.doi.org/10.1016/j.neuropharm.2006.06.011.

Kask, A., & Harro, J. (2000). Inhibition of amphetamine- and apomorphine-induced behavioural effects by neuropeptide Y Y(1) receptor antagonist BIBO 3304. *Neuropharmacology, 39*(7), 1292–1302.

Kauhanen, J., Karvonen, M. K., Pesonen, U., Koulu, M., Tuomainen, T. P., Uusitupa, M. I., et al. (2000). Neuropeptide Y polymorphism and alcohol consumption in middle-aged men. *American Journal of Medical Genetics, 93*(2), 117–121.

Kelley, S. P., Nannini, M. A., Bratt, A. M., & Hodge, C. W. (2001). Neuropeptide-Y in the paraventricular nucleus increases ethanol self-administration. *Peptides, 22*(3), 515–522.

Kerkerian-Le Goff, L., Forni, C., Samuel, D., Bloc, A., Dusticier, N., & Nieoullon, A. (1992). Intracerebroventricular administration of neuropeptide Y affects parameters of dopamine, glutamate and GABA activities in the rat striatum. *Brain Research Bulletin, 28*(2), 187–193.

Kinoshita, H., Jessop, D. S., Finn, D. P., Coventry, T. L., Roberts, D. J., Ameno, K., et al. (2000). Acute ethanol decreases NPY mRNA but not POMC mRNA in the arcuate nucleus. *Neuroreport, 11*(16), 3517–3519.

Kopp, J., Xu, Z. Q., Zhang, X., Pedrazzini, T., Herzog, H., Kresse, A., et al. (2002). Expression of the neuropeptide Y Y1 receptor in the CNS of rat and of wild-type and Y1 receptor knock-out mice. Focus on immunohistochemical localization. *Neuroscience, 111*(3), 443–532.

Kotagale, N. R., Walke, S., Shelkar, G. P., Kokare, D. M., Umekar, M. J., & Taksande, B. G. (2014). Agmatine attenuates nicotine induced conditioned place preference in mice through modulation of neuropeptide Y system. *Behavioural Brain Research, 262*, 118–124. http://dx.doi.org/10.1016/j.bbr.2014.01.004.

Kuntz-Melcavage, K. L., Brucklacher, R. M., Grigson, P. S., Freeman, W. M., & Vrana, K. E. (2009). Gene expression changes following extinction testing in a heroin behavioral incubation model. *BMC Neuroscience, 10*, 95. http://dx.doi.org/10.1186/1471-2202-10-95.

Lappalainen, J., Kranzler, H. R., Malison, R., Price, L. H., Van Dyck, C., Rosenheck, R. A., et al. (2002). A functional neuropeptide Y Leu7Pro polymorphism associated with alcohol dependence in a large population sample from the United States. *Archives of General Psychiatry, 59*(9), 825–831.

Li, M. D., Kane, J. K., Parker, S. L., McAllen, K., Matta, S. G., & Sharp, B. M. (2000). Nicotine administration enhances NPY expression in the rat hypothalamus. *Brain Research, 867*(1–2), 157–164.

Lindell, S. G., Schwandt, M. L., Sun, H., Sparenborg, J. D., Björk, K., Kasckow, J. W., et al. (2010). Functional NPY variation as a factor in stress resilience and alcohol consumption in rhesus macaques. *Archives of General Psychiatry*, *67*(4), 423–431. http://dx.doi.org/10.1001/archgenpsychiatry.2010.23.

Lovinger, D. M., & Roberto, M. (2013). Synaptic effects induced by alcohol. *Current Topics in Behavioral Neurosciences*, *13*, 31–86. http://dx.doi.org/10.1007/7854_2011_143.

Maric, T., Cantor, A., Cuccioletta, H., Tobin, S., & Shalev, U. (2009). Neuropeptide Y augments cocaine self-administration and cocaine-induced hyperlocomotion in rats. *Peptides*, *30*(4), 721–726. http://dx.doi.org/10.1016/j.peptides.2008.11.006.

Maric, T., Tobin, S., Quinn, T., & Shalev, U. (2008). Food deprivation-like effects of neuropeptide Y on heroin self-administration and reinstatement of heroin seeking in rats. *Behavioural Brain Research*, *194*(1), 39–43. http://dx.doi.org/10.1016/j.bbr.2008.06.023.

McClintick, J. N., McBride, W. J., Bell, R. L., Ding, Z. M., Liu, Y., Xuei, X., et al. (2015). Gene expression changes in serotonin, GABA-A receptors, neuropeptides and ion channels in the dorsal raphe nucleus of adolescent alcohol-preferring (P) rats following binge-like alcohol drinking. *Pharmacology, Biochemistry, and Behavior*, *129*, 87–96. http://dx.doi.org/10.1016/j.pbb.2014.12.007.

McGue, M. (1999). Phenotyping alcoholism. *Alcoholism, Clinical and Experimental Research*, *23*(5), 757–758.

Midgley, L. P., Bush, L. G., Gibb, J. W., & Hanson, G. R. (1992). Characterization of phencyclidine-induced effects on neuropeptide Y systems in the rat caudate-putamen. *Brain Research*, *593*(1), 89–96.

Midgley, L. P., Wagstaff, J. D., Singh, N. A., Bush, L. G., Gibb, J. W., & Hanson, G. R. (1994). Dynamic dopaminergic regulation of neuropeptide Y systems in discrete striatal and accumbens regions. *European Journal of Pharmacology*, *251*(2–3), 191–199.

Misra, K., & Pandey, S. C. (2006). The decreased cyclic-AMP dependent-protein kinase A function in the nucleus accumbens: A role in alcohol drinking but not in anxiety-like behaviors in rats. *Neuropsychopharmacology*, *31*(7), 1406–1419. http://dx.doi.org/10.1038/sj.npp.1300900.

Morin, S. M., & Gehlert, D. R. (2006). Distribution of NPY Y5-like immunoreactivity in the rat brain. *Journal of Molecular Neuroscience*, *29*(2), 109–114. http://dx.doi.org/10.1385/JMN:29:2:109.

Mottagui-Tabar, S., Prince, J. A., Wahlestedt, C., Zhu, G., Goldman, D., & Heilig, M. (2005). A novel single nucleotide polymorphism of the neuropeptide Y (NPY) gene associated with alcohol dependence. *Alcoholism, Clinical and Experimental Research*, *29*(5), 702–707.

Okubo, T., & Harada, S. (2001). Polymorphism of the neuropeptide Y gene: An association study with alcohol withdrawal. *Alcoholism, Clinical and Experimental Research*, *25*(6 Suppl), 59S–62S.

Palmiter, R. D., Erickson, J. C., Hollopeter, G., Baraban, S. C., & Schwartz, M. W. (1998). Life without neuropeptide Y. *Recent Progress in Hormone Research*, *53*, 163–199.

Pandey, S. C., Zhang, H., Roy, A., & Xu, T. (2005). Deficits in amygdaloid cAMP-responsive element-binding protein signaling play a role in genetic predisposition to anxiety and alcoholism. *The Journal of Clinical Investigation*, *115*(10), 2762–2773. http://dx.doi.org/10.1172/JCI24381.

Parker, R. M., & Herzog, H. (1999). Regional distribution of Y-receptor subtype mRNAs in rat brain. *The European Journal of Neuroscience*, *11*(4), 1431–1448.

Pickens, R. W., Svikis, D. S., McGue, M., Lykken, D. T., Heston, L. L., & Clayton, P. J. (1991). Heterogeneity in the inheritance of alcoholism. A study of male and female twins. *Archives of General Psychiatry*, *48*(1), 19–28.

Pleil, K. E., Rinker, J. A., Lowery-Gionta, E. G., Mazzone, C. M., McCall, N. M., Kendra, A. M., et al. (2015). NPY signaling inhibits extended amygdala CRF neurons to suppress binge alcohol drinking. *Nature Neuroscience, 18*(4), 545–552. http://dx.doi.org/10.1038/nn.3972.

Quarta, D., Leslie, C. P., Carletti, R., Valerio, E., & Caberlotto, L. (2011). Central administration of NPY or an NPY-Y5 selective agonist increase in vivo extracellular monoamine levels in mesocorticolimbic projecting areas. *Neuropharmacology, 60*(2–3), 328–335. http://dx.doi.org/10.1016/j.neuropharm.2010.09.016.

Rimondini, R., Thorsell, A., & Heilig, M. (2005). Suppression of ethanol self-administration by the neuropeptide Y (NPY) Y2 receptor antagonist BIIE0246: Evidence for sensitization in rats with a history of dependence. *Neuroscience Letters, 375*(2), 129–133. http://dx.doi.org/10.1016/j.neulet.2004.10.084.

Roberto, M., Gilpin, N. W., & Siggins, G. R. (2012). The central amygdala and alcohol: Role of γ-aminobutyric acid, glutamate, and neuropeptides. *Cold Spring Harbor Perspectives in Medicine, 2*(12), a012195. http://dx.doi.org/10.1101/cshperspect.a012195.

Roy, A., & Pandey, S. C. (2002). The decreased cellular expression of neuropeptide Y protein in rat brain structures during ethanol withdrawal after chronic ethanol exposure. *Alcoholism, Clinical and Experimental Research, 26*(6), 796–803.

Rylkova, D., Boissoneault, J., Isaac, S., Prado, M., Shah, H. P., & Bruijnzeel, A. W. (2008). Effects of NPY and the specific Y1 receptor agonist [D-His(26)]-NPY on the deficit in brain reward function and somatic signs associated with nicotine withdrawal in rats. *Neuropeptides, 42*(3), 215–227. http://dx.doi.org/10.1016/j.npep.2008.03.004.

Sah, R., Ekhator, N. N., Jefferson-Wilson, L., Horn, P. S., & Geracioti, T. D. (2014). Cerebrospinal fluid neuropeptide Y in combat veterans with and without posttraumatic stress disorder. *Psychoneuroendocrinology, 40,* 277–283. http://dx.doi.org/10.1016/j.psyneuen.2013.10.017.

Sah, R., Ekhator, N. N., Strawn, J. R., Sallee, F. R., Baker, D. G., Horn, P. S., et al. (2009). Low cerebrospinal fluid neuropeptide Y concentrations in posttraumatic stress disorder. *Biological Psychiatry, 66*(7), 705–707. http://dx.doi.org/10.1016/j.biopsych.2009.04.037.

Saurer, T. B., Ijames, S. G., & Lysle, D. T. (2006). Neuropeptide Y Y1 receptors mediate morphine-induced reductions of natural killer cell activity. *Journal of Neuroimmunology, 177*(1–2), 18–26. http://dx.doi.org/10.1016/j.jneuroim.2006.05.002.

Sørensen, G., Jensen, M., Weikop, P., Dencker, D., Christiansen, S. H., Loland, C. J., et al. (2012). Neuropeptide Y Y5 receptor antagonism attenuates cocaine-induced effects in mice. *Psychopharmacology, 222*(4), 565–577. http://dx.doi.org/10.1007/s00213-012-2651-y.

Sørensen, G., Wegener, G., Hasselstrøm, J., Hansen, T. V., Wörtwein, G., Fink-Jensen, A., et al. (2009). Neuropeptide Y infusion into the shell region of the rat nucleus accumbens increases extracellular levels of dopamine. *Neuroreport, 20*(11), 1023–1026.

Sørensen, G., & Woldbye, D. P. (2012). Mice lacking neuropeptide Y show increased sensitivity to cocaine. *Synapse, 66*(9), 840–843. http://dx.doi.org/10.1002/syn.21568.

Sørensen, G., Wörtwein, G., Fink-Jensen, A., & Woldbye, D. P. (2013). Neuropeptide Y Y5 receptor antagonism causes faster extinction and attenuates reinstatement in cocaine-induced place preference. *Pharmacology, Biochemistry, and Behavior, 105,* 151–156. http://dx.doi.org/10.1016/j.pbb.2013.02.010.

Sparrow, A. M., Lowery-Gionta, E. G., Pleil, K. E., Li, C., Sprow, G. M., Cox, B. R., et al. (2012). Central neuropeptide Y modulates binge-like ethanol drinking in C57BL/6J mice via Y1 and Y2 receptors. *Neuropsychopharmacology, 37*(6), 1409–1421. http://dx.doi.org/10.1038/npp.2011.327.

Sparta, D. R., Fee, J. R., Hayes, D. M., Knapp, D. J., MacNeil, D. J., & Thiele, T. E. (2004). Peripheral and central administration of a selective neuropeptide Y Y1 receptor antagonist suppresses ethanol intake by C57BL/6J mice. *Alcoholism, Clinical and Experimental Research, 28*(9), 1324–1330.

Stanley, B. G., Chin, A. S., & Leibowitz, S. F. (1985). Feeding and drinking elicited by central injection of neuropeptide Y: Evidence for a hypothalamic site(s) of action. *Brain Research Bulletin*, *14*(6), 521–524.

Suzuki, R., Lumeng, L., McBride, W. J., Li, T. K., & Hwang, B. H. (2004). Reduced neuropeptide Y mRNA expression in the central nucleus of amygdala of alcohol preferring (P) rats: Its potential involvement in alcohol preference and anxiety. *Brain Research*, *1014*(1–2), 251–254. http://dx.doi.org/10.1016/j.brainres.2004.04.037.

Thiele, T. E., Koh, M. T., & Pedrazzini, T. (2002). Voluntary alcohol consumption is controlled via the neuropeptide Y Y1 receptor. *The Journal of Neuroscience*, *22*(3), RC208.

Thiele, T. E., Marsh, D. J., Ste Marie, L., Bernstein, I. L., & Palmiter, R. D. (1998). Ethanol consumption and resistance are inversely related to neuropeptide Y levels. *Nature*, *396*(6709), 366–369. http://dx.doi.org/10.1038/24614.

Thiele, T. E., Miura, G. I., Marsh, D. J., Bernstein, I. L., & Palmiter, R. D. (2000). Neurobiological responses to ethanol in mutant mice lacking neuropeptide Y or the Y5 receptor. *Pharmacology, Biochemistry, and Behavior*, *67*(4), 683–691.

Thiele, T. E., Naveilhan, P., & Ernfors, P. (2004a). Assessment of ethanol consumption and water drinking by NPY Y(2) receptor knockout mice. *Peptides*, *25*(6), 975–983. http://dx.doi.org/10.1016/j.peptides.2004.03.009.

Thiele, T. E., Sparta, D. R., Hayes, D. M., & Fee, J. R. (2004b). A role for neuropeptide Y in neurobiological responses to ethanol and drugs of abuse. *Neuropeptides*, *38*(4), 235–243. http://dx.doi.org/10.1016/j.npep.2004.04.005.

Thiriet, N., Deng, X., Solinas, M., Ladenheim, B., Curtis, W., Goldberg, S. R., et al. (2005). Neuropeptide Y protects against methamphetamine-induced neuronal apoptosis in the mouse striatum. *The Journal of Neuroscience*, *25*(22), 5273–5279. http://dx.doi.org/10.1523/JNEUROSCI.4893-04.2005.

Thorsell, A., Repunte-Canonigo, V., O'Dell, L. E., Chen, S. A., King, A. R., Lekic, D., et al. (2007). Viral vector-induced amygdala NPY overexpression reverses increased alcohol intake caused by repeated deprivations in Wistar rats. *Brain*, *130*(Pt. 5), 1330–1337. http://dx.doi.org/10.1093/brain/awm033.

Thorsell, A., Rimondini, R., & Heilig, M. (2002). Blockade of central neuropeptide Y (NPY) Y2 receptors reduces ethanol self-administration in rats. *Neuroscience Letters*, *332*(1), 1–4.

Thorsell, A., Slawecki, C. J., & Ehlers, C. L. (2005). Effects of neuropeptide Y on appetitive and consummatory behaviors associated with alcohol drinking in wistar rats with a history of ethanol exposure. *Alcoholism, Clinical and Experimental Research*, *29*(4), 584–590.

Wahlestedt, C., Karoum, F., Jaskiw, G., Wyatt, R. J., Larhammar, D., Ekman, R., et al. (1991). Cocaine-induced reduction of brain neuropeptide Y synthesis dependent on medial prefrontal cortex. *Proceedings of the National Academy of Sciences of the United States of America*, *88*(6), 2078–2082.

Waller, M., McGuire, A. C., & Dobson, A. J. (2015). Alcohol use in the military: Associations with health and wellbeing. *Substance Abuse Treatment, Prevention, and Policy*, *10*, 27. http://dx.doi.org/10.1186/s13011-015-0023-4.

Westwood, S. C., & Hanson, G. R. (1999). Effects of stimulants of abuse on extrapyramidal and limbic neuropeptide Y systems. *The Journal of Pharmacology and Experimental Therapeutics*, *288*(3), 1160–1166.

Wetherill, L., Schuckit, M. A., Hesselbrock, V., Xuei, X., Liang, T., Dick, D. M., et al. (2008). Neuropeptide Y receptor genes are associated with alcohol dependence, alcohol withdrawal phenotypes, and cocaine dependence. *Alcoholism, Clinical and Experimental Research*, *32*(12), 2031–2040. http://dx.doi.org/10.1111/j.1530-0277.2008.00790.x.

Woldbye, D. P., Klemp, K., & Madsen, T. M. (1998). Neuropeptide Y attenuates naloxone-precipitated morphine withdrawal via Y5-like receptors. *The Journal of Pharmacology and Experimental Therapeutics*, *284*(2), 633–636.

Wu, C. C., & Shete, S. (2005). Analysis of genes for alcoholism using two-disease-locus models. *BMC Genetics*, *6*(Suppl. 1), S149. http://dx.doi.org/10.1186/1471-2156-6-S1-S149.

Yoshihara, T., Honma, S., Mitome, M., & Honma, K. (1996). Methamphetamine stimulates the release of neuropeptide Y and noradrenaline from the paraventricular nucleus in rats. *Brain Research*, *707*(1), 119–121.

Zhang, H., & Pandey, S. C. (2003). Effects of PKA modulation on the expression of neuropeptide Y in rat amygdaloid structures during ethanol withdrawal. *Peptides*, *24*(9), 1397–1402.

Zhu, G., Pollak, L., Mottagui-Tabar, S., Wahlestedt, C., Taubman, J., Virkkunen, M., et al. (2003). NPY Leu7Pro and alcohol dependence in Finnish and Swedish populations. *Alcoholism, Clinical and Experimental Research*, *27*(1), 19–24. http://dx.doi.org/10.1097/01.ALC.0000050642.62233.44.

Zill, P., Preuss, U. W., Koller, G., Bondy, B., & Soyka, M. (2008). Analysis of single nucleotide polymorphisms and haplotypes in the neuropeptide Y gene: No evidence for association with alcoholism in a German population sample. *Alcoholism, Clinical and Experimental Research*, *32*(3), 430–434. http://dx.doi.org/10.1111/j.1530-0277.2007.00586.x.

CHAPTER EIGHT

Orexin/Hypocretin System: Role in Food and Drug Overconsumption

Jessica R. Barson*, Sarah F. Leibowitz[†,1]
*Drexel University College of Medicine, Philadelphia, PA, United States
[†]The Rockefeller University, New York, NY, United States
[1]Corresponding author: e-mail address: leibow@rockefeller.edu

Contents

Abstract

The neuropeptide orexin/hypocretin (OX), while largely transcribed within the hypothalamus, is released throughout the brain to affect complex behaviors. Primarily through the hypothalamus itself, OX homeostatically regulates adaptive behaviors needed for survival, including food intake, sleep–wake regulation, mating, and maternal behavior. However, through extrahypothalamic limbic brain regions, OX promotes seeking and intake of rewarding substances of abuse, like palatable food, alcohol, nicotine, and cocaine. This neuropeptide, in turn, is stimulated by the intake of or early life exposure to these substances, forming a nonhomeostatic, positive feedback loop. The

International Review of Neurobiology, Volume 136
ISSN 0074-7742
http://dx.doi.org/10.1016/bs.irn.2017.06.006

specific OX receptor involved in these behaviors, whether adaptive behavior or substance seeking and intake, is dependent on the particular brain region that contributes to them. Thus, we propose that, while the primary function of OX is to maintain arousal for the performance of adaptive behaviors, this neuropeptide system is readily co-opted by rewarding substances that involve positive feedback, ultimately promoting their abuse.

1. INTRODUCTION

In 1998, the neuropeptide orexin/hypocretin (OX) was identified independently by two different laboratories, both of which suggested that it played a major role in food intake and energy homeostasis (de Lecea et al., 1998; Sakurai et al., 1998). Although the cell bodies that transcribe this neuropeptide were shown to be largely restricted to the hypothalamus that regulates food intake, these neurons were also found to send extensive projections throughout the brain, most notably to the limbic system (Peyron et al., 1998; Trivedi, Yu, MacNeil, Ploeg, & Guan, 1998). Numerous studies over the past 20 years have demonstrated that hypothalamic OX neurons and their projections have broader functions beyond homeostatic food intake, controlling a range of other adaptive behaviors such as sleep–wake regulation, mating, and maternal behavior. Of particular interest is the additional evidence that OX promotes dysregulated, nonhomeostatic intake, such as occurs with abused, rewarding substances. It stimulates excessive intake of palatable sweet and fatty foods, alcohol, nicotine, and cocaine among other substances, and in many cases, it is in turn stimulated by their intake as part of a positive feedback loop. This is the focus of the present review.

2. ANATOMY OF OREXIN/HYPOCRETIN-EXPRESSING NEURONS

The first description of OX came from de Lecea et al. (1998), who identified a gene which they termed hypocretin, given its nearly exclusive expression in the *hypo*thalamus and its similar amino acid sequence to the gut hormone se*cretin*. These authors noted that this precursor gene actually encodes two peptides, which they termed hypocretin 1 and hypocretin 2 (de Lecea et al., 1998). Soon thereafter, Sakurai et al. (1998) identified the same 130 amino acid neuropeptide gene, which they termed orexin, after the Greek word for appetite, *orexis*, and they noted that its sequence is well conserved between rats, mice, and humans. This group also identified the same encoded peptides, orexin-A (OX-A), equivalent to hypocretin 1 and composed of 33 amino acids, and orexin-B (OX-B), equivalent to hypocretin 2 and composed of 28 amino acids, and they noted that the two peptides were 46% identical in their sequence (Sakurai et al., 1998). A remarkable feature of OX is that, despite its restricted expression in a discrete set of predominantly hypothalamic nuclei, the neurons expressing this neuropeptide project widely to brain regions involved in feeding, arousal, and hormone release (Peyron et al., 1998; Trivedi et al., 1998) to influence an array of behaviors. This section will describe the anatomical locations of the OX neurons, their efferent and afferent projections, and the neurochemical properties of the OX receptors.

2.1 Cell Bodies

Neurons transcribing OX comprise a small cluster of cells that lie predominantly within the hypothalamus (Peyron et al., 1998; Sakurai et al., 1998). The subnuclei in which they have been described differ somewhat across publications, depending on the anatomical terminology used by the particular laboratory. In the rat, about half of the 6700 OX-expressing neurons can be found in the perifornical region of the hypothalamus, with the rest found across the lateral hypothalamus, dorsomedial hypothalamus, and posterior hypothalamus, as well as the subthalamus (de Lecea et al., 1998; Modirrousta, Mainville, & Jones, 2005; Peyron et al., 1998; Sakurai et al., 1998). While this distribution of OX overlaps significantly with that of another orexigenic neuropeptide, melanin-concentrating hormone, these two neuropeptides show virtually no colocalization within neurons (Hakansson, de Lecea, Sutcliffe, Yanagisawa, & Meister, 1999). OX-expressing neurons have been identified in a host of mammalian vertebrates, including rats, mice, and humans (de Lecea et al., 1998; Sakurai et al., 1998), as well as dogs (Lin et al., 1999), sheep (Archer, Findlay, Rhind, Mercer, & Adam, 2002), pigs (Dyer, Touchette, Carroll, Allee, & Matteri, 1999), and nonhuman primates (Horvath, Diano, & van den Pol, 1999). They have also been observed in nonmammalian vertebrates, such as zebrafish, where OX neurons lie in the rostral hypothalamus and preoptic area (Kaslin, Nystedt, Ostergard, Peitsaro, & Panula, 2004).

While not with melanin-concentrating hormone, OX is found to colocalize with a number of other neuropeptides and neurotransmitters.

Most notably, 94% of OX neurons also contain the opioid neuropeptide dynorphin (Chou et al., 2001), and both OX and dynorphin have even been observed within the same synaptic vesicles (Muschamp et al., 2014). There is also significant colocalization of OX with glutamate (Rosin, Weston, Sevigny, Stornetta, & Guyenet, 2003; Torrealba, Yanagisawa, & Saper, 2003) and prolactin (Risold, Griffond, Kilduff, Sutcliffe, & Fellmann, 1999) and a more limited colocalization with galanin (Hakansson et al., 1999), nociceptin/orphanin FQ (Maolood & Meister, 2010), neurotensin (Furutani et al., 2013), and nitric oxide (Cheng, Kuchiiwa, Gao, Kuchiiwa, & Nakagawa, 2003). Thus, the release of OX is likely to cooccur with these other neurochemicals, which may act on different timescales and have different effects from OX.

2.2 Receptors

The two receptors for OX have different binding affinities for the OX peptides. The orexin 1 receptor (OX1R, also called hypocretin receptor 1) primarily interacts with OX-A, requiring for a half-maximum response a concentration of 30 nM of OX-A but 2500 nM of OX-B (Sakurai et al., 1998). This is in contrast to the orexin 2 receptor (OX2R, also called hypocretin receptor 2), which interacts nearly equally with OX-A and OX-B, requiring for a half-maximum response a concentration of 34 nM of OX-A and 60 nM of OX-B (Sakurai et al., 1998). Thus, OX-A binds to both OX1R and OX2R with equal affinity, while OX-B binds almost exclusively to OX2R. Both receptors interact with Gq, Gs, and Gi proteins (Magga et al., 2006; Tang et al., 2008) and have been observed both pre- and postsynaptically (van den Pol, Gao, Obrietan, Kilduff, & Belousov, 1998).

2.3 Efferent Projections

While the OX peptides are released widely through the brain, their projections are most highly concentrated in different limbic regions involved in arousal and emotion. Thus, OX axons are particularly dense within the hypothalamus from where they follow both ascending and descending pathways (Peyron et al., 1998). The heaviest OX projection is to the locus coeruleus, but axons are also very dense in multiple regions in the forebrain (bed nucleus of the stria terminalis, paraventricular and central medial thalamus, and subthalamus) and hindbrain (substantia nigra, nucleus of the solitary tract, periaqueductal gray, dorsal raphe and raphe magus, and reticular nuclei) (Peyron et al., 1998). The OX receptors are also highly expressed in each of these regions, where in many cases only one or the other receptor is present. For example, the locus coeruleus and bed nucleus have heavy gene expression of the OX1R but no detectable OX2R, similar to the hypothalamic ventromedial nucleus which contains OX1R but not OX2R; in contrast, the central medial thalamus contains OX2R but not OX1R, and the hypothalamic paraventricular nucleus has dense expression of OX2R but not OX1R (Trivedi et al., 1998). Nuclei containing both receptors at similar levels include the paraventricular thalamus, dorsal raphe, and ventral tegmental area (Marcus et al., 2001; Trivedi et al., 1998). It is notable that brain regions not related to arousal or emotion contain fewer, if any, OX fibers and receptors, including the caudate putamen, pretectal area, and oculomotor nucleus (Peyron et al., 1998; Trivedi et al., 1998).

2.4 Afferent Inputs

In line with their projections to limbic brain regions involved in arousal and emotion, the OX cell bodies also receive afferent input from brain regions and neurochemicals involved in these behavioral functions, with the connections often being reciprocal. The heaviest input to OX neurons comes from the lateral septum and medial preoptic area, as well as the bed nucleus of the stria terminalis and posterior hypothalamus to which these neurons themselves project (Yoshida, McCormack, Espana, Crocker, & Scammell, 2006). In addition, the neurons receive input from select nuclei of the forebrain (infralimbic cortex, cingulate cortex, nucleus accumbens, and amygdala), hindbrain (locus coeruleus, ventral tegmental area, substantia nigra, periaqueductal gray, and dorsal raphe), and hypothalamus (suprachiasmatic nucleus, anterior hypothalamus, ventromedial hypothalamus, and supramammillary nucleus), but not from regions such as the caudate putamen, somatosensory cortex, or piriform cortex that are unrelated to arousal and emotion (Yoshida et al., 2006). Consistent with these diverse afferents, the OX neurons are found to express different types of receptors, in addition to their own, the OX1R (Backberg, Hervieu, Wilson, & Meister, 2002) and OX2R (Yamanaka, Tabuchi, Tsunematsu, Fukazawa, & Tominaga, 2010). These other receptors include those for serotonin (Collin, Backberg, Onnestam, & Meister, 2002), acetylcholine (Chou, Rotman, & Saper, 2004; Garcia et al., 2015), adenosine (Thakkar, Winston, & McCarley, 2002), GABA (Backberg, Collin, Ovesjo, & Meister, 2003; Backberg, Ultenius, Fritschy, & Meister, 2004), and leptin (Hakansson et al., 1999), although rarely dopamine (Bubser et al., 2005), and also those for the neuropeptides, cholecystokinin (Tsujino et al., 2005), corticotropin-releasing factor (Winsky-Sommerer et al., 2004), neurotensin (Furutani et al., 2013), neuropeptide Y (Fu, Acuna-Goycolea, & van den Pol, 2004), and galanin-like peptide (Kageyama et al., 2006), as well as the mu opioid receptor (Zhou et al., 2006). Together, these connections position OX at the center of a network regulating motivation.

3. ROLE OF OREXIN/HYPOCRETIN IN HOMEOSTATIC FEEDING AND OTHER ADAPTIVE BEHAVIORS

While OX was originally noted for its role in food intake and energy homeostasis, it soon became apparent that this neuropeptide participates in a host of other behaviors, many of which serve to promote survival. These include not only homeostatic food intake but also sleep–wake regulation, mating behavior, and maternal behavior.

3.1 Homeostatic Food Intake

Studies provide clear evidence that OX neurons act to promote food-seeking behavior, as normally occurs during a prolonged fast, restricted feeding, and hypoglycemia, with the activation of these neurons being reduced once food is consumed. Gene expression and brain peptide levels of OX have been reported to be elevated after a 48 or 72 h fast (Fujiki et al., 2001; Moriguchi, Sakurai, Nambu, Yanagisawa, & Goto, 1999; Sakurai et al., 1998). Similarly, under a restricted feeding paradigm, OX neurons are more active immediately prior to scheduled food availability (Johnstone, Fong, & Leng, 2006; Mieda et al., 2004), while mice lacking the OX gene show reduced waking and locomotor activity during the food-anticipatory period (Akiyama et al., 2004; Mieda et al., 2004). Gene expression and peptide levels as well as the activity of OX neurons are also increased by a state of hypoglycemia (Bayer et al., 2000; Griffond, Risold, Jacquemard, Colard, & Fellmann, 1999), and are corrected when glucose levels are restored by food intake (Cai et al., 2001). Although hyperglycemia does not appear to affect mRNA levels of OX (Griffond et al., 1999), glucose at physiological levels reduces the firing of OX neurons in a concentration-dependent manner (Burdakov, Gerasimenko, & Verkhratsky, 2005). This indicates that under physiological conditions OX neurons are activated and transcribe the neuropeptide when glucose levels are low, such as from fasting or restricted feeding, and they are inhibited when glucose levels are high.

The role of OX in stimulating intake of standard food appears to involve the OX1R primarily in the hypothalamic areas where the OX neurons exist, with OX likely acting on these neurons as well as on other local neuronal populations. Most notably, the intake of laboratory chow is increased by pharmacogenetic activation of OX neurons and reduced by extensive ablation of these neurons or peripheral injection of an OX1R antagonist (Haynes et al., 2000; Inutsuka et al., 2014). Through microinjection experiments, these effects of OX in promoting food intake are found to be largely due to the action of OX-A in the same hypothalamic subregions where OX itself is transcribed, namely, the perifornical area, lateral hypothalamus, and dorsomedial hypothalamus, with OX-B having little effect on feeding (Dube, Kalra, & Kalra, 1999; Sweet, Levine, Billington, & Kotz, 1999). These actions of OX-A are thus likely to be mediated by OX1R, which are located both on OX neurons themselves and on neurons expressing the feeding-promoting neuropeptide, melanin-concentrating hormone (Backberg et al., 2002). In contrast, there is little effect on food intake from OX injected in other brain areas, including the ventromedial hypothalamus, preoptic area, amygdala, ventral tegmental area, or nucleus of the solitary tract (Dube et al., 1999; Sweet et al., 1999). Thus, when promoting homeostatic food intake, OX appears to act predominantly on OX1R in the dorsal and lateral regions of hypothalamus, where it may be responding to physiological signals such as a drop in glucose levels that occurs when animals most need to seek out and consume food.

3.2 Sleep–Wake Regulation

Feeding and waking are closely associated with one another. During food deprivation, animals show shorter episodes of sleep, which decrease in frequency as the length of deprivation increases (Borbely, 1977; Dewasmes, Duchamp, & Minaire, 1989). They also exhibit enhanced locomotor activity (Jones, Bellingham, & Ward, 1990; Mabry & Campbell, 1975; Moskowitz, 1959), possibly reflecting an increase in foraging behavior. Similar to food seeking, OX is robustly tied to waking behavior in the sleep–wake cycle, with this behavior caused in part by its release in the locus coeruleus. This connection has been particularly noted in studies of narcolepsy, which in canines is caused by a disruption of the OX2R (Lin et al., 1999) and in humans is associated with nearly undetectable numbers of OX neurons in the brain and levels of OX in cerebrospinal fluid (Nishino, Ripley, Overeem, Lammers, & Mignot, 2000; Thannickal et al., 2000).

Moreover, narcolepsy is ameliorated by intranasal administration of OX-A (Baier et al., 2011). In line with these findings, manipulations of endogenous OX gene expression affect waking and the sleep cycle. Mice lacking the OX gene show behavioral state instability, having frequent and rapid transitions between wake and sleep periods (Chemelli et al., 1999; Diniz Behn, Klerman, Mochizuki, Lin, & Scammell, 2010; Mochizuki et al., 2004), and zebrafish overexpressing OX show increased arousal and reduced initiation and duration of rest periods (Prober, Rihel, Onah, Sung, & Schier, 2006). Under normal conditions, the firing of OX neurons is closely tied to the state of arousal, occurring most frequently during active waking and less frequently during quiet waking, and it is absent during slow wave sleep, is transient during rapid eye movement sleep, and increases again just prior to waking (Lee, Hassani, & Jones, 2005; Takahashi, Lin, & Sakai, 2008). Levels of OX peptide in the hypothalamus and thalamus have also been found to be higher during waking compared to sleep (Kiyashchenko et al., 2002; Yoshida et al., 2001). Perhaps the most compelling evidence that OX firing promotes waking is the finding that optogenetic stimulation of OX neurons increases the probability of a transition from sleep to wakefulness (Adamantidis, Zhang, Aravanis, Deisseroth, & de Lecea, 2007), and this appears to be largely due to changes in the activity of OX1R in the locus coeruleus (Carter et al., 2012; Hasegawa, Yanagisawa, Sakurai, & Mieda, 2014) where OX afferents are most dense (Peyron et al., 1998). Collectively, these studies confirm that OX, through both OX1R and OX2R, stimulates an aroused state, which is conducive to seeking for food.

3.3 Mating Behavior

Orexin/hypocretin has also been tied to mating behavior, with published studies examining male rats and suggesting that this occurs largely through OX1R in the hypothalamic medial preoptic area. Neurons expressing OX are activated during copulatory behavior (Muschamp, Dominguez, Sato, Shen, & Hull, 2007), and injection of OX-A into the medial preoptic area increases sexual arousal and performance (Gulia, Mallick, & Kumar, 2003) while systemic blockade of the OX1R impairs copulatory behavior (Muschamp et al., 2007). These effects may also be due, in part, to an effect on the rewarding or the motivational aspects of this behavior. This is suggested by evidence that lesions of OX neurons prevent the formation of conditioned place preference for a chamber previously paired with sexual behavior (Di Sebastiano, Wilson-Perez, Lehman, & Coolen, 2011) and that OX fibers are apposed to tyrosine hydroxylase-positive neurons in the ventral tegmental area that also show increased activation during copulation (Muschamp et al., 2007). Thus, mating is another behavior, related to arousal and survival, which is enhanced by the activity of OX, with the behavior itself affected by OX acting on OX1R in the hypothalamic medial preoptic area and its reward predominantly involving OX in the limbic ventral tegmental area.

3.4 Maternal Behavior

Studies also indicate that OX enhances maternal behavior, occurring as with mating behavior through OX1R in the hypothalamic medial preoptic area. Gene expression of OX is increased in pregnant and lactating female rats compared to nonpregnant or nonlactating rats (Kanenishi et al., 2004; Sun, Narita, Murata, Honda, & Higuchi, 2003), and the number of OX-expressing neurons is also increased during lactation (Sun et al., 2003). In accord with this, injection of OX-A into the cerebral ventricles or medial preoptic area specifically increases licking and grooming of pups but not of the self, while peripheral or medial preoptic area injections of an OX1R antagonist reduces these behaviors (D'Anna & Gammie, 2006; Rivas, Torterolo, Ferreira, & Benedetto, 2016). It is possible that the increase in OX during the perinatal period also contributes to other behaviors observed in mothers during this time, including disturbed sleep (Bei, Coo, & Trinder, 2015) and increased caloric intake (Whichelow, 1975), which are ultimately tied to care of the offspring.

4. ROLE OF OREXIN/HYPOCRETIN IN NONHOMEOSTATIC INTAKE

Whereas adaptive behaviors are generally performed in a homeostatic manner that involves negative feedback, the intake of rewarding substances of abuse is often nonhomeostatic, involving a positive feedback loop that promotes overconsumption. While the mechanisms behind these adaptive and overconsumption behaviors may seem distinct, the behaviors are known to share in common their functioning through dopamine in the nucleus accumbens shell. Levels and release of accumbal dopamine are increased in response to adaptive behaviors, including food intake (Martel & Fantino, 1996), the waking state (Lena et al., 2005), mating behavior (Pfaus et al., 1990), and maternal behavior (Champagne et al., 2004), and also prior to these behaviors (Champagne et al., 2004), with dopamine seemingly contributing to their initiation (Pisanu et al., 2015). Similarly, accumbal dopamine levels are increased by the intake of rewarding substances, including palatable foods high in fat or sugar (Rada, Avena, Barson, Hoebel, & Leibowitz, 2012; Rada, Avena, & Hoebel, 2005; Sahr et al., 2008) and drugs of abuse such as alcohol (Howard, Schier, Wetzel, Duvauchelle, & Gonzales, 2008), nicotine (Cadoni & Di Chiara, 2000; Pontieri, Tanda, Orzi, & Di Chiara, 1996), and cocaine (Pontieri, Tanda, & Di Chiara, 1995), and they are also elevated prior to the initiation of their intake (Cacciapaglia, Saddoris, Wightman, & Carelli, 2012; Doyon et al., 2003; Suto, Ecke, You, & Wise, 2010). Notably, the increase in accumbal dopamine produced by the intake of these rewarding substances is significantly greater than that induced by the performance of adaptive behaviors (Martel & Fantino, 1996), suggesting that the system governing these behaviors differs only in degree. While OX is similar to dopamine in having an important function in promoting both the adaptive behaviors and nonhomeostatic consumption of the rewarding substances, it is notable that OX acts through the hypothalamus to exert its effects on the homeostatic behaviors but acts through the extrahypothalamic limbic regions to exert its effects on the nonhomeostatic behaviors. Thus, OX increases the intake, willingness to work, seeking behavior, and the reward derived from the intake of rewarding substances, by acting through the nucleus accumbens, ventral tegmental area, and paraventricular nucleus of the thalamus, where local application of OX leads to increased extracellular levels of accumbal dopamine (Choi et al., 2012; Korotkova, Sergeeva, Eriksson, Haas, & Brown, 2003; Narita et al., 2006; Patyal, Woo, & Borgland, 2012). By working primarily outside of the hypothalamus to affect the use of rewarding substances, the role of OX in nonhomeostatic behaviors may be fundamentally different from that of homeostatic behaviors geared toward survival, promoting excess consumption that can lead to addiction.

4.1 Role of Orexin/Hypocretin in Overeating

In stimulating the intake of standard food to maintain physiological glucose levels, OX acts in a homeostatic manner. This differs greatly from the role of OX in the consumption of palatable foods high in fat or sugar, as OX is found not only to promote their intake but also to be significantly enhanced by their intake. This interaction between OX and palatable food reflects a positive feedback loop that, with a stronger OX signal occurring after each intake episode, may contribute to the overconsumption of fat and sugar.

4.1.1 Effects of Orexin/Hypocretin on Aspects of Palatable Food Intake

Similar to its role in controlling intake of standard food, endogenous OX appears to promote the consumption of palatable foods primarily through its actions on OX1R. Mice lacking the OX gene, compared to wild-type mice, consume less sucrose when available ad libitum, even after accounting for differences in their locomotor activity (Matsuo et al., 2011). A role for the OX1R in mediating intake of palatable food is supported by the findings that systemic injection of the OX1R antagonist, SB-334867, reduces operant responding for both sucrose and saccharin pellets, in a minimal work paradigm with a fixed ratio 1 schedule (Cason & Aston-Jones, 2013a, 2013b, 2014). This antagonist also suppresses binge-like intake of sucrose, saccharin, and fructose solutions under an intermittent-access home cage drinking paradigm (Alcaraz-Iborra, Carvajal, Lerma-Cabrera, Valor, & Cubero, 2014; Rorabaugh, Stratford, & Zahniser, 2014) and reduces intake of a high-fat diet under both ad libitum and limited-access feeding paradigms (Choi, Davis, Fitzgerald, & Benoit, 2010; Valdivia, Patrone, Reynaldo, & Perello, 2014; White et al., 2005). Other OX1R antagonists similarly reduce intake of sweet-fat foods whether under ad libitum or intermittent-access feeding paradigms (Piccoli et al., 2012; Steiner, Sciarretta, Pasquali, & Jenck, 2013). The one study of an OX2R antagonist, showing no effect of systemic injection of JNJ-10397049 on the consumption of Nutella® (Piccoli et al., 2012), provides support for the specific role of the OX1R but not OX2R in palatable food intake.

In contrast to the primary role of the hypothalamus in mediating OX's effect on intake of standard food, multiple extrahypothalamic limbic brain areas, including the nucleus accumbens shell, ventral tegmental area, and paraventricular nucleus of the thalamus, are suggested by central injection studies to be involved in OX's stimulatory effect on intake of palatable food. Thus, OX-A increases the intake of M&Ms® when injected into the nucleus accumbens shell (Castro, Terry, & Berridge, 2016), of a sucrose solution and high-fat diet after injection into the ventral tegmental area (Terrill et al., 2016), and of a sucrose solution after injection into the paraventricular thalamus (Barson, Ho, & Leibowitz, 2015). Conversely, the OX1R antagonist SB-334867 reduces drinking of sucrose when injected into the ventral tegmental area (Terrill et al., 2016), and knockdown of the OX1R in the paraventricular thalamus suppresses the consumption of a high-fat diet (Choi et al., 2012). A possible additional role for hypothalamic and hindbrain nuclei in the effects of OX is suggested by findings that high-fat diet consumption is stimulated by injection of OX-A into the third ventricle in the area of the hypothalamus (Clegg, Air, Woods, & Seeley, 2002), and also the fourth ventricle (Kay, Parise, Lilly, & Williams, 2014) and dorsal vagal complex (Zheng, Patterson, & Berthoud, 2005). Thus, this evidence with palatable food suggests that homeostatic intake is controlled by OX acting in the hypothalamus, while nonhomeostatic intake is controlled by OX in extrahypothalamic limbic brain areas.

In addition to intake of palatable food, OX is also found to increase an individual's willingness to work for this food, as measured using a progressive ratio task that successively increases the number of operant responses required to earn a reinforcer, with the breakpoint being the requirement at which responding ceases. The involvement of endogenous OX in the performance of this task is suggested by the findings that ventricular injection of OX-A increases the breakpoint for sucrose (Choi et al., 2010; Kay et al., 2014), and that peripheral injection of the OX1R antagonist SB-334867 reduces the breakpoint for sucrose (Cason & Aston-Jones, 2013b; Espana et al., 2010), high-fat diet (Borgland et al., 2009), and sweet-fat pellets (Choi et al., 2010). The specific brain site involved in this response has yet to be identified, with OX-A in the ventral tegmental area failing to affect the breakpoint for sucrose (Terrill et al., 2016) and knockdown of the OX1R in the paraventricular thalamus failing to affect breakpoint for high-fat diet (Choi et al., 2012). Given its strong ties with motivation (Hamid et al., 2016), the nucleus accumbens is likely to be the brain region that mediates the effect of OX on the willingness to work for palatable food.

The possibility that OX also affects the seeking of palatable food, measured by reinstatement following extinction, is suggested by some but not all reports. The OX1R antagonist SB-334867 at 30 mg/kg has been shown to inhibit cue-induced reinstatement of responding for sucrose or saccharin in male rats (Cason & Aston-Jones, 2013a, 2013b). This effect of the antagonist, however, is not evident in female rats (Cason & Aston-Jones, 2014), nor is it seen at a lower dose (10 mg/kg) in male rats with cue-induced reinstatement for sweetened condensed milk (Martin-Fardon & Weiss, 2014a). Although results clearly show that OX acts through the OX1R to promote the intake of a range of palatable foods and the willingness to work for these foods, the possibility that it affects the seeking for these foods requires further investigation.

4.1.2 Effects of Palatable Food Intake on Orexin/Hypocretin

In addition to inducing palatable food intake, endogenous OX itself is found to be highly responsive to the consumption of palatable food, with both its gene expression and peptide levels affected in a time-dependent manner. While the expression of OX is initially reduced during the first 30 min after consumption of sucrose or noncaloric saccharin compared to water (Alcaraz-Iborra et al., 2014), similar to the homeostatic inhibition of OX neuron firing induced by a physiological rise in glucose (Burdakov et al., 2005), further measurements indicate that OX peptide levels are subsequently increased by intake of palatable food. They are elevated 1 h after the drinking of saccharin compared to water, to a level higher than after standard laboratory chow (Furudono, Ando, Yamamoto, Kobashi, & Yamamoto, 2006), and also 2–4 h after the eating of sucrose pellets compared to chow (Olszewski et al., 2009). Similarly, endogenous OX is stimulated by the consumption of palatable diets high in fat content. Gene expression of OX is elevated at 2 h after a high-fat compared to low-fat meal (Gaysinskaya, Karatayev, Chang, & Leibowitz, 2007), and OX mRNA and the number of OX-A-immunoreactive neurons are higher after 3 weeks of ad libitum high-fat diet consumption (Park et al., 2004; Wortley, Chang, Davydova, & Leibowitz, 2003). It is notable that this increase in OX mRNA and peptide levels seen during the first 3 weeks of high-fat diet access is no longer evident after 8 or 12 weeks of access (Cai et al., 1999; Nobunaga et al., 2014; Xu, Wan, Tang, Wu, & Cai, 2008), similar to the temporal pattern of food intake on a high-fat diet, with hyperphagia occurring during the first few weeks of access but not later (Honors, Hargrave, & Kinzig, 2012; Vasselli & Maggio, 1990; Wang et al., 1998). These findings with foods rich in sugar and fat support the idea that OX functions in a positive feedback loop with palatable food, with OX acting at the OX1R to both promote and ultimately be increased by their intake, resulting in their overconsumption.

4.1.3 Effects of Prenatal Exposure to Palatable Food on Orexin/Hypocretin

Similar to adult intake, exposure to palatable food during the prenatal period can affect endogenous OX expression and levels. Rats exposed in utero to a high-fat diet compared to a balanced diet show in adolescence and adulthood higher gene expression and peptide levels of OX (Chang, Gaysinskaya, Karatayev, & Leibowitz, 2008; Poon, Barson, Fagan, & Leibowitz, 2012). This increase in endogenous OX in prenatally exposed animals is accompanied by an increase in the intake of calories of standard chow and also by a higher preference for the high-fat diet (Chang et al., 2008). With endogenous OX in adults shown to promote intake of a palatable fat-rich diet, this finding suggests that increased release of OX contributes to the excessive consummatory behaviors observed in animals prenatally exposed to this palatable food.

4.1.4 Clinical Evidence of a Role of Orexin/Hypocretin in Palatable Food Intake

Clinical studies also connect OX with disturbances in palatable food intake, although the relationship may not be as direct as that observed in animals. The evidence demonstrates that individuals with narcolepsy who lack OX show significantly increased palatable food intake. They consume almost four times more calories than healthy matched controls when given ad libitum access to snacks (van Holst et al., 2016), and they score significantly higher than controls on a binge eating scale (Dimitrova et al., 2011). While these results run counter to those observed in animals, which are examined primarily after short-term and reversible inhibition of the OX1R, it is possible that enhanced palatable food intake in narcoleptic patients reflects a behavioral pattern that overcompensates for the near total and chronic loss of OX function.

4.2 Role of Orexin/Hypocretin in Alcohol Drinking

The positive feedback relationship between OX and palatable food is also evident with alcohol. Since alcohol is a drug of abuse as well as a food that contains calories (7 kcal/g in ethanol, compared to 9 kcal/g in fat and 4 kcal/g in carbohydrates and protein), this substance has effects on the brain that resemble those produced by both drugs and food. As with palatable food, OX is found to promote the intake of alcohol in animals engaging in pharmacologically relevant alcohol drinking that achieves blood alcohol levels sufficient to alter their behavior, and it does this primarily through extrahypothalamic limbic areas where it also acts to enhance levels of dopamine in the nucleus accumbens. Similarly, OX levels are enhanced by alcohol intake, in a dose- and time-dependent manner, to promote further intake. Thus, OX and alcohol as with palatable food appear to interact in a positive feedback loop to contribute to substance overconsumption.

4.2.1 Effects of Orexin/Hypocretin on Aspects of Alcohol Drinking

With models of access that induce pharmacologically relevant drinking in the home cage, OX through both OX1R and OX2R has been found to promote alcohol intake, acting through a range of extrahypothalamic limbic nuclei. These high-level drinking models involve intermittent access to alcohol in the home cage and lead many subjects to binge on alcohol. The models include either the two-bottle choice procedure that provides 24-h access to water plus 10%–20% alcohol for 3 days per week (Carnicella, Ron, & Barak, 2014) or the "drinking in the dark" model that replaces the water bottle with a bottle of 20% alcohol for 2–4 h (Thiele, Crabbe, & Boehm, 2014). With these intermittent access paradigms, peripheral administration of the OX1R antagonists SB-334867 and GSK1059865 is found to reduce alcohol intake and preference in both rats and mice (Lopez, Moorman, Aston-Jones, & Becker, 2016; Moorman & Aston-Jones, 2009; Olney, Navarro, & Thiele, 2015). Similarly, SB-334867 inhibits the lower-level drinking obtained from ad libitum home cage access (Anderson, Becker, Adams, Jesudason, & Rorick-Kehn, 2014). In contrast, systemic administration of the OX2R antagonist LSN424100 reduces the high-level alcohol intake from an intermittent access paradigm but has no effect on lower-level drinking (Anderson et al., 2014), indicating that the OX2R may be recruited along with OX1R only when animals drink alcohol at sufficiently high levels. Studies with local injections provide further evidence that both OX1R and OX2R affect high-level drinking, with OX1R acting in a wider range of brain regions. The OX1R antagonists reduce drinking in the dark when injected into the nucleus accumbens shell, central nucleus of the amygdala, or ventral tegmental area (Lei et al., 2016; Olney, Navarro, & Thiele, 2017), and intermittent-access alcohol drinking when injected into the medial prefrontal cortex, but not into the anterior insular cortex and substantia nigra (Lei et al., 2016; Srinivasan et al., 2012), or the paraventricular thalamus (Barson et al., 2015). This is in contrast to the OX2R antagonist TCS OX2 29, which reduces two-bottle choice alcohol drinking when injected into the paraventricular thalamus where OX-B increases it to a greater extent than an equimolar dose of OX-A (Barson et al., 2015, 2017), but which has no effect on drinking in the dark when injected into the nucleus accumbens shell, central nucleus of the amygdala, or ventral tegmental area (Lei et al., 2016; Olney et al., 2017). This evidence supports the idea that alcohol drinking is stimulated by OX1R in multiple limbic brain regions, while OX2R is most active in the paraventricular thalamus where it specifically promotes high-level drinking. In the hypothalamus, there is one study of low-level, ad libitum drinking in the home cage, which with OX-A injection shows that this peptide may also promote alcohol intake through this structure (Schneider, Rada, Darby, Leibowitz, & Hoebel, 2007). It is possible that OX within the hypothalamus promotes the intake of alcohol for its caloric content, in contrast to the extrahypothalamic limbic regions where it promotes alcohol intake for its pharmacological effects.

Using a fixed ratio 3 schedule of reinforcement in an operant chamber, alcohol intake has also been found to be reduced by OX antagonists targeting either receptor. Thus, in various strains of rats, lever pressing for alcohol is inhibited by peripheral injection of SB-334867 (Lawrence, Cowen, Yang, Chen, & Oldfield, 2006; Moorman, James, Kilroy, & Aston-Jones, 2017; Richards et al., 2008), the dual OX antagonist almorexant (Srinivasan et al., 2012), and the OX2R antagonist JNJ-10397049 (Shoblock et al., 2011). As with home cage drinking, however, these receptor antagonists appear to exert their effects through different brain regions. This is suggested by the finding that operant self-administration of alcohol is reduced by injection of an OX1R antagonist into the shell of the nucleus accumbens (Lei et al., 2016) but by an OX2R antagonist in the core of the accumbens (Brown, Khoo, & Lawrence, 2013).

As with palatable food, OX increases not only the intake of alcohol but also an individual's willingness to work for alcohol, as measured by breakpoint in a progressive ratio task. Studies have implicated both receptor subtypes in this effect and suggest that this may be sex dependent. Specifically, the OX1R antagonist SB-334867 reduces breakpoint for alcohol in male but not female rats (Anderson et al., 2014; Jupp, Krivdic, Krstew, & Lawrence, 2011), whereas the OX2R antagonist LSN424100 reduces breakpoint in female rats (Anderson et al., 2014). Further studies are needed to determine if this reflects a sex difference in the recruitment of different brain regions during alcohol intake and the progressive ratio task.

Seeking for alcohol is also promoted by OX under several reinstatement paradigms and may be mediated by the OX1R. Cue-induced reinstatement of alcohol seeking is reduced or abolished by SB-334867 when injected systemically (Lawrence et al., 2006; Martin-Fardon & Weiss, 2014b; Moorman et al., 2017) or locally into the ventral tegmental area or medial prefrontal cortex (Brown et al., 2013). This contrasts with intracerebroventricular injection of the OX2R antagonist TCS OX2 29 which has no effect on this behavior (Brown et al., 2013). The possibility that OX participates in alcohol seeking under multiple relapse conditions is supported by the finding that systemic injection of the OX1R antagonist also inhibits yohimbine stress-induced relapse of alcohol seeking (Richards et al., 2008).

Despite the seemingly greater role of the OX1R in most aspects of alcohol drinking, the OX2R appears to be the primary receptor involved in reward from alcohol intake. In a conditioned place preference paradigm, where mice are conditioned to associate a distinct chamber with experimenter-administered alcohol, the OX2R antagonist JNJ-10397049 attenuates the acquisition, expression, and reinstatement of alcohol-conditioned place preference (Shoblock et al., 2011). This is in contrast to the OX1R antagonist SB-334867 which, with the same dose of experimenter-administered alcohol, has no effect on the acquisition of alcohol-conditioned place preference and blocks expression only of a weak place preference, but not a moderate or strong one (Voorhees & Cunningham, 2011). This suggests that reward following alcohol intake is primarily mediated by the OX2R.

4.2.2 Effects of Alcohol Drinking on Orexin/Hypocretin

In relation to consumption of alcohol, OX neuron activity, gene expression, and peptide levels vary in a time-dependent manner much as they do with the intake of palatable food. The activity of OX-expressing neurons is increased prior to voluntary alcohol drinking, consistent with the results of injection studies suggesting that this neuropeptide initiates and promotes alcohol intake. Specifically, levels of c-Fos are increased in OX neurons under conditions that induce relapse of alcohol seeking, including the presentation of an alcohol-paired cue (Dayas, McGranahan, Martin-Fardon, & Weiss, 2008), temporary inactivation of the nucleus accumbens (Millan, Furlong, & McNally, 2010), and placement into an alcohol-paired context (Moorman, James, Kilroy, & Aston-Jones, 2016). In response to alcohol intake, there is a change in OX that varies with dose and across time. While OX activity is inhibited by acute alcohol at a very high dose that promotes sleep (Sharma, Sahota, & Thakkar, 2014), similar to its inhibition after a physiological rise in glucose levels (Burdakov et al., 2005), gene expression and peptide levels of OX are significantly increased by alcohol administered at lower doses (Morganstern et al., 2010). Similarly, in response to voluntary intake, OX gene expression is stimulated during the first 30 min following the start of intake (Barson et al., 2015; Lawrence et al., 2006; Sterling, Karatayev, Chang, Algava, & Leibowitz, 2015) but is subsequently unaffected or reduced along with OX peptide levels over the next several hours

(Morganstern et al., 2010; Olney et al., 2015, 2017). This time-dependent shift in the effects of alcohol on OX, a stimulation followed by a suppression, may help to explain why individuals engage in alcohol binge sessions of a discrete duration (Carnicella et al., 2014), with drinking continuing until blood alcohol levels are sufficiently high to alter behavior and then ceasing until blood levels have significantly declined.

4.2.3 Effects of Prenatal Exposure to Alcohol on Orexin/Hypocretin

With OX promoting pharmacologically relevant alcohol drinking, it is notable that it is endogenously increased in animals prenatally exposed to alcohol. Embryonic exposure to low-to-moderate levels of alcohol leads to elevated alcohol drinking in adolescent and adult rodents and even zebrafish (Fabio, Macchione, Nizhnikov, & Pautassi, 2015; Nizhnikov, Popoola, & Cameron, 2016; Sterling, Chang, Karatayev, Chang, & Leibowitz, 2016). Notably, this exposure to alcohol also stimulates the proliferation of OX neurons, subsequently leading to their increase in gene expression and density (Chang, Karatayev, Liang, Barson, & Leibowitz, 2012; Sterling et al., 2016). This suggests that the greater number of neuropeptide neurons may contribute to elevated levels of alcohol intake.

4.2.4 Clinical Evidence of a Role for Orexin/Hypocretin in Alcohol Drinking

Individuals with OX-deficient narcolepsy show significant differences in their alcohol drinking. They have been found to be more likely to engage in frequent bouts of alcohol drinking compared to controls (Barateau et al., 2016), as they have with palatable food, possibly reflecting a compensatory mechanism for the chronic and significant loss of OX. However, this same study found these subjects to be less likely to engage in heavy alcohol drinking (Barateau et al., 2016), which is in line with a role for OX in promoting pharmacologically relevant alcohol intake. Thus, in relation to alcohol, OX may be more involved in the excessive drinking that involves extrahypothalamic limbic brain regions and feeds back to alter behavior, rather than the type of intake that involves the hypothalamus and is mainly for calories.

4.3 Role of Orexin/Hypocretin in Nicotine Intake

Literature exists that also connects OX with the use of nicotine in animals and cigarettes in humans. As with other rewarding substances, administration of OX promotes the intake of nicotine, with the specific brain regions involved yet to be determined, and endogenous OX levels are in turn enhanced by nicotine intake. This stimulatory effect of nicotine on OX has been observed at all time points examined, suggesting more clearly than with palatable food and alcohol that OX interacts with nicotine in a positive feedback loop to enhance substance overconsumption.

4.3.1 Effects of Orexin/Hypocretin on Aspects of Nicotine Intake

Evidence obtained primarily with systemic injection of OX receptor antagonists demonstrates that OX promotes not only the intake of nicotine but also the willingness to work for nicotine and the reinstatement of seeking for nicotine. Under a fixed ratio 5 schedule of reinforcement in an operant chamber, nicotine intake is reduced by the OX1R antagonist SB-334867 when injected systemically (Hollander, Lu, Cameron, Kamenecka, & Kenny, 2008; LeSage, Perry, Kotz, Shelley, & Corrigall, 2010) or into the limbic insular cortex but not the adjacent somatosensory cortex (Hollander et al., 2008). These reports suggest a function for endogenous activity at the OX1R in the insular cortex in promoting the intake of nicotine, with further research needed to determine the involvement of other limbic brain regions and the OX2R in this behavior. Analysis of the willingness to work for nicotine indicates a role for OX1R but not OX2R, with lever pressing for nicotine in a progressive ratio task found to be reduced by SB-334867 (Hollander et al., 2008), but not the OX2R antagonist 2-SORA (Uslaner et al., 2014). On the other hand, both receptors may participate in the seeking for nicotine under some reinstatement models, with intracerebroventricular injection of OX-A sufficient to reinstate previously extinguished nicotine seeking (Plaza-Zabala, Martin-Garcia, de Lecea, Maldonado, & Berrendero, 2010) and with SB-334867 and 2-SORA independently attenuating or blocking cue-induced reinstatement of nicotine seeking (Plaza-Zabala et al., 2013; Uslaner et al., 2014). These receptors do not appear to be involved in nicotine reinstatement using other models, since the OX1R antagonist has no effect on footshock-induced reinstatement of nicotine seeking (Plaza-Zabala et al., 2010) and the OX2R antagonist has no effect on nicotine-induced reinstatement (Uslaner et al., 2014). These findings, consistent with those for palatable food and alcohol, support a role for OX in multiple aspects of nicotine intake behavior, such as willingness to work and certain models of seeking, but the specific brain regions involved have yet to be determined.

As nicotine is a stimulant drug, its intake leads to both reward and anxiety, with these effects mediated in part by the OX1R (Hollander et al.,

2008; Plaza-Zabala et al., 2010). Notably, the ability of nicotine to lower intracranial self-stimulation threshold is blocked by systemic administration of SB-334867 (Hollander et al., 2008). Similarly, the ability of nicotine to increase anxiety-like behavior in an elevated plus maze is also antagonized by SB-334867 (Plaza-Zabala et al., 2010). Thus, in addition to promoting the intake of nicotine, OX also plays a role in the emotional response to this intake.

4.3.2 Effects of Nicotine Intake on Orexin/Hypocretin

In relation to the intake of nicotine, OX neuron activity, gene expression, and peptide levels are elevated in an anatomically distinct pattern, with the activity of OX-expressing neurons being higher before voluntary nicotine intake as well as after nicotine administration. Specifically, levels of c-Fos are increased in OX neurons following presentation of a cue predicting nicotine availability, in the perifornical and lateral hypothalamus but not the dorsomedial hypothalamus (Plaza-Zabala et al., 2013). This effect is also seen in the perifornical and dorsomedial but not lateral hypothalamus after acute injection of nicotine at a relatively low dose (Plaza-Zabala et al., 2010) while in all populations of OX neurons after a higher dose of nicotine (Pasumarthi & Fadel, 2008; Pasumarthi, Reznikov, & Fadel, 2006). This difference in the areas recruited by a low vs high dose of nicotine may reflect a greater sensitivity of perifornical and dorsomedial hypothalamic neurons to acetylcholine. These effects on neuronal activity may also be specific to OX neurons, as they are not detected in adjacent neurons expressing melanin-concentrating hormone (Pasumarthi et al., 2006), and they are also evident in vitro, with acute nicotine application stimulating the firing of OX neurons (Huang, Xu, & van den Pol, 2011; Zhou, Gao, & Picciotto, 2015). Repeated nicotine exposure in vivo, either through experimenter-administered injection or through voluntary oral drinking, increases the gene expression of OX and peptide levels of both OX-A and OX-B in the hypothalamus (Kane et al., 2000; Tsuneki et al., 2016). While levels of OX-A are elevated only in the dorsomedial hypothalamus, levels of OX-B are increased in both the dorsomedial and paraventricular hypothalamus (Kane et al., 2000), suggesting that some of the major effects of nicotine are mediated by the OX2R. While not measured in extrahypothalamic areas, it is likely that OX is similarly released into other limbic regions following nicotine intake, where it acts to further perpetuate this behavior.

4.3.3 Effects of Prenatal Exposure to Nicotine on Orexin/Hypocretin

With OX both promoting and being activated by nicotine intake in adult animals, it is notable that this neuropeptide and nicotine consumption are similarly increased in animals exposed in utero to nicotine. Prenatal exposure to nicotine across a range of doses is shown to elevate nicotine intake in adolescent animals (Chang, Karatayev, & Leibowitz, 2013; Levin et al., 2006). Prenatal nicotine exposure at low doses (0.15–1.5 mg/kg/day) also increases OX gene expression and peptide levels (Chang et al., 2013; Morgan, Harrod, Lacy, Stanley, & Fadel, 2013), while the number of OX neurons and their gene expression are unaffected or reduced by nicotine at high levels (6 mg/kg/day) (Boychuk, Fuller, & Hayward, 2011). Further tests are needed to determine whether endogenous OX is indeed a critical driver of the elevated nicotine intake exhibited by offspring exposed early in life.

4.3.4 Clinical Evidence of a Role for Orexin/Hypocretin in Nicotine Intake

Changes in OX function have been associated with cigarette smoking in humans, although the direction of this association is not clear. For example, an allele for the OX2R has been linked to nicotine dependence in a Japanese population (Nishizawa et al., 2015). Also, individuals with OX-deficient narcolepsy who exhibit disturbances in their patterns of alcohol drinking are found to be more likely to smoke cigarettes than matched control subjects (Barateau et al., 2016). While the precise nature of the relationship remains to be resolved, these results provide some evidence linking OX to the use of nicotine.

4.4 Role of Orexin/Hypocretin in Cocaine Use

Research also connects OX with the use of cocaine, although the available studies suggest that this relationship is less direct than that observed with palatable food, alcohol, and nicotine. Possible reasons for the stronger and more direct connections between OX and these latter substances may be the fact that OX neurons are highly responsive to glucose that is elevated by palatable food intake (Burdakov et al., 2005), express both GABA receptors known to be strongly affected by alcohol intake (Backberg et al., 2003, 2004), and also express nicotinic acetylcholine receptors strongly stimulated by nicotine intake (Garcia et al., 2015). In contrast, the relationship between OX and cocaine may be weaker or less direct due to the fact that OX neurons are found only rarely to express receptors for dopamine (Bubser et al., 2005), which is strongly elevated by cocaine intake. Thus, while able to promote various aspects of cocaine intake behavior, endogenous OX remains largely unaltered by the intake of cocaine, indicating that this drug may not directly affect OX neurons and alter their function.

4.4.1 Effects of Orexin/Hypocretin on Aspects of Cocaine Use

Through investigations of the OX1R, endogenous OX has been found to promote the intake of cocaine, although only under specific conditions. Whereas intracerebroventricular injection of OX-A and systemic injection of the OX1R antagonist SB-334867 have no effect on operant responding for standard doses of cocaine under a fixed ratio 1 paradigm (Boutrel et al., 2005; Espana et al., 2010; Hutcheson et al., 2011; Zhou et al., 2012), there is other evidence suggesting that OX becomes involved when animals must work to attain their preferred blood levels of cocaine. Thus, SB-334867 significantly reduces fixed ratio 1 operant responding for very low doses of cocaine (Brodnik, Bernstein, Prince, & Espana, 2015) and also operant responding in a discrete trials paradigm involving highly restricted periods of access (Espana et al., 2010). Further, both SB-334867 and genetic knockout of the OX1R lead animals to consume less cocaine under a high work, fixed ratio 5 paradigm (Hollander, Pham, Fowler, & Kenny, 2012). The one study investigating the involvement of specific brain regions in mediating OX's actions found cocaine self-administration to be decreased by injection of SB-334867 into the central nucleus of the amygdala (Schmeichel, Herman, Roberto, & Koob, 2017). While this suggests that the OX1R in the amygdala may participate in promoting the intake of cocaine under conditions requiring work, a possible role for other limbic brain regions and the OX2R requires further investigation.

Not surprisingly, performance in a progressive ratio task that measures how much an individual is willing to work for cocaine is mediated by the OX1R. Specifically, the breakpoint for cocaine is reduced by peripheral injection of SB-334867 but not the OX2R antagonist 4PT (Borgland et al., 2009; Brodnik et al., 2015; Prince, Rau, Yorgason, & Espana, 2015). This effect of the OX1R appears to specifically involve the ventral tegmental area where injection of SB-334867 recapitulates the lowering of the breakpoint (Espana et al., 2010) and OX-A injection increases breakpoint (Espana, Melchior, Roberts, & Jones, 2011). With these injections into the ventral tegmental area also found, respectively, to inhibit or enhance the effects of cocaine on dopamine signaling (Espana et al., 2011, 2010), it is likely that dopamine mediates these changes in the amount of effort expended to obtain this drug.

As shown with the other rewarding substances, seeking for cocaine is also promoted by OX under a number of reinstatement paradigms and is mediated by the OX1R or OX2R depending on the brain region tested. Central injection of OX itself is sufficient to induce reinstatement of cocaine seeking, with this behavior mediated by the OX1R but not OX2R in the ventral tegmental area (Wang, You, & Wise, 2009) and by the OX2R but not the OX1R in the paraventricular thalamus (Matzeu, Kerr, Weiss, & Martin-Fardon, 2016). In contrast, OX antagonists are effective at inhibiting reinstatement of cocaine seeking after presentation of a cocaine-paired cue, return to the self-administration environment, or stress or yohimbine, but not after priming with cocaine itself which has minimal effects on OX levels (see Section 4.4.2). For example, cue-induced reinstatement of cocaine seeking is reduced by SB-334867 but not the OX2R antagonist 4PT when injected systemically (Hutcheson et al., 2011; Martin-Fardon & Weiss, 2014a; Smith, See, & Aston-Jones, 2009) or into the ventral tegmental area (James et al., 2011; Mahler, Smith, & Aston-Jones, 2013). Further, systemic injection of SB-334867 inhibits cocaine seeking induced by context (Smith, Tahsili-Fahadan, & Aston-Jones, 2010), footshock stress (Boutrel et al., 2005), or yohimbine (Zhou et al., 2012), with this latter reinstatement also reduced by SB-334867 injection into the central nucleus of the amygdala (Schmeichel et al., 2017). In contrast, neither SB-334867 nor 4PT affects the reinstatement of cocaine seeking induced by cocaine itself (Mahler et al., 2013; Smith et al., 2009; Zhou et al., 2012). These results demonstrate that the reinstatement of cocaine seeking is stimulated by OX in a variety of reinstatement paradigms, but not in drug-induced reinstatement.

Studies generally support the involvement of OX in the reward from cocaine intake. In a conditioned place preference paradigm in which animals are conditioned to associate a distinct chamber with experimenter-administered cocaine, the dual OX receptor antagonist almorexant is found to reduce the expression of this place preference (Steiner, Lecourt, & Jenck, 2013), similar to the effect produced by knockout of the OX gene (Shaw et al., 2016). Conversely, direct injection of OX-A into the ventral tegmental area is sufficient to reinstate an extinguished place preference (Tung et al., 2016). Further, cocaine-induced lowering of the intracranial self-stimulation threshold is also found to be reversed by systemic injection of SB-334867 in rats (Hollander et al., 2012), although this was not found to occur using a higher dose in mice (Riday et al., 2012). Overall, these results support a role for OX in the reward from cocaine intake.

4.4.2 Effects of Cocaine Use on Orexin/Hypocretin

In relation to the intake of cocaine, OX appears to be involved in behavior that occurs prior to its intake but this neuropeptide exhibits little change in response to its consumption. As with other rewarding substances, the activity of OX-expressing neurons is increased prior to seeking for cocaine, as shown by the finding that levels of c-Fos in OX neurons are increased by a cocaine-pared cue (Martin-Fardon, Cauvi, Kerr, & Weiss, 2016) or placement in a self-administration environment (Hamlin, Clemens, & McNally, 2008). In contrast to other substances, however, repeated injections of cocaine have little effect on OX gene expression in the hypothalamus (Zhou et al., 2008), and acute injection of cocaine also fails to affect OX-A or OX-B peptide levels in the nucleus accumbens (Zhang, Mao, Liu, & Wang, 2007). These results provide further evidence that OX is involved in promoting cocaine intake but is not itself significantly affected by this behavior.

4.4.3 Clinical Evidence of a Role for Orexin/Hypocretin in Cocaine Use

Currently, there is only one study that has examined OX in relation to cocaine use in humans. Individuals with OX-deficient narcolepsy, who are found to show differences in their patterns of alcohol drinking and cigarette smoking, show no difference in their likelihood of using cocaine compared to matched control subjects (Barateau et al., 2016). This finding may be due to the low frequency of cocaine use among the sampled populations, pointing to the need for further translational research on this topic.

5. CONCLUSIONS

While OX was originally identified for its role in food intake and energy homeostasis, it has since become clear that OX participates in a range of behaviors, which are both homeostatically and nonhomeostatically regulated. Many of the behaviors promoted by OX can be classified as adaptive, including food intake, sleep–wake regulation, mating, and maternal behavior. These behaviors are largely controlled by the release of OX in the hypothalamus, with food intake involving the same regions of the hypothalamus as OX transcription (perifornical, dorsomedial, and lateral areas), waking additionally involving areas like the locus coeruleus, and mating and maternal behavior involving the medial preoptic area of the hypothalamus. The enactment of these adaptive behaviors results in feedback signals such as a rise in glucose that negatively regulate OX activity, signaling the completion of the behavioral sequence. We propose that OX in promoting these adaptive behaviors under physiological conditions has an important role in maintaining arousal to promote survival, both invigorating and motivating behavior necessary for the propagation of an individual's genes. Further studies uncover another property of OX, its ability to promote overconsumption of rewarding substances of abuse, such as palatable food, alcohol, nicotine, and cocaine, with this peptide both stimulating and being stimulated by the intake of these substances. This function of OX occurs through a different set of limbic nuclei outside of the hypothalamus, most notably the nucleus accumbens, ventral tegmental area, and paraventricular nucleus of the thalamus, in addition to the amygdala, prefrontal cortex, and insular cortex. Through these extrahypothalamic limbic brain regions, OX promotes the intake, willingness to work, seeking, and reward derived from the intake of rewarding substances. These effects are mediated by activity at the OX1R in most limbic regions, except for the paraventricular thalamus where they are mediated by OX2R. Moreover, in response to the intake or early life exposure to these rewarding substances, OX gene expression and peptide levels are generally further elevated, forming a vicious cycle involving positive feedback rather than the negative feedback characteristic of homeostatic cycles. Together, these results suggest that the OX system in a broad range of extrahypothalamic limbic brain areas can be co-opted by rewarding substances, such as palatable food, alcohol, nicotine, and cocaine, to drive intake nonhomeostatically and ultimately lead to addiction.

ACKNOWLEDGMENTS

This research was supported by the National Institute on Alcohol Abuse and Alcoholism of the National Institutes of Health under Award Numbers R00AA021782 (J.R.B.) and R01AA24798 (S.F.L.). We thank Dr. Rodrigo A. España (Drexel University College of Medicine) for his invaluable assistance in interpreting portions of the cocaine literature.

REFERENCES

Adamantidis, A. R., Zhang, F., Aravanis, A. M., Deisseroth, K., & de Lecea, L. (2007). Neural substrates of awakening probed with optogenetic control of hypocretin neurons. *Nature*, *450*(7168), 420–424. http://dx.doi.org/10.1038/nature06310.

Akiyama, M., Yuasa, T., Hayasaka, N., Horikawa, K., Sakurai, T., & Shibata, S. (2004). Reduced food anticipatory activity in genetically orexin (hypocretin) neuron-ablated mice. *The European Journal of Neuroscience*, *20*(11), 3054–3062. http://dx.doi.org/10.1111/j.1460-9568.2004.03749.x.

Alcaraz-Iborra, M., Carvajal, F., Lerma-Cabrera, J. M., Valor, L. M., & Cubero, I. (2014). Binge-like consumption of caloric and non-caloric palatable substances in ad libitum-fed C57BL/6J mice: Pharmacological and molecular evidence of orexin involvement. *Behavioural Brain Research*, *272*, 93–99. http://dx.doi.org/10.1016/j.bbr.2014.06.049.

Anderson, R. I., Becker, H. C., Adams, B. L., Jesudason, C. D., & Rorick-Kehn, L. M. (2014). Orexin-1 and orexin-2 receptor antagonists reduce ethanol self-administration in high-drinking rodent models. *Frontiers in Neuroscience*, *8*, 33. http://dx.doi.org/10.3389/fnins.2014.00033.

Archer, Z. A., Findlay, P. A., Rhind, S. M., Mercer, J. G., & Adam, C. L. (2002). Orexin gene expression and regulation by photoperiod in the sheep hypothalamus. *Regulatory Peptides*, *104*(1–3), 41–45.

Backberg, M., Collin, M., Ovesjo, M. L., & Meister, B. (2003). Chemical coding of GABA(B) receptor-immunoreactive neurones in hypothalamic regions regulating body weight. *Journal of Neuroendocrinology*, *15*(1), 1–14.

Backberg, M., Hervieu, G., Wilson, S., & Meister, B. (2002). Orexin receptor-1 (OX-R1) immunoreactivity in chemically identified neurons of the hypothalamus: Focus on orexin targets involved in control of food and water intake. *The European Journal of Neuroscience*, *15*(2), 315–328.

Backberg, M., Ultenius, C., Fritschy, J. M., & Meister, B. (2004). Cellular localization of GABA receptor alpha subunit immunoreactivity in the rat hypothalamus: Relationship with neurones containing orexigenic or anorexigenic peptides. *Journal of Neuroendocrinology*, *16*(7), 589–604. http://dx.doi.org/10.1111/j.1365-2826.2004.01207.x.

Baier, P. C., Hallschmid, M., Seeck-Hirschner, M., Weinhold, S. L., Burkert, S., Diessner, N., et al. (2011). Effects of intranasal hypocretin-1 (orexin A) on sleep in narcolepsy with cataplexy. *Sleep Medicine*, *12*(10), 941–946. http://dx.doi.org/10.1016/j.sleep.2011.06.015.

Barateau, L., Jaussent, I., Lopez, R., Boutrel, B., Leu-Semenescu, S., Arnulf, I., et al. (2016). Smoking, alcohol, drug use, abuse and dependence in narcolepsy and idiopathic hypersomnia: A case-control study. *Sleep*, *39*(3), 573–580. http://dx.doi.org/10.5665/sleep.5530.

Barson, J. R., Ho, H. T., & Leibowitz, S. F. (2015). Anterior thalamic paraventricular nucleus is involved in intermittent access ethanol drinking: Role of orexin receptor 2. *Addiction Biology*, *20*(3), 469–481. http://dx.doi.org/10.1111/adb.12139.

Barson, J. R., Poon, K., Ho, H. T., Alam, M. I., Sanzalone, L., & Leibowitz, S. F. (2017). Substance P in the anterior thalamic paraventricular nucleus: Promotion of ethanol drinking in response to orexin from the hypothalamus. *Addiction Biology*, *22*(1), 58–69. http://dx.doi.org/10.1111/adb.12288.

Bayer, L., Colard, C., Nguyen, N. U., Risold, P. Y., Fellmann, D., & Griffond, B. (2000). Alteration of the expression of the hypocretin (orexin) gene by 2-deoxyglucose in the rat lateral hypothalamic area. *Neuroreport*, *11*(3), 531–533.

Bei, B., Coo, S., & Trinder, J. (2015). Sleep and mood during pregnancy and the postpartum period. *Sleep Medicine Clinics*, *10*(1), 25–33. http://dx.doi.org/10.1016/j.jsmc.2014.11.011.

Borbely, A. A. (1977). Sleep in the rat during food deprivation and subsequent restitution of food. *Brain Research*, *124*(3), 457–471.

Borgland, S. L., Chang, S. J., Bowers, M. S., Thompson, J. L., Vittoz, N., Floresco, S. B., et al. (2009). Orexin A/hypocretin-1 selectively promotes motivation for positive reinforcers. *The Journal of Neuroscience*, *29*(36), 11215–11225. http://dx.doi.org/10.1523/JNEUROSCI.6096-08.2009.

Boutrel, B., Kenny, P. J., Specio, S. E., Martin-Fardon, R., Markou, A., Koob, G. F., et al. (2005). Role for hypocretin in mediating stress-induced reinstatement of cocaine-seeking

behavior. *Proceedings of the National Academy of Sciences of the United States of America, 102*(52), 19168–19173. http://dx.doi.org/10.1073/pnas.0507480102.
Boychuk, C. R., Fuller, D. D., & Hayward, L. F. (2011). Sex differences in heart rate variability during sleep following prenatal nicotine exposure in rat pups. *Behavioural Brain Research, 219*(1), 82–91. http://dx.doi.org/10.1016/j.bbr.2010.12.013.
Brodnik, Z. D., Bernstein, D. L., Prince, C. D., & Espana, R. A. (2015). Hypocretin receptor 1 blockade preferentially reduces high effort responding for cocaine without promoting sleep. *Behavioural Brain Research, 291*, 377–384. http://dx.doi.org/10.1016/j.bbr.2015.05.051.
Brown, R. M., Khoo, S. Y., & Lawrence, A. J. (2013). Central orexin (hypocretin) 2 receptor antagonism reduces ethanol self-administration, but not cue-conditioned ethanol-seeking, in ethanol-preferring rats. *The International Journal of Neuropsychopharmacology, 16*(9), 2067–2079. http://dx.doi.org/10.1017/S1461145713000333.
Bubser, M., Fadel, J. R., Jackson, L. L., Meador-Woodruff, J. H., Jing, D., & Deutch, A. Y. (2005). Dopaminergic regulation of orexin neurons. *The European Journal of Neuroscience, 21*(11), 2993–3001. http://dx.doi.org/10.1111/j.1460-9568.2005.04121.x.
Burdakov, D., Gerasimenko, O., & Verkhratsky, A. (2005). Physiological changes in glucose differentially modulate the excitability of hypothalamic melanin-concentrating hormone and orexin neurons in situ. *The Journal of Neuroscience, 25*(9), 2429–2433. http://dx.doi.org/10.1523/JNEUROSCI.4925-04.2005.
Cacciapaglia, F., Saddoris, M. P., Wightman, R. M., & Carelli, R. M. (2012). Differential dopamine release dynamics in the nucleus accumbens core and shell track distinct aspects of goal-directed behavior for sucrose. *Neuropharmacology, 62*(5–6), 2050–2056. http://dx.doi.org/10.1016/j.neuropharm.2011.12.027.
Cadoni, C., & Di Chiara, G. (2000). Differential changes in accumbens shell and core dopamine in behavioral sensitization to nicotine. *European Journal of Pharmacology, 387*(3), R23–R25.
Cai, X. J., Evans, M. L., Lister, C. A., Leslie, R. A., Arch, J. R., Wilson, S., et al. (2001). Hypoglycemia activates orexin neurons and selectively increases hypothalamic orexin-B levels: Responses inhibited by feeding and possibly mediated by the nucleus of the solitary tract. *Diabetes, 50*(1), 105–112.
Cai, X. J., Widdowson, P. S., Harrold, J., Wilson, S., Buckingham, R. E., Arch, J. R., et al. (1999). Hypothalamic orexin expression: Modulation by blood glucose and feeding. *Diabetes, 48*(11), 2132–2137.
Carnicella, S., Ron, D., & Barak, S. (2014). Intermittent ethanol access schedule in rats as a preclinical model of alcohol abuse. *Alcohol, 48*(3), 243–252. http://dx.doi.org/10.1016/j.alcohol.2014.01.006.
Carter, M. E., Brill, J., Bonnavion, P., Huguenard, J. R., Huerta, R., & de Lecea, L. (2012). Mechanism for hypocretin-mediated sleep-to-wake transitions. *Proceedings of the National Academy of Sciences of the United States of America, 109*(39), E2635–E2644. http://dx.doi.org/10.1073/pnas.1202526109.
Cason, A. M., & Aston-Jones, G. (2013a). Attenuation of saccharin-seeking in rats by orexin/hypocretin receptor 1 antagonist. *Psychopharmacology, 228*(3), 499–507. http://dx.doi.org/10.1007/s00213-013-3051-7.
Cason, A. M., & Aston-Jones, G. (2013b). Role of orexin/hypocretin in conditioned sucrose-seeking in rats. *Psychopharmacology, 226*(1), 155–165. http://dx.doi.org/10.1007/s00213-012-2902-y.
Cason, A. M., & Aston-Jones, G. (2014). Role of orexin/hypocretin in conditioned sucrose-seeking in female rats. *Neuropharmacology, 86*, 97–102. http://dx.doi.org/10.1016/j.neuropharm.2014.07.007.
Castro, D. C., Terry, R. A., & Berridge, K. C. (2016). Orexin in rostral hotspot of nucleus accumbens enhances sucrose "liking" and intake but scopolamine in caudal shell shifts

"liking" toward "disgust" and "fear". *Neuropsychopharmacology, 41*(8), 2101–2111. http://dx.doi.org/10.1038/npp.2016.10.

Champagne, F. A., Chretien, P., Stevenson, C. W., Zhang, T. Y., Gratton, A., & Meaney, M. J. (2004). Variations in nucleus accumbens dopamine associated with individual differences in maternal behavior in the rat. *The Journal of Neuroscience, 24*(17), 4113–4123. http://dx.doi.org/10.1523/JNEUROSCI.5322-03.2004.

Chang, G. Q., Gaysinskaya, V., Karatayev, O., & Leibowitz, S. F. (2008). Maternal high-fat diet and fetal programming: Increased proliferation of hypothalamic peptide-producing neurons that increase risk for overeating and obesity. *The Journal of Neuroscience, 28*(46), 12107–12119. http://dx.doi.org/10.1523/JNEUROSCI.2642-08.2008.

Chang, G. Q., Karatayev, O., & Leibowitz, S. F. (2013). Prenatal exposure to nicotine stimulates neurogenesis of orexigenic peptide-expressing neurons in hypothalamus and amygdala. *The Journal of Neuroscience, 33*(34), 13600–13611. http://dx.doi.org/10.1523/JNEUROSCI.5835-12.2013.

Chang, G. Q., Karatayev, O., Liang, S. C., Barson, J. R., & Leibowitz, S. F. (2012). Prenatal ethanol exposure stimulates neurogenesis in hypothalamic and limbic peptide systems: Possible mechanism for offspring ethanol overconsumption. *Neuroscience, 222*, 417–428. http://dx.doi.org/10.1016/j.neuroscience.2012.05.066.

Chemelli, R. M., Willie, J. T., Sinton, C. M., Elmquist, J. K., Scammell, T., Lee, C., et al. (1999). Narcolepsy in orexin knockout mice: Molecular genetics of sleep regulation. *Cell, 98*(4), 437–451.

Cheng, S. B., Kuchiiwa, S., Gao, H. Z., Kuchiiwa, T., & Nakagawa, S. (2003). Morphological study of orexin neurons in the hypothalamus of the Long-Evans rat, with special reference to co-expression of orexin and NADPH-diaphorase or nitric oxide synthase activities. *Neuroscience Research, 46*(1), 53–62.

Choi, D. L., Davis, J. F., Fitzgerald, M. E., & Benoit, S. C. (2010). The role of orexin-A in food motivation, reward-based feeding behavior and food-induced neuronal activation in rats. *Neuroscience, 167*(1), 11–20. http://dx.doi.org/10.1016/j.neuroscience.2010.02.002.

Choi, D. L., Davis, J. F., Magrisso, I. J., Fitzgerald, M. E., Lipton, J. W., & Benoit, S. C. (2012). Orexin signaling in the paraventricular thalamic nucleus modulates mesolimbic dopamine and hedonic feeding in the rat. *Neuroscience, 210*, 243–248. http://dx.doi.org/10.1016/j.neuroscience.2012.02.036.

Chou, T. C., Lee, C. E., Lu, J., Elmquist, J. K., Hara, J., Willie, J. T., et al. (2001). Orexin (hypocretin) neurons contain dynorphin. *The Journal of Neuroscience, 21*(19), RC168.

Chou, T. C., Rotman, S. R., & Saper, C. B. (2004). Lateral hypothalamic acetylcholinesterase-immunoreactive neurons co-express either orexin or melanin concentrating hormone. *Neuroscience Letters, 370*(2–3), 123–126. http://dx.doi.org/10.1016/j.neulet.2004.08.070.

Clegg, D. J., Air, E. L., Woods, S. C., & Seeley, R. J. (2002). Eating elicited by orexin-a, but not melanin-concentrating hormone, is opioid mediated. *Endocrinology, 143*(8), 2995–3000. http://dx.doi.org/10.1210/endo.143.8.8977.

Collin, M., Backberg, M., Onnestam, K., & Meister, B. (2002). 5-HT1A receptor immunoreactivity in hypothalamic neurons involved in body weight control. *Neuroreport, 13*(7), 945–951.

D'Anna, K. L., & Gammie, S. C. (2006). Hypocretin-1 dose-dependently modulates maternal behaviour in mice. *Journal of Neuroendocrinology, 18*(8), 553–566. http://dx.doi.org/10.1111/j.1365-2826.2006.01448.x.

Dayas, C. V., McGranahan, T. M., Martin-Fardon, R., & Weiss, F. (2008). Stimuli linked to ethanol availability activate hypothalamic CART and orexin neurons in a reinstatement model of relapse. *Biological Psychiatry, 63*(2), 152–157. http://dx.doi.org/10.1016/j.biopsych.2007.02.002.

de Lecea, L., Kilduff, T. S., Peyron, C., Gao, X., Foye, P. E., Danielson, P. E., et al. (1998). The hypocretins: Hypothalamus-specific peptides with neuroexcitatory activity. *Proceedings of the National Academy of Sciences of the United States of America, 95*(1), 322–327.

Dewasmes, G., Duchamp, C., & Minaire, Y. (1989). Sleep changes in fasting rats. *Physiology & Behavior, 46*(2), 179–184.

Di Sebastiano, A. R., Wilson-Perez, H. E., Lehman, M. N., & Coolen, L. M. (2011). Lesions of orexin neurons block conditioned place preference for sexual behavior in male rats. *Hormones and Behavior, 59*(1), 1–8. http://dx.doi.org/10.1016/j.yhbeh.2010.09.006.

Dimitrova, A., Fronczek, R., Van der Ploeg, J., Scammell, T., Gautam, S., Pascual-Leone, A., et al. (2011). Reward-seeking behavior in human narcolepsy. *Journal of Clinical Sleep Medicine, 7*(3), 293–300. http://dx.doi.org/10.5664/JCSM.1076.

Diniz Behn, C. G., Klerman, E. B., Mochizuki, T., Lin, S. C., & Scammell, T. E. (2010). Abnormal sleep/wake dynamics in orexin knockout mice. *Sleep, 33*(3), 297–306.

Doyon, W. M., York, J. L., Diaz, L. M., Samson, H. H., Czachowski, C. L., & Gonzales, R. A. (2003). Dopamine activity in the nucleus accumbens during consummatory phases of oral ethanol self-administration. *Alcoholism, Clinical and Experimental Research, 27*(10), 1573–1582. http://dx.doi.org/10.1097/01.ALC.0000089959.66222.B8.

Dube, M. G., Kalra, S. P., & Kalra, P. S. (1999). Food intake elicited by central administration of orexins/hypocretins: Identification of hypothalamic sites of action. *Brain Research, 842*(2), 473–477.

Dyer, C. J., Touchette, K. J., Carroll, J. A., Allee, G. L., & Matteri, R. L. (1999). Cloning of porcine prepro-orexin cDNA and effects of an intramuscular injection of synthetic porcine orexin-B on feed intake in young pigs. *Domestic Animal Endocrinology, 16*(3), 145–148.

Espana, R. A., Melchior, J. R., Roberts, D. C., & Jones, S. R. (2011). Hypocretin 1/orexin A in the ventral tegmental area enhances dopamine responses to cocaine and promotes cocaine self-administration. *Psychopharmacology, 214*(2), 415–426. http://dx.doi.org/10.1007/s00213-010-2048-8.

Espana, R. A., Oleson, E. B., Locke, J. L., Brookshire, B. R., Roberts, D. C., & Jones, S. R. (2010). The hypocretin-orexin system regulates cocaine self-administration via actions on the mesolimbic dopamine system. *The European Journal of Neuroscience, 31*(2), 336–348. http://dx.doi.org/10.1111/j.1460-9568.2009.07065.x.

Fabio, M. C., Macchione, A. F., Nizhnikov, M. E., & Pautassi, R. M. (2015). Prenatal ethanol increases ethanol intake throughout adolescence, alters ethanol-mediated aversive learning, and affects mu but not delta or kappa opioid receptor mRNA expression. *The European Journal of Neuroscience, 41*(12), 1569–1579. http://dx.doi.org/10.1111/ejn.12913.

Fu, L. Y., Acuna-Goycolea, C., & van den Pol, A. N. (2004). Neuropeptide Y inhibits hypocretin/orexin neurons by multiple presynaptic and postsynaptic mechanisms: Tonic depression of the hypothalamic arousal system. *The Journal of Neuroscience, 24*(40), 8741–8751. http://dx.doi.org/10.1523/JNEUROSCI.2268-04.2004.

Fujiki, N., Yoshida, Y., Ripley, B., Honda, K., Mignot, E., & Nishino, S. (2001). Changes in CSF hypocretin-1 (orexin A) levels in rats across 24 hours and in response to food deprivation. *Neuroreport, 12*(5), 993–997.

Furudono, Y., Ando, C., Yamamoto, C., Kobashi, M., & Yamamoto, T. (2006). Involvement of specific orexigenic neuropeptides in sweetener-induced overconsumption in rats. *Behavioural Brain Research, 175*(2), 241–248. http://dx.doi.org/10.1016/j.bbr.2006.08.031.

Furutani, N., Hondo, M., Kageyama, H., Tsujino, N., Mieda, M., Yanagisawa, M., et al. (2013). Neurotensin co-expressed in orexin-producing neurons in the lateral hypothalamus plays an important role in regulation of sleep/wakefulness states. *PloS One*, *8*(4), e62391. http://dx.doi.org/10.1371/journal.pone.0062391.

Garcia, A. P., Aitta-aho, T., Schaaf, L., Heeley, N., Heuschmid, L., Bai, Y., et al. (2015). Nicotinic alpha4 receptor-mediated cholinergic influences on food intake and activity patterns in hypothalamic circuits. *PloS One*, *10*(8), e0133327. http://dx.doi.org/10.1371/journal.pone.0133327.

Gaysinskaya, V. A., Karatayev, O., Chang, G. Q., & Leibowitz, S. F. (2007). Increased caloric intake after a high-fat preload: Relation to circulating triglycerides and orexigenic peptides. *Physiology & Behavior*, *91*(1), 142–153. http://dx.doi.org/10.1016/j.physbeh.2007.02.002.

Griffond, B., Risold, P. Y., Jacquemard, C., Colard, C., & Fellmann, D. (1999). Insulin-induced hypoglycemia increases preprohypocretin (orexin) mRNA in the rat lateral hypothalamic area. *Neuroscience Letters*, *262*(2), 77–80.

Gulia, K. K., Mallick, H. N., & Kumar, V. M. (2003). Orexin A (hypocretin-1) application at the medial preoptic area potentiates male sexual behavior in rats. *Neuroscience*, *116*(4), 921–923.

Hakansson, M., de Lecea, L., Sutcliffe, J. G., Yanagisawa, M., & Meister, B. (1999). Leptin receptor- and STAT3-immunoreactivities in hypocretin/orexin neurones of the lateral hypothalamus. *Journal of Neuroendocrinology*, *11*(8), 653–663.

Hamid, A. A., Pettibone, J. R., Mabrouk, O. S., Hetrick, V. L., Schmidt, R., Vander Weele, C. M., et al. (2016). Mesolimbic dopamine signals the value of work. *Nature Neuroscience*, *19*(1), 117–126. http://dx.doi.org/10.1038/nn.4173.

Hamlin, A. S., Clemens, K. J., & McNally, G. P. (2008). Renewal of extinguished cocaine-seeking. *Neuroscience*, *151*(3), 659–670. http://dx.doi.org/10.1016/j.neuroscience.2007.11.018.

Hasegawa, E., Yanagisawa, M., Sakurai, T., & Mieda, M. (2014). Orexin neurons suppress narcolepsy via 2 distinct efferent pathways. *The Journal of Clinical Investigation*, *124*(2), 604–616. http://dx.doi.org/10.1172/JCI71017.

Haynes, A. C., Jackson, B., Chapman, H., Tadayyon, M., Johns, A., Porter, R. A., et al. (2000). A selective orexin-1 receptor antagonist reduces food consumption in male and female rats. *Regulatory Peptides*, *96*(1–2), 45–51.

Hollander, J. A., Lu, Q., Cameron, M. D., Kamenecka, T. M., & Kenny, P. J. (2008). Insular hypocretin transmission regulates nicotine reward. *Proceedings of the National Academy of Sciences of the United States of America*, *105*(49), 19480–19485. http://dx.doi.org/10.1073/pnas.0808023105.

Hollander, J. A., Pham, D., Fowler, C. D., & Kenny, P. J. (2012). Hypocretin-1 receptors regulate the reinforcing and reward-enhancing effects of cocaine: Pharmacological and behavioral genetics evidence. *Frontiers in Behavioral Neuroscience*, *6*, 47. http://dx.doi.org/10.3389/fnbeh.2012.00047.

Honors, M. A., Hargrave, S. L., & Kinzig, K. P. (2012). Glucose tolerance in response to a high-fat diet is improved by a high-protein diet. *Obesity (Silver Spring)*, *20*(9), 1859–1865. http://dx.doi.org/10.1038/oby.2011.297.

Horvath, T. L., Diano, S., & van den Pol, A. N. (1999). Synaptic interaction between hypocretin (orexin) and neuropeptide Y cells in the rodent and primate hypothalamus: A novel circuit implicated in metabolic and endocrine regulations. *The Journal of Neuroscience*, *19*(3), 1072–1087.

Howard, E. C., Schier, C. J., Wetzel, J. S., Duvauchelle, C. L., & Gonzales, R. A. (2008). The shell of the nucleus accumbens has a higher dopamine response compared with the core after non-contingent intravenous ethanol administration. *Neuroscience*, *154*(3), 1042–1053. http://dx.doi.org/10.1016/j.neuroscience.2008.04.014.

Huang, H., Xu, Y., & van den Pol, A. N. (2011). Nicotine excites hypothalamic arcuate anorexigenic proopiomelanocortin neurons and orexigenic neuropeptide Y neurons:

Similarities and differences. *Journal of Neurophysiology, 106*(3), 1191–1202. http://dx.doi.org/10.1152/jn.00740.2010.

Hutcheson, D. M., Quarta, D., Halbout, B., Rigal, A., Valerio, E., & Heidbreder, C. (2011). Orexin-1 receptor antagonist SB-334867 reduces the acquisition and expression of cocaine-conditioned reinforcement and the expression of amphetamine-conditioned reward. *Behavioural Pharmacology, 22*(2), 173–181. http://dx.doi.org/10.1097/FBP.0b013e328343d761.

Inutsuka, A., Inui, A., Tabuchi, S., Tsunematsu, T., Lazarus, M., & Yamanaka, A. (2014). Concurrent and robust regulation of feeding behaviors and metabolism by orexin neurons. *Neuropharmacology, 85*, 451–460. http://dx.doi.org/10.1016/j.neuropharm.2014.06.015.

James, M. H., Charnley, J. L., Levi, E. M., Jones, E., Yeoh, J. W., Smith, D. W., et al. (2011). Orexin-1 receptor signalling within the ventral tegmental area, but not the paraventricular thalamus, is critical to regulating cue-induced reinstatement of cocaine-seeking. *The International Journal of Neuropsychopharmacology, 14*(5), 684–690. http://dx.doi.org/10.1017/S1461145711000423.

Johnstone, L. E., Fong, T. M., & Leng, G. (2006). Neuronal activation in the hypothalamus and brainstem during feeding in rats. *Cell Metabolism, 4*(4), 313–321. http://dx.doi.org/10.1016/j.cmet.2006.08.003.

Jones, L. C., Bellingham, W. P., & Ward, L. C. (1990). Sex differences in voluntary locomotor activity of food-restricted and ad libitum-fed rats. Implications for the maintenance of a body weight set-point. *Comparative Biochemistry and Physiology A, Comparative Physiology, 96*(2), 287–290.

Jupp, B., Krivdic, B., Krstew, E., & Lawrence, A. J. (2011). The orexin(1) receptor antagonist SB-334867 dissociates the motivational properties of alcohol and sucrose in rats. *Brain Research, 1391*, 54–59. http://dx.doi.org/10.1016/j.brainres.2011.03.045.

Kageyama, H., Kita, T., Toshinai, K., Guan, J. L., Date, Y., Takenoya, F., et al. (2006). Galanin-like peptide promotes feeding behaviour via activation of orexinergic neurones in the rat lateral hypothalamus. *Journal of Neuroendocrinology, 18*(1), 33–41. http://dx.doi.org/10.1111/j.1365-2826.2005.01382.x.

Kane, J. K., Parker, S. L., Matta, S. G., Fu, Y., Sharp, B. M., & Li, M. D. (2000). Nicotine up-regulates expression of orexin and its receptors in rat brain. *Endocrinology, 141*(10), 3623–3629. http://dx.doi.org/10.1210/endo.141.10.7707.

Kanenishi, K., Ueno, M., Momose, S., Kuwabara, H., Tanaka, H., Sato, C., et al. (2004). Prepro-orexin mRNA expression in the rat brain is increased during pregnancy. *Neuroscience Letters, 368*(1), 73–77. http://dx.doi.org/10.1016/j.neulet.2004.06.068.

Kaslin, J., Nystedt, J. M., Ostergard, M., Peitsaro, N., & Panula, P. (2004). The orexin/hypocretin system in zebrafish is connected to the aminergic and cholinergic systems. *The Journal of Neuroscience, 24*(11), 2678–2689. http://dx.doi.org/10.1523/JNEUROSCI.4908-03.2004.

Kay, K., Parise, E. M., Lilly, N., & Williams, D. L. (2014). Hindbrain orexin 1 receptors influence palatable food intake, operant responding for food, and food-conditioned place preference in rats. *Psychopharmacology, 231*(2), 419–427. http://dx.doi.org/10.1007/s00213-013-3248-9.

Kiyashchenko, L. I., Mileykovskiy, B. Y., Maidment, N., Lam, H. A., Wu, M. F., John, J., et al. (2002). Release of hypocretin (orexin) during waking and sleep states. *The Journal of Neuroscience: The Official Journal of the Society for Neuroscience, 22*(13), 5282–5286.

Korotkova, T. M., Sergeeva, O. A., Eriksson, K. S., Haas, H. L., & Brown, R. E. (2003). Excitation of ventral tegmental area dopaminergic and nondopaminergic neurons by orexins/hypocretins. *The Journal of Neuroscience, 23*(1), 7–11.

Lawrence, A. J., Cowen, M. S., Yang, H. J., Chen, F., & Oldfield, B. (2006). The orexin system regulates alcohol-seeking in rats. *British Journal of Pharmacology, 148*(6), 752–759. http://dx.doi.org/10.1038/sj.bjp.0706789.

Lee, M. G., Hassani, O. K., & Jones, B. E. (2005). Discharge of identified orexin/hypocretin neurons across the sleep-waking cycle. *The Journal of Neuroscience, 25*(28), 6716–6720. http://dx.doi.org/10.1523/JNEUROSCI.1887-05.2005.

Lei, K., Wegner, S. A., Yu, J. H., Mototake, A., Hu, B., & Hopf, F. W. (2016). Nucleus accumbens shell and mPFC but not insula orexin-1 receptors promote excessive alcohol drinking. *Frontiers in Neuroscience, 10*, 400. http://dx.doi.org/10.3389/fnins.2016.00400.

Lena, I., Parrot, S., Deschaux, O., Muffat-Joly, S., Sauvinet, V., Renaud, B., et al. (2005). Variations in extracellular levels of dopamine, noradrenaline, glutamate, and aspartate across the sleep–wake cycle in the medial prefrontal cortex and nucleus accumbens of freely moving rats. *Journal of Neuroscience Research, 81*(6), 891–899. http://dx.doi.org/10.1002/jnr.20602.

LeSage, M. G., Perry, J. L., Kotz, C. M., Shelley, D., & Corrigall, W. A. (2010). Nicotine self-administration in the rat: Effects of hypocretin antagonists and changes in hypocretin mRNA. *Psychopharmacology, 209*(2), 203–212. http://dx.doi.org/10.1007/s00213-010-1792-0.

Levin, E. D., Lawrence, S., Petro, A., Horton, K., Seidler, F. J., & Slotkin, T. A. (2006). Increased nicotine self-administration following prenatal exposure in female rats. *Pharmacology, Biochemistry, and Behavior, 85*(3), 669–674. http://dx.doi.org/10.1016/j.pbb.2006.11.006.

Lin, L., Faraco, J., Li, R., Kadotani, H., Rogers, W., Lin, X., et al. (1999). The sleep disorder canine narcolepsy is caused by a mutation in the hypocretin (orexin) receptor 2 gene. *Cell, 98*(3), 365–376.

Lopez, M. F., Moorman, D. E., Aston-Jones, G., & Becker, H. C. (2016). The highly selective orexin/hypocretin 1 receptor antagonist GSK1059865 potently reduces ethanol drinking in ethanol dependent mice. *Brain Research, 1636*, 74–80. http://dx.doi.org/10.1016/j.brainres.2016.01.049.

Mabry, P. D., & Campbell, B. A. (1975). Food-deprivation-induced behavioral arousal: Mediation by hypothalamus and amygdala. *Journal of Comparative and Physiological Psychology, 89*(1), 19–38.

Magga, J., Bart, G., Oker-Blom, C., Kukkonen, J. P., Akerman, K. E., & Nasman, J. (2006). Agonist potency differentiates G protein activation and Ca2+ signalling by the orexin receptor type 1. *Biochemical Pharmacology, 71*(6), 827–836. http://dx.doi.org/10.1016/j.bcp.2005.12.021.

Mahler, S. V., Smith, R. J., & Aston-Jones, G. (2013). Interactions between VTA orexin and glutamate in cue-induced reinstatement of cocaine seeking in rats. *Psychopharmacology, 226*(4), 687–698. http://dx.doi.org/10.1007/s00213-012-2681-5.

Maolood, N., & Meister, B. (2010). Nociceptin/orphanin FQ peptide in hypothalamic neurones associated with the control of feeding behaviour. *Journal of Neuroendocrinology, 22*(2), 75–82. http://dx.doi.org/10.1111/j.1365-2826.2009.01946.x.

Marcus, J. N., Aschkenasi, C. J., Lee, C. E., Chemelli, R. M., Saper, C. B., Yanagisawa, M., et al. (2001). Differential expression of orexin receptors 1 and 2 in the rat brain. *The Journal of Comparative Neurology, 435*(1), 6–25.

Martel, P., & Fantino, M. (1996). Mesolimbic dopaminergic system activity as a function of food reward: A microdialysis study. *Pharmacology, Biochemistry, and Behavior, 53*(1), 221–226.

Martin-Fardon, R., Cauvi, G., Kerr, T. M., & Weiss, F. (2016). Differential role of hypothalamic orexin/hypocretin neurons in reward seeking motivated by cocaine versus palatable food. *Addiction Biology*. http://dx.doi.org/10.1111/adb.12441, Epub ahead of print.

Martin-Fardon, R., & Weiss, F. (2014a). Blockade of hypocretin receptor-1 preferentially prevents cocaine seeking: Comparison with natural reward seeking. *Neuroreport, 25*(7), 485–488. http://dx.doi.org/10.1097/WNR.0000000000000120.

Martin-Fardon, R., & Weiss, F. (2014b). N-(2-methyl-6-benzoxazolyl)-N′-1,5-naphthyridin-4-yl urea (SB334867), a hypocretin receptor-1 antagonist, preferentially prevents ethanol seeking: Comparison with natural reward seeking. *Addiction Biology, 19*(2), 233–236. http://dx.doi.org/10.1111/j.1369-1600.2012.00480.x.

Matsuo, E., Mochizuki, A., Nakayama, K., Nakamura, S., Yamamoto, T., Shioda, S., et al. (2011). Decreased intake of sucrose solutions in orexin knockout mice. *Journal of Molecular Neuroscience, 43*(2), 217–224. http://dx.doi.org/10.1007/s12031-010-9475-1.

Matzeu, A., Kerr, T. M., Weiss, F., & Martin-Fardon, R. (2016). Orexin-A/hypocretin-1 mediates cocaine-seeking behavior in the posterior paraventricular nucleus of the thalamus via orexin/hypocretin receptor-2. *The Journal of Pharmacology and Experimental Therapeutics, 359*(2), 273–279. http://dx.doi.org/10.1124/jpet.116.235945.

Mieda, M., Williams, S. C., Sinton, C. M., Richardson, J. A., Sakurai, T., & Yanagisawa, M. (2004). Orexin neurons function in an efferent pathway of a food-entrainable circadian oscillator in eliciting food-anticipatory activity and wakefulness. *The Journal of Neuroscience, 24*(46), 10493–10501. http://dx.doi.org/10.1523/JNEUROSCI.3171-04.2004.

Millan, E. Z., Furlong, T. M., & McNally, G. P. (2010). Accumbens shell-hypothalamus interactions mediate extinction of alcohol seeking. *The Journal of Neuroscience, 30*(13), 4626–4635. http://dx.doi.org/10.1523/JNEUROSCI.4933-09.2010.

Mochizuki, T., Crocker, A., McCormack, S., Yanagisawa, M., Sakurai, T., & Scammell, T. E. (2004). Behavioral state instability in orexin knock-out mice. *The Journal of Neuroscience, 24*(28), 6291–6300. http://dx.doi.org/10.1523/JNEUROSCI.0586-04.2004.

Modirrousta, M., Mainville, L., & Jones, B. E. (2005). Orexin and MCH neurons express c-Fos differently after sleep deprivation vs. recovery and bear different adrenergic receptors. *The European Journal of Neuroscience, 21*(10), 2807–2816. http://dx.doi.org/10.1111/j.1460-9568.2005.04104.x.

Moorman, D. E., & Aston-Jones, G. (2009). Orexin-1 receptor antagonism decreases ethanol consumption and preference selectively in high-ethanol–preferring Sprague–Dawley rats. *Alcohol, 43*(5), 379–386. http://dx.doi.org/10.1016/j.alcohol.2009.07.002.

Moorman, D. E., James, M. H., Kilroy, E. A., & Aston-Jones, G. (2016). Orexin/hypocretin neuron activation is correlated with alcohol seeking and preference in a topographically specific manner. *The European Journal of Neuroscience, 43*(5), 710–720. http://dx.doi.org/10.1111/ejn.13170.

Moorman, D. E., James, M. H., Kilroy, E. A., & Aston-Jones, G. (2017). Orexin/hypocretin-1 receptor antagonism reduces ethanol self-administration and reinstatement selectively in highly-motivated rats. *Brain Research, 1654*(Pt. A), 34–42. http://dx.doi.org/10.1016/j.brainres.2016.10.018.

Morgan, A. J., Harrod, S. B., Lacy, R. T., Stanley, E. M., & Fadel, J. R. (2013). Intravenous prenatal nicotine exposure increases orexin expression in the lateral hypothalamus and orexin innervation of the ventral tegmental area in adult male rats. *Drug and Alcohol Dependence, 132*(3), 562–570. http://dx.doi.org/10.1016/j.drugalcdep.2013.04.003.

Morganstern, I., Chang, G. Q., Barson, J. R., Ye, Z., Karatayev, O., & Leibowitz, S. F. (2010). Differential effects of acute and chronic ethanol exposure on orexin expression in the perifornical lateral hypothalamus. *Alcoholism, Clinical and Experimental Research, 34*(5), 886–896. http://dx.doi.org/10.1111/j.1530-0277.2010.01161.x.

Moriguchi, T., Sakurai, T., Nambu, T., Yanagisawa, M., & Goto, K. (1999). Neurons containing orexin in the lateral hypothalamic area of the adult rat brain are activated by insulin-induced acute hypoglycemia. *Neuroscience Letters, 264*(1–3), 101–104.

Moskowitz, M. J. (1959). Running-wheel activity in the white rat as a function of combined food and water deprivation. *Journal of Comparative and Physiological Psychology, 52*, 621–625.

Muschamp, J. W., Dominguez, J. M., Sato, S. M., Shen, R. Y., & Hull, E. M. (2007). A role for hypocretin (orexin) in male sexual behavior. *The Journal of Neuroscience*, *27*(11), 2837–2845. http://dx.doi.org/10.1523/JNEUROSCI.4121-06.2007.

Muschamp, J. W., Hollander, J. A., Thompson, J. L., Voren, G., Hassinger, L. C., Onvani, S., et al. (2014). Hypocretin (orexin) facilitates reward by attenuating the antireward effects of its cotransmitter dynorphin in ventral tegmental area. *Proceedings of the National Academy of Sciences of the United States of America*, *111*(16), E1648–E1655. http://dx.doi.org/10.1073/pnas.1315542111.

Narita, M., Nagumo, Y., Hashimoto, S., Narita, M., Khotib, J., Miyatake, M., et al. (2006). Direct involvement of orexinergic systems in the activation of the mesolimbic dopamine pathway and related behaviors induced by morphine. *The Journal of Neuroscience*, *26*(2), 398–405. http://dx.doi.org/10.1523/JNEUROSCI.2761-05.2006.

Nishino, S., Ripley, B., Overeem, S., Lammers, G. J., & Mignot, E. (2000). Hypocretin (orexin) deficiency in human narcolepsy. *Lancet*, *355*(9197), 39–40. http://dx.doi.org/10.1016/S0140-6736(99)05582-8.

Nishizawa, D., Kasai, S., Hasegawa, J., Sato, N., Yamada, H., Tanioka, F., et al. (2015). Associations between the orexin (hypocretin) receptor 2 gene polymorphism Val308Ile and nicotine dependence in genome-wide and subsequent association studies. *Molecular Brain*, *8*, 50. http://dx.doi.org/10.1186/s13041-015-0142-x.

Nizhnikov, M. E., Popoola, D. O., & Cameron, N. M. (2016). Transgenerational transmission of the effect of gestational ethanol exposure on ethanol use-related behavior. *Alcoholism, Clinical and Experimental Research*, *40*(3), 497–506. http://dx.doi.org/10.1111/acer.12978.

Nobunaga, M., Obukuro, K., Kurauchi, Y., Hisatsune, A., Seki, T., Tsutsui, M., et al. (2014). High fat diet induces specific pathological changes in hypothalamic orexin neurons in mice. *Neurochemistry International*, *78*, 61–66. http://dx.doi.org/10.1016/j.neuint.2014.09.002.

Olney, J. J., Navarro, M., & Thiele, T. E. (2015). Binge-like consumption of ethanol and other salient reinforcers is blocked by orexin-1 receptor inhibition and leads to a reduction of hypothalamic orexin immunoreactivity. *Alcoholism, Clinical and Experimental Research*, *39*(1), 21–29. http://dx.doi.org/10.1111/acer.12591.

Olney, J. J., Navarro, M., & Thiele, T. E. (2017). The role of orexin signaling in the ventral tegmental area and central amygdala in modulating binge-like ethanol drinking behavior. *Alcoholism, Clinical and Experimental Research*, *41*(3), 551–561. http://dx.doi.org/10.1111/acer.13336.

Olszewski, P. K., Shaw, T. J., Grace, M. K., Hoglund, C. E., Fredriksson, R., Schioth, H. B., et al. (2009). Complexity of neural mechanisms underlying overconsumption of sugar in scheduled feeding: Involvement of opioids, orexin, oxytocin and NPY. *Peptides*, *30*(2), 226–233. http://dx.doi.org/10.1016/j.peptides.2008.10.011.

Park, E. S., Yi, S. J., Kim, J. S., Lee, H. S., Lee, I. S., Seong, J. K., et al. (2004). Changes in orexin-A and neuropeptide Y expression in the hypothalamus of the fasted and high-fat diet fed rats. *Journal of Veterinary Science*, *5*(4), 295–302.

Pasumarthi, R. K., & Fadel, J. (2008). Activation of orexin/hypocretin projections to basal forebrain and paraventricular thalamus by acute nicotine. *Brain Research Bulletin*, 77(6), 367–373. http://dx.doi.org/10.1016/j.brainresbull.2008.09.014.

Pasumarthi, R. K., Reznikov, L. R., & Fadel, J. (2006). Activation of orexin neurons by acute nicotine. *European Journal of Pharmacology*, *535*(1–3), 172–176. http://dx.doi.org/10.1016/j.ejphar.2006.02.021.

Patyal, R., Woo, E. Y., & Borgland, S. L. (2012). Local hypocretin-1 modulates terminal dopamine concentration in the nucleus accumbens shell. *Frontiers in Behavioral Neuroscience*, *6*, 82. http://dx.doi.org/10.3389/fnbeh.2012.00082.

Peyron, C., Tighe, D. K., van den Pol, A. N., de Lecea, L., Heller, H. C., Sutcliffe, J. G., et al. (1998). Neurons containing hypocretin (orexin) project to multiple neuronal systems. *The Journal of Neuroscience, 18*(23), 9996–10015.

Pfaus, J. G., Damsma, G., Nomikos, G. G., Wenkstern, D. G., Blaha, C. D., Phillips, A. G., et al. (1990). Sexual behavior enhances central dopamine transmission in the male rat. *Brain Research, 530*(2), 345–348.

Piccoli, L., Micioni Di Bonaventura, M. V., Cifani, C., Costantini, V. J., Massagrande, M., Montanari, D., et al. (2012). Role of orexin-1 receptor mechanisms on compulsive food consumption in a model of binge eating in female rats. *Neuropsychopharmacology, 37*(9), 1999–2011. http://dx.doi.org/10.1038/npp.2012.48.

Pisanu, A., Lecca, D., Valentini, V., Bahi, A., Dreyer, J. L., Cacciapaglia, F., et al. (2015). Impairment of acquisition of intravenous cocaine self-administration by RNA-interference of dopamine D1-receptors in the nucleus accumbens shell. *Neuropharmacology, 89*, 398–411. http://dx.doi.org/10.1016/j.neuropharm.2014.10.018.

Plaza-Zabala, A., Flores, A., Martin-Garcia, E., Saravia, R., Maldonado, R., & Berrendero, F. (2013). A role for hypocretin/orexin receptor-1 in cue-induced reinstatement of nicotine-seeking behavior. *Neuropsychopharmacology, 38*(9), 1724–1736. http://dx.doi.org/10.1038/npp.2013.72.

Plaza-Zabala, A., Martin-Garcia, E., de Lecea, L., Maldonado, R., & Berrendero, F. (2010). Hypocretins regulate the anxiogenic-like effects of nicotine and induce reinstatement of nicotine-seeking behavior. *The Journal of Neuroscience, 30*(6), 2300–2310. http://dx.doi.org/10.1523/JNEUROSCI.5724-09.2010.

Pontieri, F. E., Tanda, G., & Di Chiara, G. (1995). Intravenous cocaine, morphine, and amphetamine preferentially increase extracellular dopamine in the "shell" as compared with the "core" of the rat nucleus accumbens. *Proceedings of the National Academy of Sciences of the United States of America, 92*(26), 12304–12308.

Pontieri, F. E., Tanda, G., Orzi, F., & Di Chiara, G. (1996). Effects of nicotine on the nucleus accumbens and similarity to those of addictive drugs. *Nature, 382*(6588), 255–257. http://dx.doi.org/10.1038/382255a0.

Poon, K., Barson, J. R., Fagan, S. E., & Leibowitz, S. F. (2012). Developmental changes in embryonic hypothalamic neurons during prenatal fat exposure. *American Journal of Physiology. Endocrinology and Metabolism, 303*(3), E432–E441. http://dx.doi.org/10.1152/ajpendo.00238.2012.

Prince, C. D., Rau, A. R., Yorgason, J. T., & Espana, R. A. (2015). Hypocretin/orexin regulation of dopamine signaling and cocaine self-administration is mediated predominantly by hypocretin receptor 1. *ACS Chemical Neuroscience, 6*(1), 138–146. http://dx.doi.org/10.1021/cn500246j.

Prober, D. A., Rihel, J., Onah, A. A., Sung, R. J., & Schier, A. F. (2006). Hypocretin/orexin overexpression induces an insomnia-like phenotype in zebrafish. *The Journal of Neuroscience, 26*(51), 13400–13410. http://dx.doi.org/10.1523/JNEUROSCI.4332-06.2006.

Rada, P., Avena, N. M., Barson, J. R., Hoebel, B. G., & Leibowitz, S. F. (2012). A high-fat meal, or intraperitoneal administration of a fat emulsion, increases extracellular dopamine in the nucleus accumbens. *Brain Sciences, 2*(2), 242–253. http://dx.doi.org/10.3390/brainsci2020242.

Rada, P., Avena, N. M., & Hoebel, B. G. (2005). Daily bingeing on sugar repeatedly releases dopamine in the accumbens shell. *Neuroscience, 134*(3), 737–744. http://dx.doi.org/10.1016/j.neuroscience.2005.04.043.

Richards, J. K., Simms, J. A., Steensland, P., Taha, S. A., Borgland, S. L., Bonci, A., et al. (2008). Inhibition of orexin-1/hypocretin-1 receptors inhibits yohimbine-induced reinstatement of ethanol and sucrose seeking in Long-Evans rats. *Psychopharmacology, 199*(1), 109–117. http://dx.doi.org/10.1007/s00213-008-1136-5.

Riday, T. T., Fish, E. W., Robinson, J. E., Jarrett, T. M., McGuigan, M. M., & Malanga, C. J. (2012). Orexin-1 receptor antagonism does not reduce the rewarding potency of cocaine in Swiss-Webster mice. *Brain Research*, *1431*, 53–61. http://dx.doi.org/10.1016/j.brainres.2011.11.003.

Risold, P. Y., Griffond, B., Kilduff, T. S., Sutcliffe, J. G., & Fellmann, D. (1999). Preprohypocretin (orexin) and prolactin-like immunoreactivity are coexpressed by neurons of the rat lateral hypothalamic area. *Neuroscience Letters*, *259*(3), 153–156.

Rivas, M., Torterolo, P., Ferreira, A., & Benedetto, L. (2016). Hypocretinergic system in the medial preoptic area promotes maternal behavior in lactating rats. *Peptides*, *81*, 9–14. http://dx.doi.org/10.1016/j.peptides.2016.04.002.

Rorabaugh, J. M., Stratford, J. M., & Zahniser, N. R. (2014). A relationship between reduced nucleus accumbens shell and enhanced lateral hypothalamic orexin neuronal activation in long-term fructose bingeing behavior. *PloS One*, *9*(4), e95019. http://dx.doi.org/10.1371/journal.pone.0095019.

Rosin, D. L., Weston, M. C., Sevigny, C. P., Stornetta, R. L., & Guyenet, P. G. (2003). Hypothalamic orexin (hypocretin) neurons express vesicular glutamate transporters VGLUT1 or VGLUT2. *The Journal of Comparative Neurology*, *465*(4), 593–603. http://dx.doi.org/10.1002/cne.10860.

Sahr, A. E., Sindelar, D. K., Alexander-Chacko, J. T., Eastwood, B. J., Mitch, C. H., & Statnick, M. A. (2008). Activation of mesolimbic dopamine neurons during novel and daily limited access to palatable food is blocked by the opioid antagonist LY255582. *American Journal of Physiology. Regulatory, Integrative and Comparative Physiology*, *295*(2), R463–R471. http://dx.doi.org/10.1152/ajpregu.00390.2007.

Sakurai, T., Amemiya, A., Ishii, M., Matsuzaki, I., Chemelli, R. M., Tanaka, H., et al. (1998). Orexins and orexin receptors: A family of hypothalamic neuropeptides and G protein-coupled receptors that regulate feeding behavior. *Cell*, *92*(4), 573–585.

Schmeichel, B. E., Herman, M. A., Roberto, M., & Koob, G. F. (2017). Hypocretin neurotransmission within the central amygdala mediates escalated cocaine self-administration and stress-induced reinstatement in rats. *Biological Psychiatry*, *81*(7), 606–615. http://dx.doi.org/10.1016/j.biopsych.2016.06.010.

Schneider, E. R., Rada, P., Darby, R. D., Leibowitz, S. F., & Hoebel, B. G. (2007). Orexigenic peptides and alcohol intake: Differential effects of orexin, galanin, and ghrelin. *Alcoholism, Clinical and Experimental Research*, *31*(11), 1858–1865. http://dx.doi.org/10.1111/j.1530-0277.2007.00510.x.

Sharma, R., Sahota, P., & Thakkar, M. M. (2014). Role of adenosine and the orexinergic perifornical hypothalamus in sleep-promoting effects of ethanol. *Sleep*, *37*(3), 525–533. http://dx.doi.org/10.5665/sleep.3490.

Shaw, J. K., Ferris, M. J., Locke, J. L., Brodnik, Z. D., Jones, S. R., & Espana, R. A. (2016). Hypocretin/orexin knock-out mice display disrupted behavioral and dopamine responses to cocaine. *Addiction Biology*. http://dx.doi.org/10.1111/adb.12432, Epub ahead of print.

Shoblock, J. R., Welty, N., Aluisio, L., Fraser, I., Motley, S. T., Morton, K., et al. (2011). Selective blockade of the orexin-2 receptor attenuates ethanol self-administration, place preference, and reinstatement. *Psychopharmacology*, *215*(1), 191–203. http://dx.doi.org/10.1007/s00213-010-2127-x.

Smith, R. J., See, R. E., & Aston-Jones, G. (2009). Orexin/hypocretin signaling at the orexin 1 receptor regulates cue-elicited cocaine-seeking. *The European Journal of Neuroscience*, *30*(3), 493–503. http://dx.doi.org/10.1111/j.1460-9568.2009.06844.x.

Smith, R. J., Tahsili-Fahadan, P., & Aston-Jones, G. (2010). Orexin/hypocretin is necessary for context-driven cocaine-seeking. *Neuropharmacology*, *58*(1), 179–184. http://dx.doi.org/10.1016/j.neuropharm.2009.06.042.

Srinivasan, S., Simms, J. A., Nielsen, C. K., Lieske, S. P., Bito-Onon, J. J., Yi, H., et al. (2012). The dual orexin/hypocretin receptor antagonist, almorexant, in the ventral tegmental area attenuates ethanol self-administration. *PloS One*, 7(9), e44726. http://dx.doi.org/10.1371/journal.pone.0044726.

Steiner, M. A., Lecourt, H., & Jenck, F. (2013a). The dual orexin receptor antagonist almorexant, alone and in combination with morphine, cocaine and amphetamine, on conditioned place preference and locomotor sensitization in the rat. *The International Journal of Neuropsychopharmacology*, *16*(2), 417–432. http://dx.doi.org/10.1017/S1461145712000193.

Steiner, M. A., Sciarretta, C., Pasquali, A., & Jenck, F. (2013b). The selective orexin receptor 1 antagonist ACT-335827 in a rat model of diet-induced obesity associated with metabolic syndrome. *Frontiers in Pharmacology*, *4*, 165. http://dx.doi.org/10.3389/fphar.2013.00165.

Sterling, M. E., Chang, G. Q., Karatayev, O., Chang, S. Y., & Leibowitz, S. F. (2016). Effects of embryonic ethanol exposure at low doses on neuronal development, voluntary ethanol consumption and related behaviors in larval and adult zebrafish: Role of hypothalamic orexigenic peptides. *Behavioural Brain Research*, *304*, 125–138. http://dx.doi.org/10.1016/j.bbr.2016.01.013.

Sterling, M. E., Karatayev, O., Chang, G. Q., Algava, D. B., & Leibowitz, S. F. (2015). Model of voluntary ethanol intake in zebrafish: Effect on behavior and hypothalamic orexigenic peptides. *Behavioural Brain Research*, *278*, 29–39. http://dx.doi.org/10.1016/j.bbr.2014.09.024.

Sun, G., Narita, K., Murata, T., Honda, K., & Higuchi, T. (2003). Orexin-A immunoreactivity and prepro-orexin mRNA expression in hyperphagic rats induced by hypothalamic lesions and lactation. *Journal of Neuroendocrinology*, *15*(1), 51–60.

Suto, N., Ecke, L. E., You, Z. B., & Wise, R. A. (2010). Extracellular fluctuations of dopamine and glutamate in the nucleus accumbens core and shell associated with lever-pressing during cocaine self-administration, extinction, and yoked cocaine administration. *Psychopharmacology*, *211*(3), 267–275. http://dx.doi.org/10.1007/s00213-010-1890-z.

Sweet, D. C., Levine, A. S., Billington, C. J., & Kotz, C. M. (1999). Feeding response to central orexins. *Brain Research*, *821*(2), 535–538.

Takahashi, K., Lin, J. S., & Sakai, K. (2008). Neuronal activity of orexin and non-orexin waking-active neurons during wake-sleep states in the mouse. *Neuroscience*, *153*(3), 860–870. http://dx.doi.org/10.1016/j.neuroscience.2008.02.058.

Tang, J., Chen, J., Ramanjaneya, M., Punn, A., Conner, A. C., & Randeva, H. S. (2008). The signalling profile of recombinant human orexin-2 receptor. *Cellular Signalling*, *20*(9), 1651–1661. http://dx.doi.org/10.1016/j.cellsig.2008.05.010.

Terrill, S. J., Hyde, K. M., Kay, K. E., Greene, H. E., Maske, C. B., Knierim, A. E., et al. (2016). Ventral tegmental area orexin 1 receptors promote palatable food intake and oppose postingestive negative feedback. *American Journal of Physiology. Regulatory, Integrative and Comparative Physiology*, *311*(3), R592–R599. http://dx.doi.org/10.1152/ajpregu.00097.2016.

Thakkar, M. M., Winston, S., & McCarley, R. W. (2002). Orexin neurons of the hypothalamus express adenosine A1 receptors. *Brain Research*, *944*(1–2), 190–194.

Thannickal, T. C., Moore, R. Y., Nienhuis, R., Ramanathan, L., Gulyani, S., Aldrich, M., et al. (2000). Reduced number of hypocretin neurons in human narcolepsy. *Neuron*, *27*(3), 469–474.

Thiele, T. E., Crabbe, J. C., & Boehm, S. L., 2nd (2014). "Drinking in the Dark" (DID): A simple mouse model of binge-like alcohol intake. *Current Protocols in Neuroscience*, *68*. http://dx.doi.org/10.1002/0471142301.ns0949s68. 9 49 1–12.

Torrealba, F., Yanagisawa, M., & Saper, C. B. (2003). Colocalization of orexin a and glutamate immunoreactivity in axon terminals in the tuberomammillary nucleus in rats. *Neuroscience*, *119*(4), 1033–1044.

Trivedi, P., Yu, H., MacNeil, D. J., Van der Ploeg, L. H., & Guan, X. M. (1998). Distribution of orexin receptor mRNA in the rat brain. *FEBS Letters*, *438*(1–2), 71–75.

Tsujino, N., Yamanaka, A., Ichiki, K., Muraki, Y., Kilduff, T. S., Yagami, K., et al. (2005). Cholecystokinin activates orexin/hypocretin neurons through the cholecystokinin A receptor. *The Journal of Neuroscience*, *25*(32), 7459–7469. http://dx.doi.org/10.1523/JNEUROSCI.1193-05.2005.

Tsuneki, H., Nagata, T., Fujita, M., Kon, K., Wu, N., Takatsuki, M., et al. (2016). Nighttime administration of nicotine improves hepatic glucose metabolism via the hypothalamic orexin system in mice. *Endocrinology*, *157*(1), 195–206. http://dx.doi.org/10.1210/en.2015-1488.

Tung, L. W., Lu, G. L., Lee, Y. H., Yu, L., Lee, H. J., Leishman, E., et al. (2016). Orexins contribute to restraint stress-induced cocaine relapse by endocannabinoid-mediated disinhibition of dopaminergic neurons. *Nature Communications*, 7, 12199. http://dx.doi.org/10.1038/ncomms12199.

Uslaner, J. M., Winrow, C. J., Gotter, A. L., Roecker, A. J., Coleman, P. J., Hutson, P. H., et al. (2014). Selective orexin 2 receptor antagonism blocks cue-induced reinstatement, but not nicotine self-administration or nicotine-induced reinstatement. *Behavioural Brain Research*, *269*, 61–65. http://dx.doi.org/10.1016/j.bbr.2014.04.012.

Valdivia, S., Patrone, A., Reynaldo, M., & Perello, M. (2014). Acute high fat diet consumption activates the mesolimbic circuit and requires orexin signaling in a mouse model. *PloS One*, *9*(1), e87478. http://dx.doi.org/10.1371/journal.pone.0087478.

van den Pol, A. N., Gao, X. B., Obrietan, K., Kilduff, T. S., & Belousov, A. B. (1998). Presynaptic and postsynaptic actions and modulation of neuroendocrine neurons by a new hypothalamic peptide, hypocretin/orexin. *The Journal of Neuroscience*, *18*(19), 7962–7971.

van Holst, R. J., van der Cruijsen, L., van Mierlo, P., Lammers, G. J., Cools, R., Overeem, S., et al. (2016). Aberrant food choices after satiation in human orexin-deficient narcolepsy type 1. *Sleep*, *39*(11), 1951–1959. http://dx.doi.org/10.5665/sleep.6222.

Vasselli, J. R., & Maggio, C. A. (1990). Diet composition determines course of hyperphagia in developing Zucker obese rats. *Physiology & Behavior*, *48*(6), 805–811.

Voorhees, C. M., & Cunningham, C. L. (2011). Involvement of the orexin/hypocretin system in ethanol conditioned place preference. *Psychopharmacology*, *214*(4), 805–818. http://dx.doi.org/10.1007/s00213-010-2082-6.

Wang, J., Akabayashi, A., Yu, H. J., Dourmashkin, J., Alexander, J. T., Silva, I., et al. (1998). Hypothalamic galanin: Control by signals of fat metabolism. *Brain Research*, *804*(1), 7–20.

Wang, B., You, Z. B., & Wise, R. A. (2009). Reinstatement of cocaine seeking by hypocretin (orexin) in the ventral tegmental area: Independence from the local corticotropin-releasing factor network. *Biological Psychiatry*, *65*(10), 857–862. http://dx.doi.org/10.1016/j.biopsych.2009.01.018.

Whichelow, M. J. (1975). Letter: Calorie requirements for successful breast feeding. *Archives of Disease in Childhood*, *50*(8), 669.

White, C. L., Ishii, Y., Mendoza, T., Upton, N., Stasi, L. P., Bray, G. A., et al. (2005). Effect of a selective OX1R antagonist on food intake and body weight in two strains of rats that differ in susceptibility to dietary-induced obesity. *Peptides*, *26*(11), 2331–2338. http://dx.doi.org/10.1016/j.peptides.2005.03.042.

Winsky-Sommerer, R., Yamanaka, A., Diano, S., Borok, E., Roberts, A. J., Sakurai, T., et al. (2004). Interaction between the corticotropin-releasing factor system and

hypocretins (orexins): A novel circuit mediating stress response. *The Journal of Neuroscience, 24*(50), 11439–11448. http://dx.doi.org/10.1523/JNEUROSCI.3459-04.2004.
Wortley, K. E., Chang, G. Q., Davydova, Z., & Leibowitz, S. F. (2003). Peptides that regulate food intake: Orexin gene expression is increased during states of hypertriglyceridemia. *American Journal of Physiology. Regulatory, Integrative and Comparative Physiology, 284*(6), R1454–R1465. http://dx.doi.org/10.1152/ajpregu.00286.2002.
Xu, R. Y., Wan, Y. P., Tang, Q. Y., Wu, J., & Cai, W. (2008). The effects of high fat on central appetite genes in Wistar rats: A microarray analysis. *Clinica Chimica Acta, 397*(1–2), 96–100. http://dx.doi.org/10.1016/j.cca.2008.07.027.
Yamanaka, A., Tabuchi, S., Tsunematsu, T., Fukazawa, Y., & Tominaga, M. (2010). Orexin directly excites orexin neurons through orexin 2 receptor. *The Journal of Neuroscience, 30*(38), 12642–12652. http://dx.doi.org/10.1523/JNEUROSCI.2120-10.2010.
Yoshida, Y., Fujiki, N., Nakajima, T., Ripley, B., Matsumura, H., Yoneda, H., et al. (2001). Fluctuation of extracellular hypocretin-1 (orexin A) levels in the rat in relation to the light-dark cycle and sleep-wake activities. *The European Journal of Neuroscience, 14*(7), 1075–1081.
Yoshida, K., McCormack, S., Espana, R. A., Crocker, A., & Scammell, T. E. (2006). Afferents to the orexin neurons of the rat brain. *The Journal of Comparative Neurology, 494*(5), 845–861. http://dx.doi.org/10.1002/cne.20859.
Zhang, G. C., Mao, L. M., Liu, X. Y., & Wang, J. Q. (2007). Long-lasting up-regulation of orexin receptor type 2 protein levels in the rat nucleus accumbens after chronic cocaine administration. *Journal of Neurochemistry, 103*(1), 400–407. http://dx.doi.org/10.1111/j.1471-4159.2007.04748.x.
Zheng, H., Patterson, L. M., & Berthoud, H. R. (2005). Orexin-A projections to the caudal medulla and orexin-induced c-Fos expression, food intake, and autonomic function. *The Journal of Comparative Neurology, 485*(2), 127–142. http://dx.doi.org/10.1002/cne.20515.
Zhou, Y., Bendor, J., Hofmann, L., Randesi, M., Ho, A., & Kreek, M. J. (2006). Mu opioid receptor and orexin/hypocretin mRNA levels in the lateral hypothalamus and striatum are enhanced by morphine withdrawal. *The Journal of Endocrinology, 191*(1), 137–145. http://dx.doi.org/10.1677/joe.1.06960.
Zhou, Y., Cui, C. L., Schlussman, S. D., Choi, J. C., Ho, A., Han, J. S., et al. (2008). Effects of cocaine place conditioning, chronic escalating-dose "binge" pattern cocaine administration and acute withdrawal on orexin/hypocretin and preprodynorphin gene expressions in lateral hypothalamus of Fischer and Sprague-Dawley rats. *Neuroscience, 153*(4), 1225–1234. http://dx.doi.org/10.1016/j.neuroscience.2008.03.023.
Zhou, W. L., Gao, X. B., & Picciotto, M. R. (2015). Acetylcholine acts through nicotinic receptors to enhance the firing rate of a subset of Hypocretin neurons in the mouse hypothalamus through distinct presynaptic and postsynaptic mechanisms. *eNeuro, 2*(1). http://dx.doi.org/10.1523/ENEURO.0052-14.2015.
Zhou, L., Ghee, S. M., Chan, C., Lin, L., Cameron, M. D., Kenny, P. J., et al. (2012). Orexin-1 receptor mediation of cocaine seeking in male and female rats. *The Journal of Pharmacology and Experimental Therapeutics, 340*(3), 801–809. http://dx.doi.org/10.1124/jpet.111.187567.

CHAPTER NINE

Oxytocin, Tolerance, and the Dark Side of Addiction

Cort A. Pedersen[1]
The University of North Carolina at Chapel Hill, Chapel Hill, NC, United States
[1]Corresponding author: e-mail address: cort_pedersen@med.unc.edu

Contents

Abstract

Substance use disorders blight the lives of millions of people and inflict a heavy financial burden on society. There is a compelling need for new pharmacological treatments as current drugs have limited efficacy and other major drawbacks. A substantial number of animal and recent clinical studies indicate that the neuropeptide, oxytocin, is a particularly promising therapeutic agent for human addictions, especially alcohol use disorders. In preliminary trials, we found that oxytocin administered by the intranasal route,

International Review of Neurobiology, Volume 136
ISSN 0074-7742
http://dx.doi.org/10.1016/bs.irn.2017.08.003

which produces some neuropeptide penetration into the CNS, potently blocked withdrawal and reduced alcohol consumption in heavy drinkers. A considerable body of earlier animal studies demonstrated that oxytocin inhibits tolerance to alcohol, opioids, and stimulants as well as withdrawal from alcohol and opioids. Based on these preclinical findings and our clinical results, we hypothesize that oxytocin may exert therapeutic effects in substance dependence by the novel mechanism of diminishing established tolerance. A newer wave of studies has almost unanimously found that oxytocin decreases self-administration of a number of addictive substances in several animal models of addiction. Reduction of established tolerance should be included among the potential explanations of oxytocin effects in these studies and changes in tolerance should be examined in future studies in relationship to oxytocin influences on acquisition and reinstatement of self-administration as well as extinction of drug seeking. Oxytocin efficacy in reducing anxiety and stress responses as well as established tolerance suggests it may be uniquely effective in reducing negative reinforcement (Koob's "dark side" of addiction) that maintains chronic substance use.

Addiction encompasses a distressingly prevalent set of disorders that not only heavily burden society but also blight the lives of many millions of individuals (Barrio, Reynolds, Garcia-Altés, Gual, & Anderson, 2017; Surgeon General's Report, 2016). Alcohol use disorders are a major public health problem in the United States. The most recent 12-month/lifetime prevalence figures are, respectively, 13.9% and 29.1% (17.6% and 36.0% in men) (Grant et al., 2015). The Center for Disease Control places alcohol as the number three cause of preventable deaths in the United States (Mokdad, Marks, Stroup, & Gerberding, 2004). The World Health Organization estimates that alcohol consumption causes an annual worldwide death toll of 3.3 million (5.9% of all deaths) (Gulland, 2014). The economic costs of alcohol misuse in the United States are estimated at about $223.5 billion per year (Bouchery, Harwood, Sacks, Simon, & Brewer, 2011). Three new agents for the treatment of alcohol use disorders have received FDA approval in the past 20 years (oral naltrexone, long-acting intramuscular naltrexone, and acamprosate). However, these medications have only demonstrated small effect sizes when studied in the broad population of all patients with alcohol use disorders (Kranzler & Van Kirk, 2001; Maisel, Blodgett, Wilbourne, Humphreys, & Finney, 2013) and their full benefits require detoxification or a period of abstinence prior to initiating treatment (Maisel et al., 2013). Their utilization in clinical practice has been limited (Mark, Kassed, Vandivort-Warren, Levit, & Kranzler, 2009). Opioid use disorders have emerged over the past 10 years as a public health crisis in the United States, and opioid overdose is now the leading cause of death for individuals between 25 and 44 years of age. Opioid misuse and lethal overdoses, fueled by easy access to prescription opioids and heroin, have reached epidemic levels in the United States (Kanouse & Compton, 2015; Kolodny et al., 2015). Overall, more than 52,000 individuals died from drug/alcohol overdoses in 2015, an 11.4% increase from 2014, with the majority of those deaths occurring due to opioid misuse (Rudd, Seth, David, & Scholl, 2016). Methadone, and, more recently, buprenorphine–naloxone opioid replacement treatments are highly effective in reducing illicit opioid use in individuals with opioid use disorder (Fullerton et al., 2014; Salsitz & Wiegand, 2016; Weiss & Rao, 2017). However, these medications simply replace illegal and diverted opioids for medically managed opioid treatment and, therefore, maintain opioid dependence. Weaning patients from buprenorphine–naloxone results in very high rates of relapse (Weiss & Rao, 2017). Effective medications to truly eliminate opioid dependence remain elusive. Stimulant use disorders are also prevalent, but the search for effective medication treatments is in its infancy (Kiluk et al., 2016; Stoops & Rush, 2014). Cannabinoids, most frequently cannabis, are the most widely used illicit psychoactive substances in the United States, but there are no effective pharmacological treatments (Laprevote, Schwan, Schwitzer, Rolland, & Thome, 2015). To move the treatment of substance use disorders forward there is a compelling need to find new medications that are both effective and well tolerated.

The neuropeptide, oxytocin, is emerging as a particularly promising new treatment for substance use disorders in general, especially alcohol, opioid, and stimulant use disorders. There is now an impressive preclinical literature on oxytocin and addictive substances. This literature is composed of: (1) an early seminal body of studies demonstrating that oxytocin inhibits tolerance formation to alcohol, opioids, and stimulants in nontolerant animals as well as attenuates rates of self-administration and withdrawal in tolerant animal (Kovács, Sarnyai, & Szabó, 1998; Sarnyai & Kovács, 1994); and (2) a recent explosion of studies employing methods used widely in more current preclinical addiction research, such as conditioned place preference (CPP), two-bottle choice, and drinking in the dark paradigms to study oxytocin effects on the initial acquisition of addictive substance self-administration, maintenance of established rats of self-administration, and relapse of self-administration after periods of deprivation. Despite the large body of promising preclinical work, studies of oxytocin effects in human addictions are few in number. To date, only four small clinical trials (two published), some very promising, have been conducted. This chapter is not a review of this burgeoning area of investigation. The reader is referred to Lee, Rohn, Tanda, and Leggio (2016) who provide an excellent, comprehensive summary of animal and human studies relevant to the topic of oxytocin and dependence on/consumption of a wide variety of addictive substances. This chapter is primarily theoretical, first developing the hypothesis that oxytocin may exert therapeutic effects in substance use disorders by a unique mechanism: rapid inhibition of neuroadaptations underlying established tolerance. This notion is based on: (1) the results of Phase IIa randomized, placebo-controlled clinical trials conducted by my colleagues and me that demonstrate that daily intranasal administration of oxytocin potently blocks withdrawal symptoms and consumption of alcohol in very heavy drinkers; and (2) the early, ground-breaking preclinical studies, referred to above, demonstrating that oxytocin inhibits tolerance formation to addictive substances in nontolerant animals and reduces rates of self-administration and withdrawal in highly tolerant animals. The chapter will begin with our findings on oxytocin treatment of withdrawal and drinking in heavy drinkers and the implications of those findings. Immediately following will be a summary of the small number of published human studies of oxytocin effects in subjects with substance dependence disorders. I will then discuss in some detail the important inferences from the earlier seminal work on oxytocin effects on tolerance formation and established tolerance, which is usually glossed over briefly in newer reports on oxytocin and addiction, to add depth to the hypothesis that oxytocin exerts antiaddictive effects by inhibiting established tolerance.

Thereafter, I will provide an overview of the newer generation of preclinical studies of oxytocin effects in animal models of addiction and make the case that their findings generally support the hypothesis that oxytocin inhibits established tolerance, a viewpoint that is not mentioned in the discussion sections of recent publications but is compatible with the interpretations of the reported results. Some recent animal studies have examined neurochemical effects of oxytocin that may provide insights into mechanisms whereby oxytocin may reverse neuroadaptations underlying established tolerance. Oxytocin is well known to attenuate stress effects such as heightened anxiety and hypothalamic–pituitary–adrenal axis activation, phenomena that occur during chronic addictive substance use, withdrawal, and protracted abstinence (Cardoso, Kingdon, & Ellenbogen, 2014; Heinrichs, Baumgartner, Kirschbaum, & Ehlert, 2003; Koob, 2015; Koob et al., 2014; MacDonald & Feifel, 2014). The potential contributions of these effects on oxytocin influences in newer generation models of addiction-like behavior will be examined. Koob has persuasively made the case in numerous influential publications that the motivational processes leading to addiction progress from "impulsivity (positive reinforcement) to compulsivity (negative reinforcement)" (Koob, 2009, 2013a, 2013b, 2014, 2015; Koob et al., 2014; Koob & Le Moal, 2008; Koob & Mason, 2016; Koob & Volkow, 2016). In other words, while the rewarding, even euphoric, effects of addictive substances motivate use early on, with sustained, frequent use and the formation of increasing tolerance and eventually dependence, the primary motivation to use shifts toward avoiding the negative effects of abstinence such as dysphoria and even physical withdrawal (the "dark side" of addiction). I suggest that the hypothesized inhibitory effect of oxytocin on established tolerance and the well-established antistress effects of oxytocin indicate it might be an effective treatment for the "dark side" negative motivation that Koob and colleagues argue maintain addiction.

1. INTRANASAL OXYTOCIN TREATMENT OF ALCOHOL WITHDRAWAL

Direct and indirect evidence indicates that administration in a nasal spray results in some oxytocin penetration into the brain. In humans and monkeys intranasal administration significantly increases CSF concentrations of oxytocin (Dal Monte, Noble, Turchi, Cummins, & Averbeck, 2014; Freeman et al., 2016; Lee et al., 2017; Striepens et al., 2013), and intranasal administration in rodents significantly raises oxytocin concentrations in CSF and deep brain structures (Neumann, Maloumby, Beiderbeck, Lukas, & Landgraf, 2013). The Lee et al. (2017) study is of particular importance because it employed unique methods to demonstrate that intranasal administration of oxytocin does not increase CSF concentrations by stimulating release of endogenous oxytocin within the CNS. Several fMRI studies have shown activity changes in specific brain regions (e.g., the central amygdala) after intranasal oxytocin administration (Domes et al., 2007, 2010; Kirsch et al., 2005; Zink & Meyer-Lindenberg, 2012), findings that strongly support CNS penetration of oxytocin.

Single doses of intranasal oxytocin have been reported by numerous investigators to exert prosocial and antistress effects (enhanced trust, cooperation, social cognition, and reduction in stressor-induced anxiety and HPA axis activation; Williams, Chang, & Pedersen, 2014; Young & Flanagan-Cato, 2012) akin to results found in animal studies (Carter,

Lederhendler, & Kirkpatrick, 1997; Lee, Macbeth, Pagani, & Young, 2009; Pedersen, Caldwell, Jirikowski, & Insel, 1992). Currently there are at least 32 published human placebo-controlled, randomized clinical trials (RCTs) testing the clinical efficacy of repeated daily dosing with intranasal oxytocin over periods ranging from 3 days to several months including: (1) five trials in borderline personality disorder (Amad, Thomas, & Perez-Rodriguez, 2015), (2) seven trials in autism (Preti et al., 2014), (3) a 13-week trial of twice-daily doses of 40 IU oxytocin in chronic constipation (Ohlsson et al., 2005), (4) an 8-week trial of four times daily doses of 24 IU oxytocin in obesity (Zhang et al., 2013), and (5) a 12-week trial of twice-daily doses of 24 IU in schizophrenia (Jarskog et al., 2017). Prolonged intranasal oxytocin appears to be safe (MacDonald et al., 2011): no serious adverse events have occurred in the numerous published clinical trials.

Based on evidence that oxytocin pretreatment diminished picrotoxin-induced withdrawal symptoms (seizures) in alcohol-dependent mice (Szabó, Kovács, & Telegdy, 1987) and other evidence that oxytocin reduces effects of addictive substances in tolerant animals (see below), we conducted a 3-day placebo-controlled RCT testing the efficacy of intranasal oxytocin compared to placebo treatment on alcohol withdrawal symptoms in heavy drinking human subjects studied in a clinical research unit. Inclusion in the trial required a history of consuming 8–30 standard drinks/day (a standard drink contains 12 g of ethanol, approximately the same amount of ethanol as a 12-oz beer) for at least 4 weeks before study entry and a history of having withdrawal (excluding those with history of delirium tremens or withdrawal seizures) upon stopping or cutting down on alcohol consumption. Subjects were not required to stop drinking before entering the trial. Starting shortly after admission to the research unit, subjects received twice-daily intranasal doses (six 0.1 mL insufflations) containing 24 IU of oxytocin per dose (Syntocinon Spray, Novartis) or placebo solution containing the same ingredients as Syntocinon Spray except for oxytocin. Clinical Institute Withdrawal Assessment for Alcohol scores (CIWA, Sullivan, Sykora, Schneiderman, Naranjo, & Sellers, 1989), modified to include vital signs in the scores, was obtained every 4 h. Lorazepam (2 mg) was given when CIWA scores were ≥ 12 and repeated at 1-h intervals until CIWA scores dropped below 10. In oxytocin recipients ($N = 7$) compared to placebo recipients ($N = 4$), mean CIWA scores on inpatient days 1 and 2 were much lower (Fig. 1). Also, oxytocin recipients required far fewer cumulative milligrams of lorazepam than did placebo recipients to complete detoxification

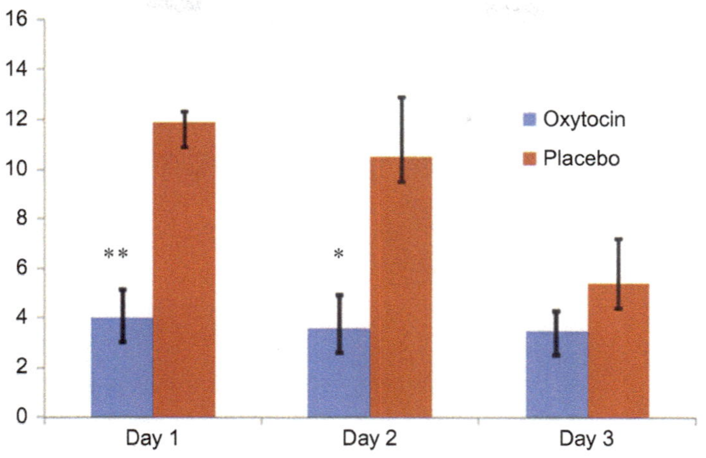

Fig. 1 Comparison of oxytocin and placebo groups on mean CIWA scores (±SEs) on research unit admission days 1–3. *$P=0.0015$; **$P<0.0001$.

during the 3-day period on the research unit [3.4 mg ± 4.7 (SD) vs 16.5 mg ± 4.4; $P=0.0015$]. Self-rating on the Profile of Mood States Tension/Anxiety subscale (McNair, Lorr, & Droppleman, 1971) was also significantly lower in subjects who received oxytocin compared to placebo on day 2 on the research unit ($P=0.007$). This was the first published RCT testing repeated daily doses of oxytocin for any substance use disorder (Pedersen et al., 2013).

We have replicated our original results in a substantially different cohort: low SES heavy drinking individuals seeking treatment at a community inpatient detoxification center. Study inclusion required an average of at least 3 heavy drinking days/week (men ≥5; women ≥4 standard drinks/day) during the 4 weeks prior to admission and meeting criteria for DSM-IV-TR (First, Spitzer, Gibbon, & Williams, 2002) for current alcohol dependence and prior withdrawal. Intranasal test treatments (24 IU oxytocin or placebo three times daily) were initiated within the first 36 h of their inpatient stay in individuals who had not received more than a total of 6 mg of lorazepam to control withdrawal symptoms since admission. CIWA ratings were obtained every 4 h in patients admitted for alcohol detoxification. During the 48 h after initiation of randomly assigned intranasal test treatments, mean CIWA scores declined in oxytocin recipients ($N=4$) but increased in placebo recipients ($N=4$) ($P=0.025$, Fig. 2). The sum of lorazepam doses (each 2 mg) given for elevated CIWA scores during the 48-h period trended toward being lower in oxytocin recipients ($P=0.076$).

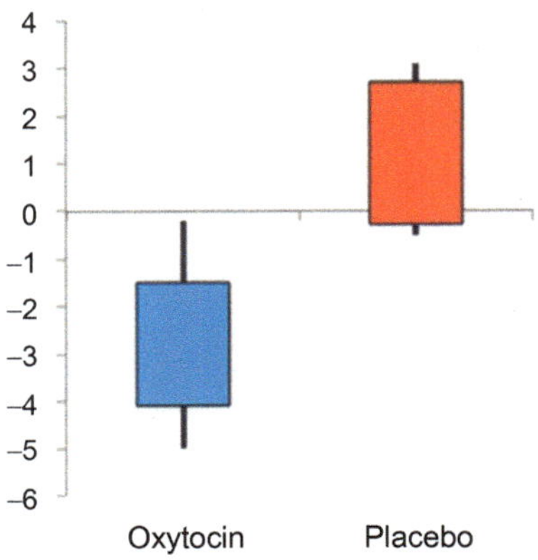

Fig. 2 Mean CIWA scores (±SE) decreased in the oxytocin group but increased in the placebo group during the 48h after initiation of three times daily intranasal treatment ($P = 0.025$).

2. INTRANASAL OXYTOCIN TREATMENT EFFECTS ON ALCOHOL CONSUMPTION IN HEAVY DRINKERS

We recently completed an NIAAA-funded 12-week, double blind, RCT comparing twice-daily intranasal oxytocin and placebo on alcohol consumption, craving, and anxiety in subjects who met DSM-IV-TR criteria (First et al., 2002) for alcohol dependence and were heavy drinkers (≥35 standard drinks/week for men, ≥28 standard drinks/week for women) for at least 4 weeks prior to study entry. Subjects were recruited from individuals who responded to radio and social media ads. Screening procedures assured that participants were medically and psychiatrically stable. After training in intranasal self-administration technique and research staff monitoring of self-administration of the first dose of the treatment to which they were randomly assigned, subjects took 60-mL bottles containing 50mL of intranasal test solution home and self-administered intranasal doses (40IU of oxytocin/dose or placebo) three times daily for the first 2 days and then twice daily for the remainder of the 12-week treatment period. New spray bottles were exchanged for used bottles every 2 weeks. Participants were evaluated at clinic visits at 1, 2, 3, 4, 6, 8, 10, and 12 weeks after starting intranasal test treatments. Primary outcome measures included: (1) heavy drinking days (≥5 standard drinks for men, ≥4 drinks for women), abstinent days, and drinks/drinking day as determined by the Timeline Followback interview (TLFB, Sobell, Sobell, Leo, & Cancilla, 1988);

(2) subject self-ratings on the Penn Alcohol Craving Scale (Flannery, Volpicelli, & Pettinati, 1999); and (3) subject self-ratings on the Spielberger State-Trait Anxiety Inventory (Spielberger, Gorsuch, & Lushene, 1969). Blood and urine laboratory tests were obtained during initial screening as well as at the 4-, 8-, and 12-week clinic visits. The heavy drinking requirement for entry into this study as well as the 12-week duration of the trial was chosen to decrease the magnitude of the placebo effect that can confound detecting a test treatment effect (Maisel et al., 2013).

22 subjects (13 men, 9 women) screened into the study: 12 were randomly assigned to intranasal oxytocin and 10 were assigned to placebo. 10 of the subjects assigned to oxytocin and 5 assigned to placebo completed the entire 12-week trial. There were no significant differences between treatment groups on baseline measures of drinking, craving, or anxiety. Nor were there significant group differences on age, gender, education, and other demographic factors.

Statistical analyses revealed that the oxytocin compared to the placebo treatment group had significantly fewer heavy drinking days (Fig. 3; $P=0.047$; about 15 fewer heavy drinking days during the 12-week trial, $d=0.97$) and drank very significantly fewer drinks/drinking day (Fig. 4;

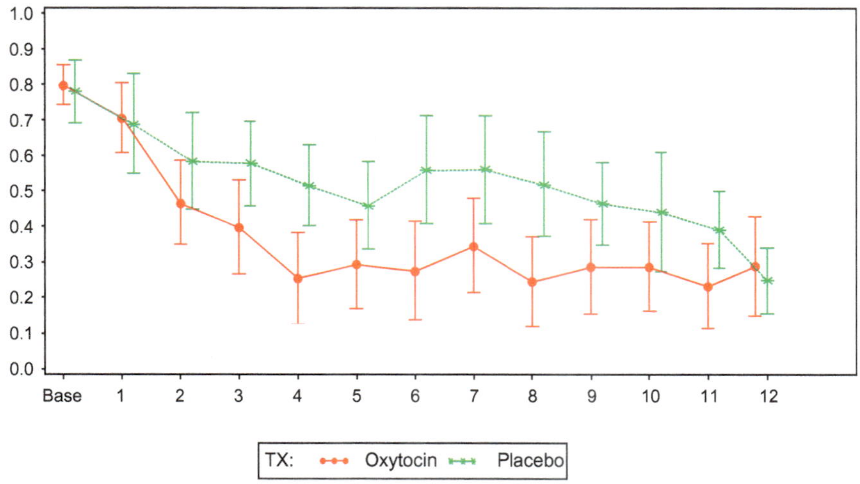

Fig. 3 Comparison of oxytocin and placebo on mean (±SE) weekly proportion of heavy drinking days during the 12-week treatment trial ($P=0.047$). Base values are the mean (±SE) proportion of heavy drinking days during the 90 days prior to study entry derived from the Timeline Followback interview conducted during screening to determine study eligibility.

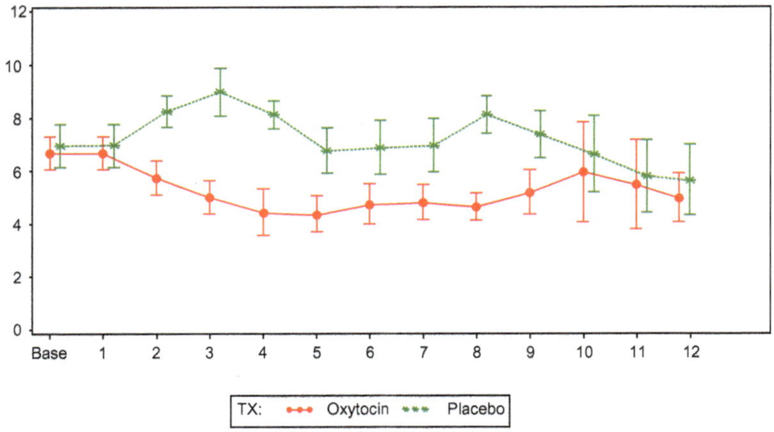

Fig. 4 Comparison of oxytocin and placebo on mean (±SE) weekly drinks/drinking day during the 12-week treatment trial ($P = 0.0008$). Base values are the mean (±SE) weekly drinks/drinking day during the 90 days prior to study entry derived from the Timeline Followback interview conducted during screening to determine study eligibility.

$P = 0.0008$; about 2.6 fewer drinks/drinking day during the trial, $d = 1.83$). No significant oxytocin effect on abstinent days was noted. Spielberger state anxiety self-ratings trended weakly toward being lower in the oxytocin compared to the placebo group ($P = 0.098$), corresponding to an on-average 3.5-point difference. There were no group differences in alcohol craving scores. No severe adverse events occurred. Also, vital signs remained stable and laboratory measures remained within normal limits in all subjects except for transient, asymptomatic, and mild hyponatremia in three oxytocin subjects.

There are substantial limitations to our preliminary clinical trials: each study cohort was small; because the studies only included very heavy drinkers, it remains unclear if the beneficial effects observed would be found in moderate drinkers who comprise the vast majority of individuals with alcohol use disorders. Larger trials including moderate drinkers, as well as heavy drinkers, are needed.

3. IMPLICATIONS OF OUR PRELIMINARY FINDINGS

Based on our preliminary studies, intranasal oxytocin treatment has the potential to be effective both in reducing alcohol withdrawal symptoms during detoxification and reducing drinking in individuals with alcohol use disorder living in the community. The availability of a treatment that decreases acute withdrawal and is effective in reducing drinking without requiring a period of abstinence would provide a continuum of medical care that is now lacking.

The strong efficacy of intranasal oxytocin in inhibiting alcohol withdrawal in our initial preliminary studies suggests that attenuation of withdrawal symptoms may have contributed to the significantly lower drinking in the dependent subjects receiving oxytocin treatment in our 12-week clinical trial. Besides reducing withdrawal symptoms, oxytocin has other affects that might also reduce drinking. Oxytocin is potently anxiolytic in animal and human studies (MacDonald & Feifel, 2014). Indeed, during our initial study of oxytocin treatment of alcohol withdrawal, oxytocin significantly reduced anxiety on the second day after dependent subjects stopped drinking. Oxytocin also reduces stress activation of the hypothalamic–pituitary–adrenal (HPA) axis in animals by inhibiting CRF release from hypothalamic paraventricular nucleus neurons (Slattery & Neumann, 2008; Smith & Wang, 2012). Numerous human studies have found that intranasal oxytocin administration decreases stress-induced cortisol release (Cardoso et al., 2014). Addiction is characterized by heightened HPA axis activation and anxiety, factors that contribute to negative reinforcement to continue to drink in dependent individuals (Koob, 2015; Koob et al., 2014). The anxiolytic and antistress effects of oxytocin may contribute to reduction of drinking in heavy drinkers. Oxytocin treatment also blocks alcohol-induced dopamine release in the nucleus accumbens (Peters, Bowen, Bohrer, McGregor, & Neumann, 2017), suggesting that oxytocin may also decrease alcohol consumption in heavy drinkers by reducing the positive reinforcing effects of alcohol.

Perhaps the most important implication of our evidence that oxytocin blocks alcohol withdrawal and reduces drinking is that oxytocin may exert therapeutic effects by rapidly inhibiting or reversing neuroadaptations underlying established tolerance to alcohol. This possibility is supported by early findings that oxytocin treatment attenuated the development of tolerance to alcohol (and other addictive substances) in nontolerant animals and diminished established tolerance. This would be a totally unique anti-addiction mechanism. If this hypothesis is correct, subjects who received oxytocin in our outpatient study may have decreased their drinking, in part, because oxytocin inhibition of tolerance resulted in less alcohol consumption being needed to achieve the "effect" that participants were seeking when drinking. Studies should be conducted to directly test this hypothesis. While there are numerous published papers on the neurochemical effects of addictive substance administration, this work has not been linked to changes in tolerance so there are few inferences in the literature about the neurochemical correlates of tolerance formation. Evidence that oxytocin exerts antiaddictive effects by reducing established tolerance in humans would add to the translational significance of previous evidence in animals (discussed in more detail below) that endogenous oxytocin plays a role in the formation and maintenance of tolerance. Perhaps future animal and human oxytocin studies will advance our understanding of the neurobiological mechanisms of tolerance formation.

4. ADDITIONAL HUMAN STUDIES

To date, only one other RCT has been published testing the efficacy of daily intranasal oxytocin over a sustained period in subjects with substance use disorders. Stauffer et al. (2016) studied 22 cocaine-dependent subjects who were receiving methadone maintenance treatment for opioid use disorder. Over a 2-week period, subjects self-administered either intranasal oxytocin (40 IU/dose) or placebo twice daily. Measures were obtained at baseline, the last day of intranasal treatment (day 15), and after a washout period (day 21). Both the oxytocin and placebo groups had very significant reductions in craving for and urges to use cocaine at day 15 that persisted on day 21. Although oxytocin recipients had greater overall reductions in craving and urges, there were no significant treatment group differences. Urges to use heroin trended over the trial toward significantly increasing in the placebo group, but not the oxytocin group. Both groups had a reduction in self-reported use of cocaine, but the reduction was not as great in the oxytocin group. However, the validity of group comparisons of self-reported use is questionable: comparisons with urine cocaine metabolite measures found that self-reports of cocaine use were significantly more accurate in oxytocin recipients. On a modified implicit association task, the oxytocin but not the placebo group underwent a significant switch by day 21 from associating cocaine images with "self" words to associating cocaine images with "other" words. The oxytocin group had a trend level lower self-rating for "do you feel any drug effects" on daily visual analog scales. The authors argue that this indicates oxytocin was not reducing heroin tolerance in the setting of methadone maintenance therapy. The validity of this conclusion is supported by the trend level increase in this study in the urge to use heroin by the placebo but not the oxytocin group. However, this conclusion does not take into consideration that lower self-ratings for "do you feel any drug effects" in the oxytocin group could be the consequence of oxytocin reducing withdrawal from cocaine, symptoms of which can occur in daily users because of the short half-life of this drug. This interpretation is supported by the study finding that the oxytocin group had lower craving for and urges to use cocaine (albeit insignificant).

Unfortunately, the Stauffer et al. (2016) study had a large placebo effect on craving and urges to use cocaine that may have masked a potential oxytocin effect. Group comparisons on self-reported cocaine use were confounded by evidence that the oxytocin group more accurately reported their use. To the extent that it is valid, the nonsignificant effects of oxytocin on cocaine and heroin craving in the Stauffer et al. (2016) study are consistent with the lack of effect of oxytocin on alcohol craving in our 12-week RCT summarized in heavy drinkers.

Other studies have tested the effects of single doses of intranasal oxytocin in humans with substance use disorders. Mitchell, Arcuni, Weinstein, and Wooley (2016) examined intranasal oxytocin (40 IU) compared to placebo in a crossover design on social perception, craving, and approach behavior in a cohort of 32 subjects with alcohol use disorder. Because attachment style has been reported to moderate intranasal oxytocin effects in human studies (Bartz et al., 2010; De Dreu, 2012), attachment style was assessed in subjects by administering the Experiences in Close Relationships: Relationship Structures Questionnaire (Fraley, Heffeman, Vicary, & Brumbaugh, 2011). Oxytocin had no effect on cue-induced craving in the entire sample but was significantly more effective in reducing craving in subjects who were more anxiously attached. This finding suggests that differences among subjects in attachment style may have contributed to the wide variability of alcohol craving self-ratings and lack of treatment group differences on this measure in our 12-week intranasal oxytocin treatment trial in heavy drinkers described above.

Lee et al. (2014) conducted a double blind, crossover comparison of single doses of intranasal oxytocin (24 IU) compared to placebo in 23 cocaine-dependent subjects who had been abstinent for sustained periods during inpatient admission. Oxytocin produced significantly greater baseline desire to use cocaine and cue-induced (viewing videoclips of cocaine-related scenes) excitability but not cue-induced desire to use. These results should be considered in light of preclinical evidence that oxytocin treatment increased cocaine sensitization in animals that had previously received cocaine doses on 5 consecutive days but that oxytocin treatment also reduced cue-induced reinstatement of cocaine CPP after extinction in animals that had previously acquired high rates of self-administration over 5–9 day periods (see below).

Wooley et al. (2016) tested a single dose of intranasal oxytocin (40 IU) or placebo in a randomized, double blind, crossover study in 36 male former IV heroin users on methadone replacement therapy and 34 healthy controls. Oxytocin had no significant effects on cue-induced craving or performance on an implicit association task with opioid-related and neutral images.

Using a within subject, placebo-controlled design in cigarette smokers who had abstained from smoking for 12 h ($N=17$), Miller, Bershad, King, and de Wit (2016) compared spontaneous cigarette craving at 20-min intervals after intranasal administration of an initial dose of oxytocin (20 IU) or placebo and then craving at intervals before and following exposure to cigarette cues after a second intranasal dose of the same test substance taken 75 min after the first dose. Cue-induced cigarette craving, but not spontaneous craving, was significantly lower after oxytocin compared to placebo treatment. The design of this study begs the question of whether differences in the cumulative dose of oxytocin taken prior to testing (40 IU vs 20 IU) resulted in significant reduction of cue-elicited craving but not spontaneous craving. In a cohort of 16 participants meeting DSM-IV criteria for marijuana dependence, McRae-Clark, Baker, Moran-Santa Maria, and Brady (2013) found that Marijuana Craving Questionnaire scores were significantly lower shortly after completion of the Trier Social Stress Test in subjects who self-administered intranasal oxytocin ($N=8$) 45 min before the start of the stress test but not in subjects who took intranasal placebo ($N=8$).

Relationships of oxytocin plasma levels or oxytocin gene variants with addictive substance use have been examined in two studies. Gerra et al. (2017) reported higher oxytocin plasma concentrations in abstinent heroin addicts ($N=18$) compared to 18 matched normal control subjects. Concentrations in the abstinent addicts correlated positively with psychiatric symptoms, aggressiveness, and mother neglect scores but not with heroin exposure extent or dosages. Vaht, Kurrikoff, Laas, Veidebaum, and Harro (2016) found, in a cohort that initially enrolled 593 fifteen-year-old participants, that males (but not females) homozygous for the A allele of the oxytocin receptor gene variant, rs53576 (suggested to be indicative of less efficient oxytocin functioning), compared to G allele carriers were more frequently alcohol consumers at ages 15 and 18, and also more likely to have had alcohol abuse or addiction by age 25.

Clearly human investigation of oxytocin and addiction is in its infancy.

5. PIONEERING STUDIES OF OXYTOCIN EFFECTS ON TOLERANCE FORMATION AND ESTABLISHED TOLERANCE TO ALCOHOL, OPIOIDS, AND COCAINE

Our findings that intranasal oxytocin potently reduces withdrawal and decreases alcohol consumption in very heavy drinkers lead us to the hypothesis that oxytocin exerts therapeutic effects by rapidly inhibiting established alcohol tolerance. This hypothesis is strongly supported by and now adds translational significance to an earlier seminal body of animal studies that explicitly studied exogenous and endogenous oxytocin effects on the formation of tolerance and established tolerance to addictive substances. Most work in this area focused on oxytocin and opioid tolerance, but a smaller number of studies examined oxytocin effects on alcohol and cocaine tolerance. The key finds from this body of work are summarized below.

5.1 Peripheral or Intracerebroventricular Administration of a Wide Range of Doses of Oxytocin Inhibits Tolerance Formation to Addictive Substances (Reduction in the Magnitude of Effects With Repeated Administration)

Oxytocin given peripherally or centrally 10 min to 2 h before each daily dose of addictive substance [2–5 mg/kg of alcohol intraperitoneal (IP) in mice; 7.5 mg/kg cocaine subcutaneous (SC) in rats] over 3–5 days significantly inhibited tolerance formation to substance effects: hypothermia in all alcohol studies (Jodogne, Tirelli, Klingbiel, & Legros, 1991; Pucilowski, Kotowski, & Trzaskowa, 1985; Szabó, Kovács, Székeki, & Telegdy, 1985; Szabó, Kovács, & Telegdy, 1989) but also akinesia, myorelaxation, and sedation (Jodogne et al., 1991; Pucilowski et al., 1985); and cocaine-induced stereotyped sniffing in rats (Sarnyai, Biró, et al., 1992; Sarnyai, Szabó, Kovács, & Telegdy, 1992). Studies of oxytocin and tolerance formation to opioids were conducted in mice and employed the tail flick test of nociception to quantify opioid effects. Several opioid studies employed models of acute or chronic morphine-induced tolerance: the analgesic effect of a test dose of morphine (5 mg/kg SC), a μ opioid receptor ligand, given 5 h after injection with a tolerance-inducing dose of morphine (100 mg/kg SC) or 48 h after SC placement of a morphine pellet (Kovács, Horváth, Sarnyai, Faludi, & Telegdy, 1985; Kovács, Izbéki, Horváth, & Telegdy, 1984). Peripheral or intracerebroventricular (ICV) oxytocin given 1–2 h (and even 24 h) before administration of the tolerance-inducing doses of morphine significantly increased the analgesic effect of the subsequently administered 5 mg/kg test dose of morphine. In an analogous acute opioid tolerance study, ICV injection of 5.0 μg of D-Met^2-Pro^5-enkephalinamide, a synthetic analogue of the endogenous opioid met-enkephalin that is a δ receptor ligand, significantly reduced the analgesic effect of a 0.5 μg test dose of D-Met^2-Pro^5-enkephalinamide given 330 min after the 5.0 μg tolerance-inducing dose. SC oxytocin administration 1 h prior to the 5.0 μg dose significantly increased the analgesic effect of the 0.5 μg test dose of D-Met^2-Pro^5-enkephalinamide (Schwartzberg, Kovács, Penke, & Telegdy, 1986). These findings indicate that oxytocin pretreatment inhibits opioid tolerance formation in these models.

As in the alcohol and cocaine studies mentioned above, daily SC oxytocin pretreatment has similar attenuating effects on tolerance formation to the analgesic effects of daily SC doses of heroin (Kovács, Faludi, & Telegdy, 1985) that, like morphine, is a μ opioid receptor ligand as well as tolerance formation to daily ICV doses of the endogenous opioid, β-endorphin, that is a ε opioid receptor ligand (Kovács & Telegdy, 1987). Heroin administration twice daily for 4 days produced cross-tolerance to a test dose of D-Met^2-Pro^5-enkephalinamide. Peripheral and ICV oxytocin administration prior to each heroin dose significantly decreased cross-tolerance to the met-enkephalin analogue (Kriván, Szabó, Sarnyai, Kovács, & Telegdy, 1992).

In summary, oxytocin inhibits tolerance formation to multiple addictive substances including a number of opioids that act at different opioid receptors.

5.2 Oxytocin Enhances the Development of Cocaine Sensitization

In seeming contrast to the considerable evidence summarized above that oxytocin inhibits tolerance formation, Sarnyai, Szabó, et al. (1992) reported that oxytocin increases sensitization to cocaine-induced locomotor behavior. In mice treated for 5 days with twice-daily SC doses of 7.5 mg of cocaine, a test dose of 7.5 mg cocaine given 3 days later stimulated more locomotor activity than the test dose given to animals that were not previously cocaine treated. Oxytocin (the highest of several doses) given prior to each daily cocaine injection resulted in significantly greater locomotion after the subsequent test dose of cocaine, indicating that oxytocin enhanced the sensitization effect of repeated daily doses of cocaine.

5.3 Oxytocin Inhibits Withdrawal From Addictive Substances

Kovács, Horváth, et al. (1985) reported that SC doses of oxytocin given 1 h before SC implantation of a morphine pellet in mice significantly delayed the onset of withdrawal symptoms (stereotyped jumping) precipitated by a dose of naloxone given 72 h after pellet placement. Szabó et al. (1987) rendered mice alcohol dependent with daily alternating doses of IP *tert*-butanol and ethanol for 4 days. Oxytocin or placebo was given SC 2 h before the first-day dose of alcohol and daily thereafter. On day 5, withdrawal convulsions were induced by a dose of picrotoxin. The number of seizures was unaffected by oxytocin treatments but, at the highest dose of oxytocin, the mortality rate was lower compared to controls that received no oxytocin, albeit not significantly. Evidence for oxytocin inhibition of alcohol withdrawal is weak in the Szabó et al. (1987) paper. Our demonstration of the effects of intranasal oxytocin in dependent human subjects undergoing detoxification summarized above is by far the strongest evidence that oxytocin does indeed block alcohol withdrawal (Pedersen et al., 2013).

5.4 Oxytocin Treatment Decreases Rates of Acquisition of Heroin Self-Administration and Established Rates of Heroin Self-Administration in Tolerant Rats

Kovács, Borthaiser, and Telegdy (1985) investigated oxytocin effects on intravenous (IV) heroin self-administration in rats. Some rats were rendered heroin tolerant by 7 days of twice-daily IP injections of escalating heroin doses. Subsequently the acquisition of jugular vein IV self-administration of heroin was then compared between the heroin-tolerant rats and heroin-naïve rats over an 8-day period. Animals received oxytocin or placebo doses SC 1 h before daily 90-min heroin self-administration sessions. In heroin-naïve rats, rates of self-administration of heroin increased over the test period, but the rates at which animals lever-pressed to obtain IV infusions of heroin did not differ between the oxytocin and placebo treatment groups. Placebo-treated heroin-tolerant rats acquired IV heroin self-administration faster and to significantly higher rates than heroin-naïve rats. In heroin-tolerant rats, oxytocin compared to placebo treatment before self-administration sessions significantly reduced acquisition of heroin self-administration to a rate comparable to heroin-naïve animals. Other heroin-tolerant animals that did not receive injections before daily IV heroin self-administration acquired high rates of self-administration over 8 days.

In these highly tolerant rats, pretreatment with a single SC dose of oxytocin compared to placebo significantly reduced the amount of heroin consumed during a self-administration session. In a separate report, Kovács and Van Ree (1985) also found that an SC dose of 1 μg of oxytocin preceding a heroin self-administration test conducted after a 7-day period of acquisition of IV self-administration significantly reduced lever pressing for IV doses of heroin.

5.5 Endogenous Oxytocin Regulates Tolerance Formation

SC injection of a single 60 mg/kg dose of morphine in mice produces acute tolerance as evidenced by a significant reduction in the antinociceptive effect of a 5 mg/kg test dose of morphine given SC 5 h after the 60 mg/kg dose. Kovács, Sarnyai, et al. (1987) found that ICV administration of two selective oxytocin antagonists prior to injection with the tolerance-inducing 60 mg/kg dose of morphine each further reduced the antinociceptive effect of the 5 mg/kg test dose of morphine 5 h later. In an analogous study, Sarnyai et al. (1988) produced mild acute tolerance with a single SC dose of 30 mg/kg morphine that reduced the analgesic effect of a 5 mg/kg dose of morphine given 5 h later. Central injections of a low dose of oxytocin antagonist (1 pg) into specific brain sites (see Section 5.6) further decreased the analgesic effect of the 5 mg/kg test dose of morphine. The results of these studies indicate that endogenous central oxytocin attenuates tolerance formation to morphine.

5.6 Brain Sites and Mechanisms of Oxytocin Attenuation of Tolerance and Substance Self-Administration

Several brain sites have been implicated in oxytocin effects on tolerance formation to and self-administration of addictive substances. In mice, oxytocin injection into the nucleus accumbens or dorsal hippocampus prior to pellet implantation, but not into the caudate nucleus, ventral tegmental area, or external cortical surface, significantly attenuated tolerance formation tested 48 h after SC morphine pellet implantation as well as significantly delayed naloxone-precipitated withdrawal (Kovács et al., 1984). Also in mice, injection of a low dose of oxytocin antagonist into the posterior olfactory nucleus, central amygdala nucleus, and ventral hippocampus, but not into the caudate nucleus or ICV, increased morphine tolerance (reduced the analgesic effects of 5 mg of morphine given 5 h after injection of 30 mg/kg of morphine; Sarnyai et al., 1988). Surprisingly, oxytocin antagonist injection into the nucleus accumbers and ventral tegmental area was not tested in this study. Ibragimov, Kovács, Szabó, and Telegdy (1987) reported that oxytocin administration into the anterodorsal nucleus accumbens or into the ventral hippocampus decreased self-administration of heroin in rats that had acquired a stable rate of jugular vein self-administration. Tolerance formation to cocaine-induced stereotyped sniffing in rats was attenuated by oxytocin injection into the ventral hippocampus (Sarnyai, Biró, et al., 1992). Oxytocin infusions into other sites were not tested in the Ibragimov et al. (1987) and the Sarnyai, Biró, et al. (1992) studies.

Administration of addictive substances alters oxytocin content in specific brain areas. In experiments conducted by Kovács, Laczi, et al. (1987) in mice, oxytocin content in the hippocampus, amygdala, and "basal forebrain" (not defined) increased significantly 30 min after a single SC dose of morphine (5 mg/kg). These effects were reversed by naloxone administration before morphine injection. 72 h after SC placement of a morphine pellet, oxytocin content was significantly lower in the hippocampus but not amygdala or basal forebrain. Thirty minutes after naloxone administration that precipitated withdrawal in pellet-implanted animals oxytocin content increased significantly in the hippocampus and basal forebrain but not in the amygdala. Sarnyai, Biró, et al. (1992) found that oxytocin concentrations in plasma, the hypothalamus, and the hippocampus in rats increased significantly after treatment with twice-daily doses of cocaine for 5 days, a regimen that produced tolerance to cocaine-induced stereotyped sniffing.

Interactions with central dopaminergic neurotransmission may play roles in oxytocin effects on tolerance formation and withdrawal from morphine and cocaine. Early on, Kovács et al. (1984) found that 6-hydroxydopamine lesions of the NA partially decreased oxytocin attenuation of tolerance formation to and naloxone-precipitated withdrawal from morphine. Kovács, Horváth, et al. (1985) examined in mice the influences of oxytocin on the effects of low and high apomorphine doses and the effects of high-dose apomorphine after 10 days of haloperidol treatment and concluded that oxytocin downregulates dopamine neurotransmission at postsynaptic receptors. Other studies found that oxytocin treatment decreased cocaine stimulation of locomotor activity, and this effect may be mediated by oxytocin blockade of cocaine-induced dopamine release in the nucleus accumbens but not in the caudate nucleus (Kovács, Sarnyai, Babarczy, Szabó, & Telegdy, 1990; Sarnyai, Szabó, Kovács, & Telegdy, 1990).

6. NEWER ANIMAL STUDIES ON OXYTOCIN AND ADDICTION

A new era of preclinical investigations of oxytocin and addiction began late in the first decade of this millennium and the literature in this area has expanded rapidly. Most of the studies have focused on stimulants but there has been a recent surge in alcohol studies. Opioid studies are more limited in number.

Most of the newer preclinical investigations have employed the CPP paradigm to examine oxytocin effects on self-administration of addictive substances. CPP paradigms are described in detail in Napier, Herrold, and de Wit (2013) and Tzschentke (2007). Briefly stated, the paradigm generally employs a three-chamber apparatus. Initially animals are permitted free access of all three chambers over a series of daily habituation sessions. During subsequent daily test sessions, when individual animals first enter a specific chamber, they are confined to that chamber for a specific period of time during which they receive by intravenous administration or other routes a standard dose of addictive substance when they press one of two available levers (the rewarding lever). When the rewarding lever is pressed, specific cues (sound, light, both) are activated. The cues remain on for a short period of time during which further pressing of the rewarding lever does not result in drug delivery. Pressing the rewarding lever results in delivery of another dose of drug only after cue activation ends. Pressing the other (nonrewarding) lever never results in drug delivery. After a number of consecutive days of conditioning in this way (commonly 5–7 in the literature on oxytocin effects), animals are again allowed free access to all three chambers for the entire test session. CPP is exhibited during this session if animals spend a significantly higher proportion of the test session time in the chamber where they had received lever press-activated self-administered doses of addictive substance. This outcome is referred to as *acquisition* of CPP and is thought to be a measure of the rewarding properties of the addictive substance. Stable rates of daily self-administration established during the acquisition procedure are referred to as *expression* of acquired CPP. After acquisition of a stable rate of daily self-administration of addictive drug, subsequent daily sessions are often run during which pressing of the formerly rewarding lever does not result in drug delivery. The proportion of time spent in the conditioned chamber and the rate of lever pressing gradually decline to baseline levels. This process is called *extinction*. After preference for the conditioned chamber has dropped to a stable low rate during extinction, proportion of time spent in the formerly conditioned chamber during test sessions and rates of formerly rewarding lever pressing can be rapidly increased by subjecting animals to a stressor, a dose of the addictive substance, or simply exposing animals to the cues that had previously been activated when reward lever pressing had resulted in delivery of a dose of the drug. This process is referred to as *reinstatement* of CPP and is thought to be a model of relapse.

Other procedures are employed to produce stable rates of substance self-administration. A classic and simple example is the two-bottle choice method of achieving stable rates of alcohol consumption. There are different variations but the basic approach is to introduce two drinking bottles in the animal cage, one with regular drinking water and the other with sweetener flavored water. Once animals favor drinking from the sweet fluid bottle, alcohol at steadily increasing concentrations is added to that bottle. Sweetener is phased out at some point during acceleration in alcohol concentrations. Continued consumption of a higher proportion of fluid intake from the alcohol-containing bottle compared to the drinking water-only bottle is an expression of alcohol preference. Drinking in the dark is a model of binge drinking. Initially animals are given unlimited access to drink from a cylinder containing alcohol solution starting a few hours into the dark cycle. After consumption rates stabilize, the period during the dark when the cylinder is available is shortened. When consumption rates have again stabilized, the period of cylinder availability is shortened again. Repeating this process leads to rapid rates of alcohol consumption during brief periods.

With remarkable consistency, a large number of studies employing CPP and other paradigms in alcohol- or stimulant (cocaine, methamphetamine)-treated animals have found that peripheral or central oxytocin administration (or other manipulations that enhance central oxytocin activity): (1) lower the rate of increase of self-administration of addictive substances if given during conditioning, (2) lower expression of acquired rates of self-administration after conditioning, and (3) inhibit reinstatement of self-administration after periods of extinction (Bahi, 2015; Baracz, Everett, & Cornish, 2015; Baracz, Everett, McGregor, & Cornish, 2016; Baracz et al., 2012; Carson, Cornish, Guastella, Hunt, & McGregor, 2010; Cox et al., 2017; Cox, Young, See, & Reichel, 2013; Ferland, Reichel, & McGinty, 2016; Georgiou et al., 2015; Han et al., 2014; King et al., 2017; Leong, Zhou, Ghee, See, & Reichel, 2016; MacFayden et al., 2016; Morales-Rivera et al., 2014; Peters et al., 2017; Peters, Slattery,

Flor, Neumann, & Reber, 2013; Qi et al., 2009; Zanos et al., 2014; Zhou et al., 2014). Some of these interventions also accelerated the rate of extinction of alcohol and stimulant CPP (Bahi, 2015; Qi et al., 2009). The effects of oxytocin on self-administration of addictive substances have been localized to specific brain sites: the nucleus accumbens, subthalamic nucleus, medial prefrontal cortex, and dorsal hippocampus (Baracz & Cornish, 2016; Han et al., 2014). Effects in most brain sites were blocked by coadministration of a selective oxytocin receptor antagonist. However, there is one notable exception. While site injection of oxytocin into the medial prefrontal cortex and dorsal hippocampus reduced restraint stress-induced reinstatement of methamphetamine CPP, a selective oxytocin antagonist blocked the oxytocin effect in the medial prefronal cortex but not in the dorsal hippocampus (Han et al., 2014). This indicates that oxytocin effects on reinstatement in some brain sites are mediated by a mechanism that does not require oxytocin receptor activation.

Most of the studies cited above were conducted in males. Substantial sex differences have been found in animal models of addiction (Becker & Koob, 2016). Remarkably, in the few oxytocin studies that included females, no sex differences were found for the effects of oxytocin treatment on acquisition and reinstatement of CPP for stimulants (Cox et al., 2017, 2013; Leong et al., 2016). The one exception is that Cox et al. (2013) initially reported that oxytocin reduced cue-induced reinstatement of methamphetamine seeking only in females, although there were no sex differences in oxytocin reduction of methamphetamine- and stressor (yohimbine dosing)-induced reinstatement. However, Cox et al. (2017) and Leong et al. (2016) found no sex differences in OT inhibition of methamphetamine and cocaine seeking in rats, including cue-induced reinstatement. In the recent literature, there is only one study of oxytocin effects on opioid CPP. In contrast to the results of the numerous alcohol and stimulant studies cited above, Moaddab, Hyland, and Brown (2015) found that oxytocin treatment increased the expression of morphine CPP tested the day after completion of the acquisition period. Perhaps oxytocin effects on acquisition of opioid CPP are truly different than effects on acquisition of alcohol and stimulant CPP. However, the acquisition period in the Moaddab et al. (2015) study was unusually short (only 3 days of morphine reward in the conditioned chamber in contrast to 5–7 days in alcohol and stimulant CPP studies in which oxytocin effects were investigated). Additional studies are required to clarify the effects of oxytocin on opioid CPP.

Several studies have reported negative emotional states in formerly tolerant animals well after access to addictive substances has been terminated, even in the absence of physical withdrawal symptoms (Georgiou, Zanos, Garcia-Carmona, et al., 2016; Georgiou, Zanos, Hourani, Kitchen, & Bailey, 2016; Zanos et al., 2014). After extinction of CPP, stressors rapidly reinstate preference for the formerly conditioned location including forced swimming, restraint stress, yohimbine treatment, and exposure to predator odor that produce dysphoria and fear (Cox et al., 2013; Ferland et al., 2016; Han et al., 2014; Qi et al., 2009; Zanos et al., 2014). Even reexposure to mild cues that reinstate addictive substance CPP [lights and tones previously activated by reward lever pressing during acquisition in CPP test chambers (Cox et al., 2017, 2013; Morales-Rivera et al., 2014; Zhou et al., 2014)] is anxiogenic (Morales-Rivera et al., 2014). This body of animal findings models the prolonged dysphoria and heightened stress vulnerability experienced by substance-dependent people during periods of abstinence that increase their risk for relapse (Koob, 2015; Koob et al., 2014). Cocaine and methamphetamine abstinence for extended periods in tolerant mice caused increased anxiety and anhedonia that were associated with increased oxytocin receptor binding in the piriform cortex, medial septum, and amygdala (cocaine) as well as the lateral septum (cocaine, methamphetamine) (Georgiou, Zanos, Garcia-Carmona, et al., 2016; Georgiou, Zanos, Hourani, et al., 2016), possibly the result of diminished oxytocin release in those sites. Oxytocin treatment reduces anxiety produced by cues that reinstate cocaine-seeking behavior (Morales-Rivera et al., 2014), and eliminates anxiety-like and depression-like behavior as well as deficits in social behavior exhibited by morphine-tolerant mice after 7 days of withdrawal (Zanos et al., 2014). These findings suggest that oxytocin reduction of abstinence-induced dysphoria may contribute to oxytocin inhibition of addictive substance self-administration in tolerant animals and humans.

Addictive substances alter oxytocin systems and oxytocin reverses some neurochemical effects of substances. Chronic alcohol, stimulant, or opioid treatment or consumption generally decreases oxytocin content in plasma and some brain areas as well as upregulates oxytocin receptor concentrations in specific brain areas; acute stimulant administration increases oxytocin content in some brain regions. The details of where changes occur in the brain are summarized in Lee et al. (2016). Conversely, oxytocin treatment reverses a number of neurochemical effects of addictive substances. These include acute methamphetamine: increased dopamine turnover in the striatum and the nucleus accumbens (Qi et al., 2008); increased glutamate extracellular concentrations as determined by microdialysis in the medial prefrontal cortex and reduced glutamate in the dorsal hippocampus (Qi et al., 2012); reduced extracellular GABA in the dorsal hippocampus (Qi et al., 2012); increased glutamatergic receptor NR1 subunit levels in the prefrontal cortex (Qi et al., 2012); and increased Fos expression in the subthalamic nucleus and nucleus accumbens (Carson, Hunt, et al., 2010). Oxytocin pretreatment diminished the enhancement of intracellular signaling processes that increase glutamate receptor subunit expression produced by restraint stress-induced reinstatement of methamphetamine CPP and cocaine-primed reinstatement of cocaine CPP (Han et al., 2014; Zhou et al., 2014). However, the observation by Qi et al. (2009) that oxytocin reduces restraint stress-induced glutamate release raises questions about whether the findings of Han et al. (2014) reflect oxytocin influences on neurochemical effects of restraint stress or reinstatement. The findings summarized here suggest that oxytocin attenuates addictive substance dysregulation of systems that have been identified as components of the neurocircuitry of addiction (Koob & Volkow, 2016). Pursuing the mechanism underlying their finding that ICV oxytocin administration attenuated acute alcohol-induced sedation and ataxia, Bowen et al. (2015) made the exciting discovery that oxytocin completely blocks ethanol-enhanced activity at δ subunit-containing recombinant $GABA_{(A)}$ receptors but not other recombinant $GABA_{(A)}$ receptors expressed in Xenopus oocytes. The recombinant oocytes constructs were devoid of oxytocin receptors so selective inhibitory effects of oxytocin at δ subunit-containing $GABA_{(A)}$ receptors were exerted independent of oxytocin receptor activation. Clearly there is a need for more studies of neurochemical mechanisms mediating oxytocin reduction of addictive substance self-administration and oxytocin neurochemical effects relevant to addiction.

7. COMMENTARY ON THE INTERPRETATION OF RECENT ANIMAL STUDY FINDINGS ON OXYTOCIN AND ADDICTION

In the recent literature on oxytocin reductions of acquisition and reinstatement of self-administration in animal models of addiction, oxytocin effects are generally interpreted as the results of inhibition of dopamine-mediated reward triggered by addictive substances, especially within the nucleus accumbens. Older as well as newer studies cited above have generated strong evidence supporting this hypothesis. This viewpoint is reiterated in the discussion sections of all of the new generation of papers in this area. The anxiolytic effects of oxytocin are also frequently, and probably correctly, speculated as contributing to oxytocin reduction of self-administration of addictive substances, especially during reinstatement.

I suggest that many of the inhibitory effects of oxytocin on acquisition, maintenance, extinction, and reinstatement of CPP for addictive substances and substance self-administration in other models of addictive behavior could also be the consequence of reversal of neuroadaptive changes underlying established tolerance. As argued above, our findings that oxytocin is highly efficacious both in blocking withdrawal and reducing alcohol consumption in heavy drinking human subjects strongly support the hypothesis that oxytocin inhibits established tolerance. Surprisingly, few of the more recent preclinical papers on oxytocin effects in addiction models cite earlier studies demonstrating that oxytocin regulates tolerance. None of the reports include reduction of tolerance as a possible mechanism contributing to oxytocin reduction of addictive substance self-administration. I believe our recent clinical findings, especially the potent oxytocin reduction of alcohol withdrawal, require a serious reconsideration of reversal of neuroadaptive changes underlying tolerance formation as a mechanism whereby oxytocin exerts antiaddictive effects in animal models. The validity of this hypothesis could be readily tested by measuring oxytocin effects on tolerance to addictive substances (and sensitization in the case of stimulants) in parallel with testing oxytocin effects on acquisition, extinction, and reinstatement of addictive substance self-administration in animal models. For instance, it would be valuable to investigate whether tolerance declines in parallel with extinction and then rapidly increases upon reinstatement. Would oxytocin acceleration of extinction be accompanied by oxytocin acceleration of reduction in tolerance? Would oxytocin inhibition of reinstatement also be accompanied by a parallel reduction in tolerance?

In human alcohol dependence, this hypothesis could be tested using the computer-assisted alcohol infusion system (CAIS) developed by O'Connor and associates that rapidly produces and maintains (clamps) the same target blood alcohol concentration in all subjects (O'Connor, Morzorati, Christian, & Li, 1998; O'Connor, Ramchandani, & Li, 2000; Ramchandani, Bolane, Li, & O'Connor, 1999). Alcohol-dependent subjects should have relatively less impairment on neurocognitive and motor tasks conducted shortly after CAIS has clamped blood alcohol at a concentration (e.g., 0.08 g/dL) that would be significantly impairing in nondependent individuals. Our hypothesis would be supported if intranasal oxytocin compared to placebo taken prior to and perhaps during alcohol infusion significantly worsened performance in alcohol-dependent individuals.

8. OXYTOCIN, SOCIAL BONDS, AND ADDICTION

Addictive substance use has profoundly negative effects on social relationships (Chermack et al., 2008; Hawley, Halle, Drasin, & Thomas, 1995; Inciardi, 1994; Johnson et al., 2002; Kaestner, 1995; Langevin, Paitich, Orchard, Handy, & Russon, 1982; Lejuez, Bornovalova, Daughters, & Curtin, 2005; Testa, Livingston, & Leonard, 2003). Social support decreases the odds of developing substance use disorders as well as reduces risk of relapse in recovering addicts (Bassuk, Hanson, Greene, Richard, & Laudet, 2016; Bell, Forthun, & Sun, 2000; Ellickson, Collins, & Bell, 1999; Garmendia, Alvarado, Montenegro, & Pino, 2008; Kendler et al., 2000; Kosten, Jalali, Steidl, & Kleber, 1987; Walter et al., 2006). Similar reciprocal relationships between social bonds and addictive behaviors have been demonstrated in animal models (McGregor & Bowen, 2012; McGregor, Callaghan, & Hunt, 2008; Young, Gobrogge, & Wang, 2011). In monogamous prairie voles (*Microtus ochrogaster*), amphetamine treatment or methamphetamine consumption blocked the development of mating or proximity-induced partner preference (Hostetler, Phillips, & Ryabinin, 2016; Liu et al., 2010). Conversely, amphetamine conditioning failed to establish CPP in pair-bonded males but did produce CPP in sexually naïve males (Liu, Young, Curtis, Aragona, & Wang, 2011). Voluntary alcohol consumption inhibits partner preference formation in male but not female prairie voles (Anacker et al., 2014). As in other species, prairie voles that are deprived for 72 h following an extended period of unlimited access to alcohol exhibit increased consumption when access is resumed. However, housing animals with a familiar same-sex social partner after the deprivation period prevented increased alcohol consumption in this model of relapse (Hostetler & Ryabinin, 2014).

Oxytocin has been implicated in the reciprocal relationships between social bonding and addictive behavior in prairie voles. Amphetamine exposure that inhibited the formation of partner preference reduced oxytocin receptor immunoreactivity in the medial prefrontal cortex and the nucleus accumbens (Young, Liu, Gobrogge, Wang, & Wang, 2014). Methamphetamine consumption that inhibited pair bond formation significantly reduced oxytocin content in the paraventricular nucleus of the hypothalamus

(Hostetler et al., 2016). Oxytocin infusion into the medial prefrontal cortex restored partner preferences in amphetamine-treated animals (Young et al., 2014).

It has been speculated that love is like an addictive disorder, and numerous similarities between the neural substrates of affiliative behavior and addictive substance reward have been described (Burkett & Young, 2012; Insel, 2003; McGregor & Bowen, 2012). However, the roles of oxytocin in the neurobiology of social bonds and addiction differ. Oxytocin stimulates and oxytocin antagonist administration inhibits the onset of maternal behavior and the formation of monogamous pair bonds, effects that have been localized to the ventral tegmental area and the nucleus accumbens suggesting oxytocin facilitates these attachments by releasing dopamine within reward pathways (Burkett & Young, 2012; Pedersen, 2013). In contrast, oxytocin inhibits acquisition of addictive substance self-administration, an effect that may be linked to oxytocin blockade of substance-induced release of dopamine in the nucleus accumbens (Kovács et al., 1984, 1990; Peters et al., 2017; Qi et al., 2008; Sarnyai et al., 1990). Once maternal behavior is established, blockade of central oxytocin receptors has no effect on maintenance of mothering (Fahrbach, Morrell, & Pfaff, 1985). In contrast, as reviewed above, after acquisition of addictive substance self-administration, oxytocin treatment significantly reduces rates of self-administration. A possible exception to oxytocin inhibition of self-administration of addictive substances is evidence that oxytocin release may be associated with the prosocial and prosexual effects of party drugs like ±3,4-methylenedioxymethamphetamine (Ecstasy) and gamma-hydroxybutyrate (Broadbear, Kabel, & Mak, 2014; Kirkpatrick, Francis, Lee, de Wit, & Jacob, 2014; McGregor & Bowen, 2012; McGregor et al., 2008).

Long-term addictive substance use causes persistent social deficits that are sustained well after abstinence has been achieved (Burkett & Young, 2012; McGregor & Bowen, 2012; McGregor et al., 2008), effects that may be related to chronic substance-induced changes in central oxytocin systems (Lee et al., 2016). Indeed, treatment with the oxytocin agonist, carbetocin, significantly reversed social impairment in morphine-tolerant mice after 7 days of withdrawal (Zanos et al., 2014).

9. FINAL COMMENTS

As stated above, Koob has repeatedly and convincingly argued that sustained substance use leads to addiction because of a shift from positive rewarding motivation to negative motivation to avoid the emotional and physical discomfort brought on by abstinence in dependent individuals. Oxytocin has a combination of effects suggesting that this neuropeptide may be particularly effective in reducing the "dark side" negative motivation maintaining addiction. These include blocking physical withdrawal, reducing negative emotional consequences of abstinence, diminishing vulnerability to anxiety, stressor as well as cue-induced relapse, and potentially bolstering protective social bonds. I also argue that oxytocin may exert anti-addiction effects by the unique mechanism of reversing established tolerance to dependence-inducing substances. Testing this hypothesis is not only imperative; it may also provide much needed insight into mechanisms underlying tolerance formation. Identifying medications that can effectively correct the maladaptive neuroadaptations underlying the "dark side" of addiction is the holy grail of pharmacological treatment of addiction (Koob & Mason, 2016).

REFERENCES

Amad, A., Thomas, P., & Perez-Rodriguez, M. M. (2015). Borderline personality disorder and oxytocin: Review of clinical trials and future directions. *Current Pharmaceutical Design*, *21*(23), 3311–3316.

Anacker, A. M., Ahem, T. H., Hostetler, C. M., Dufour, B. D., Smith, M. L., Cocking, D. L., et al. (2014). Drinking alcohol has sex-dependent effects on pair bond formation in prairie voles. *Proceedings of the National Academy of Sciences of the United States of America*, *111*(16), 6052–6057.

Bahi, A. (2015). The oxytocin receptor impairs ethanol reward in mice. *Physiology and Behavior*, *139*, 321–327.

Baracz, S. J., & Cornish, J. L. (2016). The neurocircuitry involved in oxytocin modulation of methamphetamine addiction. *Frontiers in Neuroendocrinology*, *43*, 1–18.

Baracz, S. J., Everett, N. A., & Cornish, J. L. (2015). The involvement of oxytocin in the subthalamic nucleus on relapse to methamphetamine-seeking behaviour. *PLoS One*, *10*(8). https://doi.org/10.1371/journal.pone.0136132.

Baracz, S. J., Everett, N. A., McGregor, I. S., & Cornish, J. L. (2016). Oxytocin in the nucleus accumbens core reduces reinstatement of methamphetamine-seeking behaviour in rats. *Addiction Biology*, *21*(2), 316–325.

Baracz, S. J., Rourke, P. I., Pardey, M. C., Hunt, G. E., McGregor, I. S., & Cornish, J. L. (2012). Oxytocin directly administered into the nucleus accumbens core or subthalamic nucleus attenuates methamphetamine-induced conditioned place preference. *Behavioral Brain Research*, *228*(1), 185–193.

Barrio, P., Reynolds, J., Garcia-Altés, A., Gual, A., & Anderson, P. (2017). Social costs of illegal drugs, alcohol and tobacco in the European Union: A systematic review. *Drug and Alcohol Review*, *36*(5), 578–588. https://doi.org/10.1111/dar.12504.

Bartz, J. A., Zaki, J., Ochsner, K. N., Bolger, N., Kolevzon, A., Ludwig, N., et al. (2010). Effects of oxytocin on recollections of maternal care and closeness. *Proceedings of the National Academy of Sciences of the United States of America*, *107*(50), 21371–21375.

Bassuk, E. L., Hanson, J., Greene, R. N., Richard, M., & Laudet, A. (2016). Peer-delivered recovery support services for addictions in the United States: A systematic review. *Journal of Substance Abuse Treatment*, *63*, 1–9.

Becker, J. B., & Koob, G. F. (2016). Sex differences in animal models: Focus on addiction. *Pharmacological Reviews*, *68*(2), 242–263.

Bell, N. J., Forthun, L. F., & Sun, S. W. (2000). Attachment, adolescent competencies, and substance use: Developmental considerations in the study of risk behaviors. *Substance Use and Misuse*, *35*(9), 1177–1206.

Bouchery, E. E., Harwood, H. J., Sacks, J. J., Simon, C. J., & Brewer, R. D. (2011). Economic costs of excessive alcohol consumption in the U.S., 2006. *American Journal of Preventative Medicine*, *41*, 516–524.

Bowen, M. T., Peters, S. T., Absalom, N., Chebib, M., Neumann, I. D., & McGregor, I. S. (2015). Oxytocin prevents ethanol actions at δ subunit-containing $GABA_A$ receptors and attenuates ethanol-induced motor impairment in rats. *Proceedings of the National Academy of Sciences of theUnited States of America*, *112*, 3104–3109.

Broadbear, J. H., Kabel, D., & Mak, T. P. (2014). Review: Oxytocin regulation of endogenous as well as drug induced mood. *Pharmacology, Biochemistry and Behavior*, *119*, 61–71.

Burkett, J. P., & Young, L. J. (2012). The behavioral, anatomical and pharmacological parallels between social attachment, love and addiction. *Psychopharmacology*, *224*(1), 1–26.

Cardoso, C., Kingdon, D., & Ellenbogen, M. A. (2014). A meta-analytic review of the impact of intranasal oxytocin administration on cortisol concentrations during laboratory tests: Moderation by method and mental health. *Psychoneuroendocrinology*, *49*, 161–170.

Carson, D. S., Cornish, J. L., Guastella, A. J., Hunt, G. E., & McGregor, I. S. (2010). Oxytocin decreases methamphetamine self-administration, methamphetamine hyperactivity, and relapse to methamphetamine-seeking behaviour in rats. *Neuropharmacology*, *58*(1), 38–43.

Carson, D. S., Hunt, G. E., Guastella, A. J., Barber, L., Cornish, J. L., Arnold, J. C., et al. (2010). Systemically administered oxytocin decreases methamphetamine activation of the subthalamic nucleus and accumbens core and stimulates oxytocinergic neurons in the hypothalamus. *Addiction Biology*, *15*(4), 448–463.

Carter, C. S., Lederhendler, I., & Kirkpatrick, B. (1997). The integrative neurobiology of affiliation. *New York Academy of Sciences*, *807*.

Chermack, S. T., Murray, R. L., Walton, M. A., Booth, B. A., Wryobeck, J., & Blow, F. C. (2008). Partner aggression among men and women in substance use disorder treatment: Correlates of psychological and physical aggression and injury. *Drug and Alcohol Dependence*, *98*(1–2), 35–44.

Cox, B. M., Bentzley, B. S., Regen-Tuero, H., See, R. E., Reichel, C. M., & Aston-Jones,-G. (2017). Oxytocin acts in nucleus accumbens to attenuate methamphetamine seeking and demand. *Biological Psychiatry*, *81*(11), 949–958.

Cox, B. M., Young, A. B., See, R. E., & Reichel, C. M. (2013). Sex differences in methamphetamine seeking in rats: Impact of oxytocin. *Psychoneuroendocrinology*, *38*, 2343–2353.

Dal Monte, O., Noble, P. L., Turchi, J., Cummins, A., & Averbeck, B. B. (2014). CSF and blood oxytocin concentration changes following intranasal delivery in macaque. *PLoS One*, *9*(8) . e103677.

De Dreu, C. K. (2012). Oxytocin modulates the link between adult attachment and cooperation through reduced betrayal aversion. *Psychoneuroendocrinology*, *37*(7), 871–880.

Domes, G., Heinrichs, M., Glascher, J., Buchel, C., Braus, D. F., & Herpertz, S. C. (2007). Oxytocin attenuates amygdala responses to emotional faces regardless of valence. *Biological Psychiatry*, *62*, 1187–1190.

Domes, G., Lischke, A., Berger, C., Grossmann, A., Hauenstein, K., Heinrichs, M., et al. (2010). Effects of intranasal oxytocin on emotional face processing in women. *Psychoneuroendocrinology*, *35*, 83–93.

Ellickson, P. L., Collins, R. L., & Bell, R. M. (1999). Adolescent use of illicit drugs other than marijuana: How important is social bonding and for which ethnic groups? *Substance Use and Misuse, 34*(3), 317–346.

Fahrbach, S. E., Morrell, J. I., & Pfaff, D. W. (1985). Possible role for endogenous oxytocin in estrogen-facilitated maternal behavior in rats. *Neuroendocrinology, 40*, 526–532.

Ferland, C. L., Reichel, C. M., & McGinty, J. E. (2016). Effects of oxytocin on methamphetamine-seeking exacerbated by predator odor pre-exposure. *Psychopharmacology, 233*(6), 1015–1024.

First, M. B., Spitzer, R. L., Gibbon, M., & Williams, J. B. W. (2002). *Structured clinical interview for DSM-IV-TR axis I disorders, research version, patient edition, (SCID-I/P).* New York: Biometrics Research, New York State Psychiatric Institute.

Flannery, B. A., Volpicelli, J. R., & Pettinati, H. M. (1999). Psychometric properties of the Penn Alcohol Craving Scale. *Alcoholism: Clinical and Experimental Research, 23*, 1289–1295.

Fraley, R. D., Heffeman, M. E., Vicary, A. M., & Brumbaugh, C. C. (2011). The experiences in close relationships-relationship structures questionnaire: A method for assessing attachment orientations across relationships. *Psychological Assessment, 23*(3), 615–625.

Freeman, S. M., Samineni, S., Allen, P. C., Stockinger, D., Bales, K. L., Hwa, G. G., et al. (2016). Plasma and CSF oxytocin levels after intranasal and intravenous oxytocin in awake macaques. *Psychoneuroendocrinology, 66*, 185–194.

Fullerton, C. A., Kim, M., Thomas, C. P., Lyman, D. R., Montejano, L. B., Dougherty, R. H., et al. (2014). Medication-assisted treatment with methadone: Assessing the evidence. *Psychiatric Services, 65*(2), 146–157.

Garmendia, M. L., Alvarado, M. E., Montenegro, M., & Pino, P. (2008). Social support as a protective factor of recurrence after drug addiction treatment. *Revista Medica De Chile, 136*, 169–178.

Georgiou, P., Zanos, P., Garcia-Carmona, J. S., Hourani, S., Kitchen, I., Kieffer, B. L., et al. (2015). The oxytocin analogue carbetocin prevents priming-induced reinstatement of morphine-seeking: Involvement of dopaminergic, noradrenergic and MOPr systems. *European Journal of Neuropsychopharmacology, 25*(12), 2459–2464.

Georgiou, P., Zanos, P., Garcia-Carmona, J. S., Hourani, S., Kitchen, I., Laorden, M. L., et al. (2016). Methamphetamine abstinence induces changes in μ-opioid receptor, oxytocin and CRF systems: Association with an anxiogenic phenotype. *Neuropsychopharmacology, 105*, 520–532.

Georgiou, P., Zanos, P., Hourani, S., Kitchen, I., & Bailey, A. (2016). Cocaine abstinence induces emotional impairment and brain region-specific upregulation of the oxytocin receptor binding. *European Journal of Neuroscience, 44*(7), 2446–2454.

Gerra, L. M., Gerra, G., Mercolini, L., Manfredini, M., Somaini, L., Pieri, C. M., et al. (2017). Increased oxytocin levels among abstinent heroin addicts: Association with aggressiveness, psychiatric symptoms and perceived childhood neglect. *Progress in Neuropsychopharmacology and Biological Psychiatry, 75*, 70–76.

Grant, B. F., Goldstein, R. B., Saha, T. D., Chou, P., Jung, J., Zhang, H., et al. (2015). Epidemiology of DSM-5 alcohol use disorder: Results from the National Epidemiologic Survey on alcohol and related conditions III. *JAMA Psychiatry, 72*, 757–766.

Gulland, A. (2014). Alcohol consumption kills 3.3 million people a year, WHO calculates. *British Medical Journal, 348* . g3335.

Han, W.-Y., Du, P., Fu, S.-Y., Wang, F., Song, M., Wu, C.-F., et al. (2014). Oxytocin via its receptor affects restraint stress-induced methamphetamine CPP reinstatement in mice: Involvement of the medial prefrontal cortex and dorsal hippocampus glutamatergic system. *Pharmacology, Biochemistry and Behavior, 119*, 80–87.

Hawley, T. L., Halle, T. G., Drasin, R. E., & Thomas, N. G. (1995). Children of addicted mothers: Effects of the "crack epidemic" on the caregiving environment and the development of preschoolers. *American Journal of Orthopsychiatry, 65*(3), 364–379.

Heinrichs, M., Baumgartner, T., Kirschbaum, C., & Ehlert, U. (2003). Social support and oxytocin interact to suppress cortisol and subjective responses to psychosocial stress. *Biological Psychiatry, 54,* 1389–1398.

Hostetler, C. M., Phillips, T. J., & Ryabinin, A. E. (2016). Methamphetamine consumption inhibits pair bonding and hypothalamic oxytocin in prairie voles. *PLoS One, 11*(7). . e0158178https://doi.org/10.1371/journal.pone.0158178.

Hostetler, C. M., & Ryabinin, A. E. (2014). Social partners prevent alcohol relapse behavior in prairie voles. *Psychoneuroendocrinology, 39,* 152–157.

Ibragimov, R., Kovács, G. L., Szabó, G., & Telegdy, G. (1987). Microinjection of oxytocin into limbic-mesolimbic brain structures disrupts heroin self-administration behavior: A receptor-mediated event? *Life Sciences, 41,* 1265–1271.

Inciardi, J. A. (1994). HIV/AIDS risks among male, heterosexual noninjecting drug users who exchange crack for sex. *NIDA Research Monograph, 143,* 26–40.

Insel, T. R. (2003). Is social attachment an addictive disorder? *Physiology and Behavior, 79*(3), 251–257.

Jarskog, L. F., Pedersen, C. A., Johnson, J. L., Hamer, R. M., Rau, S. W., Elliott, T., et al. (2017). A 12-week randomized controlled trial of twice-daily intranasal oxytocin for social cognitive deficits in people with schizophrenia. *Schizophrenia Research, 185,* 88–95.

Jodogne, C., Tirelli, E., Klingbiel, P., & Legros, J. J. (1991). Oxytocin attenuates tolerance not only to the hypothermic but also to the myorelaxant and akinesic effects of ethanol in mice. *Pharmacology, Biochemistry, and Behavior, 40,* 261–265.

Johnson, A. L., Morrow, C. E., Accornero, V. H., Xue, L., Anthony, J. C., & Bandstra, E. S. (2002). Maternal cocaine use: Estimated effects on mother-child play interactions in the preschool period. *Journal of Developmental & Behavioral Pediatrics, 23*(4), 191–202.

Kaestner, R. (1995). *The effects of cocaine and marijuana use on marriage and marital stability.* National Bureau of Economic Research, Working Paper, No. 5038.

Kanouse, A. B., & Compton, P. (2015). The epidemic of prescription opioid abuse, the subsequent rising prevalence of heroin use, and the federal response. *Journal of Pain & Palliative Care Pharmacotherapy, 29*(2), 102–114.

Kendler, K. S., Bulik, C. M., Silberg, J., Hettema, J. M., Myers, J., & Prescott, C. A. (2000). Childhood sexual abuse and adult psychiatric and substance use disorders in women: An epidemiological and cotwin control analysis. *Archives of General Psychiatry, 57*(10), 953–959.

Kiluk, B. D., Carrol, K. M., Duhig, A., Falk, D. E., Kampman, K., Lai, S., et al. (2016). Measures of outcome for stimulant trials: ACTTION recommendations and research agenda. *Drug and Alcohol Dependence, 158,* 1–7.

King, C. E., Griffin, W. C., Luderman, L. N., Kates, M. M., McGinty, J. F., & Becker, H. C. (2017). Oxytocin reduces ethanol self-administration in mice. *Alcoholism: Clinical and Experimental Research, 41*(5), 955–964.

Kirkpatrick, M. G., Francis, S. M., Lee, R., de Wit, H., & Jacob, S. (2014). Plasma oxytocin concentrations following MDMA or intranasal oxytocin in humans. *Psychoneuroendocrinology, 46,* 23–31.

Kirsch, P., Esslinger, C., Chen, Q., Mier, D., Lis, S., Siddhanti, S., et al. (2005). Oxytocin modulates neural circuitry for social cognition and fear in humans. *Journal of Neuroscience, 25*(49), 11489–11493.

Kolodny, A., Courtwright, D. T., Hwang, C. S., Kreiner, P., Eadie, J. L., Clark, T. W., et al. (2015). The prescription opioid and heroin crisis: A public health approach to an epidemic of addiction. *Annual Review of Public Health, 36,* 559–574.

Koob, G. F. (2009). Neurobiological substrates for the dark side of compulsivity in addiction. *Neuropharmacology, 56*(Suppl. 1), 18–31.

Koob, G. F. (2013a). Theoretical frameworks and mechanistic aspects of alcohol addiction: Alcohol addiction as a reward deficit disorder. *Current Topics in Behavioral Neurosciences, 13,* 3–30.

Koob, G. F. (2013b). Addiction is a reward deficit and stress surfeit disorder. *Frontiers in Psychiatry*, 4(article 72), 1–18.
Koob, G. F. (2014). Neurocircuitry of alcohol addiction: Synthesis from animal models. *Handbook of Clinical Neurology*, *125*, 33–54.
Koob, G. F. (2015). The dark side of emotion: The addiction perspective. *European Journal of Pharmacology*, *753*, 73–87.
Koob, G. F., Buck, C. L., Cohen, A., Edwards, S., Park, P. E., Schlosburg, J. E., et al. (2014). Addiction as a stress surfeit disorder. *Neuropharmacology*, 76(Pt. B), 370–382.
Koob, G. F., & Le Moal, M. (2008). Review: Neurobiological mechanisms for opponent motivational processes in addiction. *Philosophical Transactions of the Royal Society of London Series B, Biological Sciences*, *363*(1507), 3113–3123.
Koob, G. F., & Mason, B. J. (2016). Existing and future drugs for the treatment of the dark side of addiction. *Annual Review of Pharmacology and Toxicology*, *56*, 299–322.
Koob, G. F., & Volkow, N. D. (2016). Neurobiology of addiction: A neurocircuitry analysis. *Lancet Psychiatry*, *3*(8), 760–773.
Kosten, T. R., Jalali, B., Steidl, J. H., & Kleber, H. D. (1987). Relationship of marital structure and interactions to opiate abuse relapse. *The American Journal of Drug and Alcohol Abuse*, *13*(4), 387–399.
Kovács, G. L., Borthaiser, Z., & Telegdy, G. (1985). Oxytocin reduces intravenous heroin self-administration in heroin-tolerant rats. *Life Sciences*, *37*, 17–26.
Kovács, G. L., Faludi, M., & Telegdy, G. (1985). Oxytocin diminishes heroin tolerance in mice. *Psychopharmacology*, *86*, 377–379.
Kovács, G. L., Horváth, Z., Sarnyai, Z., Faludi, M., & Telegdy, G. (1985). Oxytocin and a C-terminal derivative (Z-prolyl-D-leucine) attenuate tolerance to and dependence on morphine and interact with dopaminergic neurotransmission in the mouse brain. *Neuropharmacology*, *24*(5), 413–419.
Kovács, G. L., Izbéki, F., Horváth, Z., & Telegdy, G. (1984). Effects of oxytocin and a derivative (Z-prolyl-D-leucine) on morphine tolerance/withdrawal are mediated by the limbic system. *Behavioural Brain Research*, *14*, 1–8.
Kovács, G. L., Laczi, F., Vecserrnyés, M., Hódi, K., Telegdy, G., & László, F. A. (1987). Limbic oxytocin and arginine^{8}-vasopressin in morphine tolerance and dependence. *Experimental Brain Research*, *65*, 307–311.
Kovács, G. L., Sarnyai, Z., Babarczy, E., Szabó, G., & Telegdy, G. (1990). The role of oxytocin–dopamine interactions in cocaine-induced locomotor hyperactivity. *Neuropharmacology*, *29*, 365–368.
Kovács, G. L., Sarnyai, Z., Izbéki, F., Szabó, G., Telegdy, G., Barth, T., et al. (1987). Effects of oxytocin-related peptides on acute morphine tolerance: Opposite actions by oxytocin and its receptor antagonists. *The Journal of Pharmacology and Experimental Therapeutics*, *241*(2), 269–274.
Kovács, G. L., Sarnyai, Z., & Szabó, G. (1998). Oxytocin and addiction: A review. *Psychoneuroendocrinology*, *23*, 945–962.
Kovács, G. L., & Telegdy, G. (1987). Endorphin tolerance is inhibited by oxytocin. *Pharmacology, Biochemistry and Behavior*, *26*, 57–60.
Kovács, G. L., & Van Ree, J. M. (1985). Behaviourally active oxytocin fragments simultaneously attenuate heroin self-administration and tolerance in rats. *Life Sciences*, *37*, 1895–1900.
Kranzler, H. R., & Van Kirk, J. (2001). Efficacy of naltrexone and acamprosate for alcoholism treatment: A meta-analysis. *Alcoholism: Clinical and Experimental Research*, *25*, 1335–1341.
Kriván, M., Szabó, G., Sarnyai, Z., Kovács, G. L., & Telegdy, G. (1992). Oxytocin blocks the development of heroin-enkephalin cross-tolerance in mice. *Pharmacology, Biochemistry and Behavior*, *43*(1), 187–192.
Langevin, R., Paitich, D., Orchard, B., Handy, L., & Russon, A. (1982). The role of alcohol, drugs, suicide attempts and situational strains in homicide committed by offenders seen

for psychiatric assessment. A controlled study. *Acta Psychiatrica Scandinavica*, *66*(3), 229–242.

Laprevote, V., Schwan, R., Schwitzer, T., Rolland, B., & Thome, J. (2015). Is there a place for off-label pharmacotherapy in cannabis use disorder? A review of efficacy and safety. *Current Pharmaceutical Design*, *21*(23), 3298–3305.

Lee, M. R., Glassman, M., King-Casas, B., Kelly, D. L., Stein, E. A., Schroeder, J., et al. (2014). Complexity of oxytocin's effects in a chronic cocaine dependent population. *European Neuropsychopharmacology*, *24*(9), 1483–1491.

Lee, H.-J., Macbeth, A. H., Pagani, J. H., & Young, W. S., 3rd (2009). Oxytocin: The great facilitator of life. *Progress in Neurobiology*, *88*, 127–151.

Lee, M. R., Rohn, M. C., Tanda, G., & Leggio, L. (2016). Targeting the oxytocin system to treat addictive disorders: Rationale and progress to date. *CNS Drugs*, *30*(2), 109–123.

Lee, M., Scheidweiler, K. B., Diao, X. X., Akhlaghi, F., Cummins, A., Huestis, M. A., et al. (2017). Oxytocin by intranasal and intravenous routes reaches the cerebrospinal fluid in rhesus macaques: Determination using a novel oxytocin assay. *Molecular Psychiatry* (epub ahead of print)https://doi.org/10.1038/mp.2017.27.

Lejuez, C. W., Bornovalova, M. A., Daughters, S. B., & Curtin, J. J. (2005). Differences in impulsivity and sexual risk behavior among inner-city crack/cocaine users and heroin users. *Drug and Alcohol Dependence*, 77(2), 169–175.

Leong, K. C., Zhou, L., Ghee, S. M., See, R. E., & Reichel, C. M. (2016). Oxytocin decreases cocaine taking, cocaine seeking, and locomotor activity in female rats. *Experimental and Clinical Psychopharmacology*, *24*(1), 55–64.

Liu, Y., Aragona, B. J., Young, K. A., Dietz, D. M., Kabbaj, M., Mazei-Robison, M., et al. (2010). Nucleus accumbens dopamine mediates amphetamine-induced impairment of social bonding in a monogamous rodent species. *Proceedings of the National Academy of Sciences of the United States of America*, *107*(3), 1217–1222.

Liu, Y., Young, K. A., Curtis, J. T., Aragona, B. J., & Wang, Z. (2011). Social bonding decreases the rewarding properties of amphetamine through a dopamine D1 receptor mediated mechanism. *Journal of Neuroscience*, *31*(22), 7960–7966.

MacDonald, E., Dadds, M. R., Brennan, J. L., Williams, K., Levy, F., & Cauchi, A. J. (2011). A review of safety, side-effects and subjective reactions to intranasal oxytocin in human research. *Psychoneuroendocrinology*, *36*, 1114–1126.

MacDonald, K., & Feifel, D. (2014). Oxytocin's role in anxiety: A critical appraisal. *Brain Research*, *1580*, 22–56.

MacFayden, K., Loveless, R., DeLucca, B., Wardley, K., Deogan, S., Thomas, C., et al. (2016). Peripheral oxytocin administration reduces ethanol consumption in rats. *Pharmacology, Biochemistry, and Behavior*, *140*, 27–32.

Maisel, N. C., Blodgett, J. C., Wilbourne, P. L., Humphreys, K., & Finney, W. F. (2013). Meta-analysis of naltrexone and acamprosate for treating alcohol use disorders: When are these medications most helpful? *Addiction*, *108*(2), 275–293.

Mark, T. L., Kassed, C. A., Vandivort-Warren, R., Levit, K. R., & Kranzler, H. R. (2009). Alcohol and opioid dependence medications: Prescription trends, overall and by physician specialty. *Drug and Alcohol Dependence*, *99*, 345–349.

McGregor, I. S., & Bowen, M. T. (2012). Breaking the loop: Oxytocin as a potential treatment for drug addiction. *Hormones and Behavior*, *61*, 331–339.

McGregor, I. S., Callaghan, P. D., & Hunt, G. E. (2008). From ultrasocial to antisocial: A role for oxytocin in the acute reinforcing effects and long-term adverse consequences of drug use. *British Journal of Pharmacology*, *154*(2), 358–368.

McNair, D. M., Lorr, M., & Droppleman, L. F. (1971). *EDITS manual for the profile of mood states*. Educational and Industrial Testing Service: San Diego, CA.

McRae-Clark, A. L., Baker, N. L., Moran-Santa Maria, M., & Brady, K. T. (2013). Effects of oxytocin on craving and stress response in marijuana-dependent individual: A pilot study. *Psychopharmacology*, *228*(4), 623–631.

Miller, M. A., Bershad, A., King, A. C., & de Wit, H. (2016). Intranasal oxytocin dampens cue-elicited cigarette craving in daily smokers: A pilot study. *Behavioural Pharmacology, 27*(8), 697–703.

Mitchell, J. M., Arcuni, P. A., Weinstein, D., & Wooley, J. D. (2016). Intranasal oxytocin selectively modulates social perception, craving, and approach behavior in subjects with alcohol use disorder. *Journal of Addiction Medicine, 10*(3), 182–189.

Moaddab, M., Hyland, B. I., & Brown, C. H. (2015). Oxytocin enhances the expression of morphine-induced conditioned place preference in rats. *Psychoneuroendocrinology, 53*, 159–169.

Mokdad, A., Marks, J., Stroup, D., & Gerberding, J. (2004). Actual causes of death in the United States, 2000. *Journal of the American Medical Association, 291*, 1238–1245.

Morales-Rivera, A., Hernández-Burgos, M. M., Martinez-Rivera, A., Pérez-Colón, J., Rivera, R., Montalvo, J., et al. (2014). Anxiolytic effects of oxytocin in cue-induced cocaine seeking behavior in rats. *Psychopharmacology, 231*(21), 4145–4155.

Napier, T. C., Herrold, A. A., & de Wit, H. (2013). Using conditioned place preference to identify relapse prevention medications. *Neuroscience and Biobehavioral Reviews, 37*(9 Pt. A), 2081–2086.

Neumann, I. D., Maloumby, R., Beiderbeck, D. I., Lukas, M., & Landgraf, R. (2013). Increased brain and plasma oxytocin after nasal and peripheral administration in rats and mice. *Psychoneuroendocrinology, 38*(10), 1985–1993.

O'Connor, S., Morzorati, S., Christian, J., & Li, T. K. (1998). Clamping breath alcohol concentration reduces experimental variance: Application to the study of acute tolerance to alcohol and alcohol elimination rate. *Alcoholism: Clinical and Experimental Research, 22*, 202–210.

O'Connor, S., Ramchandani, V. A., & Li, T. K. (2000). PBPK modeling as a basis for achieving a steady BrAC of 60 +/- 5 mg% within ten minutes. *Alcoholism: Clinical and Experimental Research, 24*, 426–427.

Ohlsson, B., Truedsson, M., Bengtsson, M., Torstenson, R., Sjolund, K., Bjornsson, E. S., et al. (2005). Effects of long-term treatment with oxytocin in chronic constipation; a double blind, placebo-controlled pilot trial. *Neurogastroenterology and Motility, 17*, 697–704.

Pedersen, C. A. (2013). Oxytocin regulation of maternal behavior: From rodents to humans. In E. Choleris, D. W. Pfaff, & M. Kavaliers (Eds.), *Oxytocin, vasopressin, and related peptides in the regulation of behavior* (pp. 148–182). New York: Cambridge University Press.

Pedersen, C. A., Caldwell, J. D., Jirikowski, G. F., & Insel, T. R. (1992). Oxytocin in maternal, social, and sexual behaviors. *New York Academy of Sciences, 652.*

Pedersen, C. A., Smedley, K. L., Leserman, J., Jarskog, L. F., Rau, S. W., Kampov-Polevoi,-A., et al. (2013). Intranasal oxytocin blocks alcohol withdrawal in human subjects. *Alcoholism: Clinical and Experimental Research, 37*, 484–489.

Peters, S. T., Bowen, M. T., Bohrer, K., McGregor, I. S., & Neumann, I. D. (2017). Oxytocin inhibits ethanol consumption and ethanol-induced dopamine release in the nucleus accumbens. *Addiction Biology, 22*(3), 702–711.

Peters, S., Slattery, D. A., Flor, P. J., Neumann, I. D., & Reber, S. O. (2013). Differential effects of baclofen and oxytocin on the increased ethanol consumption following chronic psychosocial stress in mice. *Addiction Biology, 18*(1), 66–77.

Preti, A., Melis, M., Siddi, S., Vellante, M., Doneddu, G., & Fadda, R. (2014). Oxytocin and autism: A systematic review of randomized controlled trials. *Journal of Child and Adolescent Psychopharmacology, 24*(2), 54–68.

Pucilowski, O., Kotowski, W., & Trzaskowa, E. (1985). The effect of oxytocin and fragment (MIF-1) on the development of tolerance to hypothermic and hypnotic action of ethanol in the rat. *Peptides, 6*, 7–10.

Qi, J., Han, W. Y., Yang, J. Y., Wang, L. H., Dong, Y. X., Wang, F., et al. (2012). Oxytocin regulates changes of extracellular glutamate and GABA levels induced by methamphetamine in the mouse brain. *Addiction Biology*, *17*(4), 758–769.

Qi, J., Yang, J. Y., Song, M., Li, Y., Wang, F., & Wu, C. F. (2008). Inhibition by oxytocin of methamphetamine-induced hyperactivity related to dopamine turnover in the mesolimbic region in mice. *Naunyn-Schmiedeberg's Archives of Pharmacology*, *376*(6), 441–448.

Qi, J., Yang, J. Y., Wang, F., Zhao, Y. N., Song, M., & Wu, C. F. (2009). Effects of oxytocin on methamphetamine-induced conditioned place preference and the possible role of glutamatergic neurotransmission in the medial prefrontal cortex of mice in reinstatement. *Neuropharmacology*, *56*(5), 856–865.

Ramchandani, V. A., Bolane, J., Li, T. K., & O'Connor, S. (1999). A physiologically-based pharmacokinetic (PBPK) model for alcohol facilitates rapid BrAC clamping. *Alcoholism: Clinical and Experimental Research*, *23*, 617–623.

Rudd, R. A., Seth, P., David, F., & Scholl, L. (2016). Increases in drug and opioid-involved overdose deaths—United States, 2010–2015. *Morbidity and Mortality Weekly Report*, *65*, 1445–1452.

Salsitz, E., & Wiegand, T. (2016). Pharmacotherapy of opioid addiction: "Putting a real face on a false demon" *Journal of Medical Toxicology*, *12*(1), 58–63.

Sarnyai, Z., Biró, E., Babarczy, E., Vecsernyés, M., Laczi, F., Szabó, G., et al. (1992). Oxytocin modulates behavioural adaptation to repeated treatment with cocaine in rats. *Neuropharmacology*, *31*, 593–598.

Sarnyai, Z., & Kovács, G. L. (1994). Review: Role of oxytocin in the neuroadaptation to drugs of abuse. *Psychoneuroendocrinology*, *19*(1), 85–117.

Sarnyai, Z., Szabó, G., Kovács, G. L., & Telegdy, G. (1990). Oxytocin attenuates the cocaine-induced exploratory hyperactivity in mice. *Neuroreport*, *1*, 200–202.

Sarnyai, Z., Szabó, G., Kovács, G. L., & Telegdy, G. (1992). Opposite actions of oxytocin and vasopressin in the development of cocaine-induced behavioral sensitization in mice. *Pharmacology, Biochemistry and Behavior*, *43*, 491–494.

Sarnyai, Z., Viski, S., Kriván, M., Szabó, G., Kovács, G. L., & Telegdy, G. (1988). Endogenous oxytocin inhibits morphine tolerance through limbic forebrain receptors. *Brain Research*, *463*, 284–288.

Schwartzberg, H., Kovács, G. L., Penke, B., & Telegdy, G. (1986). Effect of oxytocin on acute enkephalin tolerance in mice. *Neuropeptides*, *7*, 247–250.

Slattery, D. A., & Neumann, I. D. (2008). No stress please! Mechanisms of stress hyporesponsiveness of the maternal brain. *Journal of Physiology*, *586*(2), 377–385.

Smith, A. S., & Wang, Z. (2012). Salubrious effects of oxytocin on social stress-induced deficits. *Hormones and Behavior*, *61*(3), 320–330.

Sobell, L. C., Sobell, M. B., Leo, G. L., & Cancilla, A. (1988). Reliability of a timeline followback method: Assessing normal drinkers' reports of recent drinking and a comparative evaluation across several populations. *British Journal of Addiction*, *83*, 393–402.

Spielberger, C. D., Gorsuch, R. L., & Lushene, R. E. (1969). *The State Trait Anxiety Inventory manual*. Palo Alto: Consulting Psychologists Press.

Stauffer, C. S., Musinipally, V., Suen, V., Lynch, K. L., Shapiro, B., & Wooley, J. D. (2016). A two-week pilot study of intranasal oxytocin for cocaine-dependent individuals receiving methadone maintenance treatment for opioid use disorder. *Addiction Research & Theory*, *24*(6), 490–498.

Stoops, W. W., & Rush, C. R. (2014). Combination pharmacotherapies for stimulant use disorder: A review of clinical findings and recommendations for future research. *Expert Review of Clinical Pharmacology*, *7*(3), 363–374.

Striepens, N., Kendrick, K. M., Hanking, V., Landgraf, R., Wullner, U., Maier, W., et al. (2013). Elevated cerebrospinal fluid and blood concentration of oxytocin following its intranasal administration in humans. *Scientific Reports*, *3*, 3440. https://doi.org/10.1038/srep03440.

Sullivan, J. T., Sykora, K., Schneiderman, J., Naranjo, C. A., & Sellers, E. M. (1989). Assessment of alcohol withdrawal: The revised Clinical Institute Withdrawal Assessment of Alcohol Scale (CIWA-Ar). *British Journal of Addiction*, *84*, 1353–1357.

Surgeon General's Report (2016) Facing addiction in America: The Surgeon General's report on alcohol, drugs, and health (online).

Szabó, G., Kovács, G. L., Székeki, S., & Telegdy, G. (1985). The effects of neurohypophyseal hormones on tolerance to the hypothermic effect of ethanol. *Alcohol*, *2*, 567–574.

Szabó, G., Kovács, G. L., & Telegdy, G. (1987). Effects of neurohypophyseal peptide hormones on alcohol dependence and withdrawal. *Alcohol and Alcoholism*, *22*, 71–74.

Szabó, G., Kovács, G. L., & Telegdy, G. (1989). Intraventricular administration of neurohypophyseal hormones interferes with the development of tolerance to ethanol. *Acta Physiologica Hungarica*, *73*, 97–103.

Testa, M., Livingston, J. A., & Leonard, K. E. (2003). Women's substance use and experiences of intimate partner violence: A longitudinal investigation among a community sample. *Addictive Behaviors*, *28*(9), 1649–1664.

Tzschentke, M. (2007). Measuring reward with the conditioned place preference (CPP) paradigm: Update of the last decade. *Addiction Biology*, *12*(3–4), 227–462.

Vaht, M., Kurrikoff, T., Laas, K., Veidebaum, T., & Harro, J. (2016). Oxytocin receptor gene variation rs53576 and alcohol abuse in a longitudinal population representative study. *Psychoneuroendocrinology*, *74*, 333–341.

Walter, M., Gerhard, U., Duersteler-MacFarland, K. M., Weijers, H. G., Boening, J., & Wiesbeck, G. A. (2006). Social factors but not stress-coping styles predict relapse in detoxified alcoholics. *Neuropsychobiology*, *54*, 100–106.

Weiss, R. D., & Rao, V. (2017). The prescription opioid addiction treatment study: What have we learned. *Drug and Alcohol Dependence*, 173(Suppl. 1), S48–S54.

Williams, C. L., Chang, S. W. C., & Pedersen, C. A. (2014). Oxytocin in human social behavior and psychopathology. *Brain Research*, *1580*(special issue).

Wooley, J. D., Arcuni, P. A., Stauffer, C. S., Fulford, D., Carson, D. S., Batki, S., et al. (2016). The effects of intranasal oxytocin in opioid-dependent individuals and healthy control subjects: A pilot study. *Psychopharmacology*, *233*, 2571–2580.

Young, L. J., & Flanagan-Cato, L. M. (2012). Oxytocin, vasopressin and social behavior. *Hormones and Behavior*, *61*(3)(special issue).

Young, K. A., Gobrogge, K. L., & Wang, Z. (2011). The role of mesocorticolimbic dopamine in regulating interactions between drugs of abuse and social behavior. *Neuroscience and Biobehavioral Reviews*, *35*(3), 498–515.

Young, K. A., Liu, Y., Gobrogge, K. L., Wang, H., & Wang, Z. (2014). Oxytocin reverses amphetamine-induced deficits in social bonding: Evidence for an interaction with nucleus accumbens dopamine. *Journal of Neuroscience*, *34*(25), 8499–8506.

Zanos, P., Georgiou, P., Wright, S. R., Hourani, S. M., Kitchen, I., Winsky-Sommerer, R., et al. (2014). The oxytocin analogue carbetocin prevents emotional impairment and stress-induced reinstatement of opioid-seeking in morphine-abstinent mice. *Neuropsychopharmacology*, *39*(4), 855–865.

Zhang, H., Wu, C., Chen, Q., Chen, X., Xu, Z., Wu, J., et al. (2013). Treatment of obesity and diabetes using oxytocin or analogs in patients and mouse models. *PLoS One*, *8*(5). . e61477. https://doi.org/10.1371/journal.pone.0061477(online publication).

Zhou, L., Sun, W-L., Young, A. B., Lee, K., McGinty, J. F., & See, R. E. (2014). Oxytocin reduces cocaine seeking and reverses chronic cocaine-induced changes in glutamate receptor function. *International Journal of Neuropsychopharmacology*, 1–11. https://doi.org/10.1093/ijnp/pyu009 (online publication).

Zink, C. F., & Meyer-Lindenberg, A. (2012). Human neuroimaging of oxytocin and vasopressin in social cognition. *Hormones and Behavior*, *61*, 400–409.

CHAPTER TEN

Contribution of Urocortin to the Development of Excessive Drinking

Andrey E. Ryabinin[*,1], **William J. Giardino**[†]
[*]Oregon Health & Science University, Portland, OR, United States
[†]Stanford University, Stanford, CA, United States
[1]Corresponding author: e-mail address: ryabinin@ohsu.edu

Contents

Abstract

The corticotropin-releasing factor (CRF) system plays a role in alcohol consumption, and its dysregulation can contribute to alcohol use disorder. This system includes four peptide ligands: CRF, urocortin (Ucn)1, Ucn2, and Ucn3. Historically, attention focused on CRF, however, Ucn1 also plays a critical role in excessive alcohol use. This review covers evidence for this contribution and contrasts the role of Ucn1 with CRF. While CRF can promote binge consumption, this regulation occurs through generalized mechanisms that are not specific for alcohol. In contrast, inhibition of Ucn1 action specifically blunts escalation of alcohol drinking. Lesions, genetic knockout, and RNA interference experiments indicate that the centrally projecting Edinger-Westphal nucleus is the neuroanatomical source of Ucn1 critical for alcohol drinking. We propose that the contributions of Ucn1 to excessive drinking likely occur through enhancing rewarding properties of alcohol and symptoms of alcohol withdrawal, whereas CRF drives dependence-induced drinking at later stages of alcohol use. The transition from occasional binge drinking to dependence intricately depends on CRF system plasticity and coordination of CRF and Ucn1.

International Review of Neurobiology, Volume 136
ISSN 0074-7742
http://dx.doi.org/10.1016/bs.irn.2017.06.007

1. INTRODUCTION: THE PROMISE OF THE CRF SYSTEM

Moderate consumption of alcoholic beverages is popular across the world and can have some health benefits. On the other hand, excessive alcohol consumption and dependence have a tremendous negative impact on health and social well-being of individuals. Therefore, it is extremely important to understand the mechanisms leading to excessive alcohol consumption.

Increased allostatic/stress load has been put forth as one of the most important mechanisms in the development of alcohol and drug dependence (Koob & Le Moal, 2001). Since the corticotropin-releasing factor (CRF) system is a major regulator of the stress-related allostatic response, it has received significant attention in research on the neurobiology of addiction. Indeed, pharmacological inhibition of the CRF receptors can reduce alcohol self-administration in animals that have become forcibly dependent on alcohol, but not in nondependent control animals (Heilig & Koob, 2007; Koob, 2008, 2010). These findings led to clinical trials testing CRF antagonists in alcohol-dependent patients. Unfortunately, these trials were not successful (Schwandt et al., 2016). While this negative outcome contributed to a certain degree of skepticism about CRF as a target for addiction treatment, the failure most likely reflects a lack of attention to the details of the CRF system and their specific contributions to regulation of alcohol consumption (Giardino & Ryabinin, 2011; Ryabinin et al., 2012; Schank, Ryabinin, Giardino, Ciccocioppo, & Heilig, 2012). In fact, several studies indicated that global inhibition of actions of CRF can increase stress-induced alcohol consumption (Molander et al., 2012; Sillaber et al., 2002) and alter consummatory behaviors in general (Giardino & Ryabinin, 2013), suggesting cautionary approach to such global inhibition. The findings, showing that manipulation of the global CRF system can result in bidirectional regulation of alcohol consumption (inhibition in some studies and increase in others), indicate that the complexity of the CRF system's components needs to be taken into account when designing treatments for alcoholism (Giardino & Ryabinin, 2013; Kaur, Li, Stenzel-Poore, & Ryabinin, 2012; Lowery & Thiele, 2010; Molander et al., 2012; Sillaber et al., 2002). In particular, it is important to remember that CRF receptors serve as targets not only for CRF but also for the urocortin peptides.

2. THE CRF/UROCORTIN SYSTEM

Throughout evolution, the components that comprise the CRF system (endogenous peptide ligands, receptors, and soluble-binding proteins) changed dramatically across species (including their anatomical locations and functional significance) (Lovejoy & de Lannoy, 2013). The mammalian CRF system consists of four ligands and four or five protein targets (Bale & Vale, 2004; Henckens, Deussing, & Chen, 2016). Specifically, CRF was the first ligand in this system that was discovered. Later studies discovered three related endogenous peptides: urocortin (Ucn)1, Ucn2, and Ucn3. The three best-described targets of these ligands are CRF receptor (CRFR) 1, CRFR2, and CRF-binding protein (CRF-BP). CRFRs are G-protein-coupled receptors that are both positively coupled to cAMP production. CRFR1 and CRFR2 proteins are encoded by CRHR1 and CRHR2 genes. Alternative splicing of CRHR2 can result in a truncated soluble protein (Chen et al., 2005), which functions to modulate the actions of CRFR1 ligands. More recent studies suggest that CRF, and possibly Ucn1, can also directly bind to the gamma secretase (Park et al., 2015), modulating its actions.

CRF has high affinity for CRFR1, but not CRFR2 receptors. In contrast, Ucn2 and Ucn3 have high affinity for CRFR2, but not CRFR1. Ucn1 is unique among the ligands of this family, in that it has highest affinity toward both CRFR1 and CRFR2 receptors (Bale & Vale, 2004). As such, the name CRF system is a confusing term stemming from the fact that CRF has been discovered prior to urocortins. Based on these higher affinities, this peptide system should have been named the Ucn system.

A substantial number of previous reviews have described the distribution of CRF and CRF receptors through the brain (Henckens et al., 2016; Van Pett et al., 2000). Peripheral expression of urocortins has also been described elsewhere (Bamberger & Bamberger, 2000; Boorse & Denver, 2006). The urocortins vary in their distribution in the mammalian CNS (Deussing, Ryabinin, Zuniga, & Kozicz, 2016). The expression of Ucn1 is limited to only a few brain regions. The predominant neural site of Ucn1 expression is the centrally projecting Edinger-Westphal nucleus (EWcp). This nucleus is distinct from the preganglionic Edinger-Westphal nucleus (EWpg) involved in oculomotor functions (Kozicz et al., 2011; May, Reiner, & Ryabinin, 2008; Ryabinin, Tsivkovskaia, & Ryabinin, 2005). Besides Ucn1, EWcp expresses a number of peptides and receptors involved in regulation of food consumption, including the cocaine- and amphetamine-regulated transcript, cholecystokinin, nesfatin-1, and growth hormone secretagogue receptor 1 (Giardino, Cote, Li, & Ryabinin, 2012; Giardino et al., 2017; Kaur & Ryabinin, 2010). The lateral superior olive is a lesser site of Ucn1 expression. Some studies have also described expression of Ucn1 in the hypothalamic supraoptic nucleus and dorsolateral lemniscus, but this expression appears to be species and strain dependent (Weitemier & Ryabinin, 2005a; Weitemier, Tsivkovskaia, & Ryabinin, 2005). Ucn1 fibers are widely distributed throughout the brain. Some of the prominent fiber sites in the forebrain and midbrain are the lateral septum, diagonal band, lateral geniculate, subthalamic nucleus, pretectal nucleus, mammilary nucleus, inferior and superior colliculus, periaqueductal gray, interpeduncular nucleus, laterodorsal tegmental area, and the dorsal raphe. Lesion and tracing studies confirmed that Ucn1 fibers in the lateral septum, lateral hypothalamus, bed nucleus of stria terminalis (BNST), central nucleus of amygdala (CeA), and the dorsal raphe originate in the EWcp (Bachtell, Weitemier, & Ryabinin, 2004; Bittencourt et al., 1999; Dos Santos Junior et al., 2015; Weitemier & Ryabinin, 2005a, 2005b; Weitemier et al., 2005).

The distribution of Ucn2 in the brain has been less characterized than Ucn1, but Ucn2 mRNA has been detected in the paraventricular nucleus of hypothalamus (PVN), the arcuate nucleus, and several brainstem nuclei (Reyes et al., 2001; Tanaka et al., 2003). Ucn3 is expressed in the PVN and the nearby perifornical area, the preoptic region, BNST, medial amygdala, and the paraolivary and parabrachial nuclei. Ucn3 fibers are dense in the hypothalamus, the lateral septum, BNST, and medial amygdala (Cavalcante, Sita, Mascaro, Bittencourt, & Elias, 2006; Deussing et al., 2010; Lewis et al., 2001; Li, Vaughan, Sawchenko, & Vale, 2002). Taken together, the brain is densely but intricately innervated by the Ucn peptides, which are likely to regulate varied behaviors in a site-specific manner.

3. COMPLICATIONS OF STUDYING ACTIONS OF CRF AND Ucns

The varied and partially overlapping distribution of ligands of the CRF system complicates the interpretation of studies of this system. Thus, pharmacological studies using CRF, nonspecific CRF antagonists, CRFR1-specific antagonists, or sometimes even CRFR2-specific antagonists have been traditionally interpreted as testing the actions of endogenous

CRF (Giardino & Ryabinin, 2011). These interpretations do not take into account the fact that Ucn1 has equal or higher affinity to CRFR1 receptor than CRF itself and that all Ucns have higher affinity than CRF at CRFR2 receptors. Using intracranial site-specific instead of systemic injections does not overcome this complication, as multiple peptides of the CRF system innervate the same brain regions.

Interpretation of studies using global knockout or transgenic approaches is also complicated by developmental compensations that have been reported in this system. Only a few studies combined pharmacological and transgenic approaches in studying the CRF system, but the number of such studies is growing. Perhaps even more revealing would be studies using viral approaches to regulate population-specific expression of these peptides.

4. ALCOHOL SENSITIVITY OF THE Ucn1 NEUROPEPTIDE SYSTEM

The Ucn1-expressing neurons of the EWcp are extremely sensitive to alcohol. Thus, this is the only identified population of neurons that shows consistent induction of the activity marker c-Fos following voluntary alcohol drinking across a variety of paradigms in mice, rats, and prairie voles (Anacker, Loftis, Kaur, & Ryabinin, 2011; Bachtell, Tsivkovskaia, & Ryabinin, 2002; Spangler, Cote, Anacker, Mark, & Ryabinin, 2009; Weitemier, Woerner, Backstrom, Hyytia, & Ryabinin, 2001). This response is ethanol specific, as such induction is not observed following the consumption of other palatable fluids, such as sucrose or saccharine. Induction of c-Fos in this nucleus is also observed following forced administration of alcohol and several other drugs of abuse, including morphine, cocaine, and amphetamine. However, while the induction of c-Fos following voluntary drinking occurs almost exclusively in EWcp, forced alcohol administration (e.g., following injections or vapor inhalation) is accompanied by induction of c-Fos in multiple other brain regions, most likely due to the stress/novelty associated with these methods. In fact, induction of c-Fos in EWcp is substantially more sensitive to alcohol vs various stressors (Spangler et al., 2009). This idea is in agreement with the recent finding that EWcp and the CeA are the only two brain regions in which an ethanol injections lead to induction of c-Fos under isoflurane anesthesia (Smith, Li, Cote, & Ryabinin, 2016). Since EWcp has been shown to project to CeA, it is possible that activation of CeA neurons by ethanol is a consequence of ethanol's actions on EWcp.

c-Fos is also induced in EWcp during withdrawal from alcohol. Increased activity of EWcp during withdrawal is also confirmed by increased expression of a large number of transcripts (including Ucn1), in this nucleus during withdrawal (Giardino et al., 2017). Thus, transcriptional activity of EWcp is elevated both during alcohol self-administration and withdrawal. It is tempting to speculate that induction of activity during both of these processes leads to escalation of its neural activity and contributes to development of dependence. Further confirmation of this idea has been obtained in studies addressing the causal contribution of the EWcp Ucn1 system to alcohol self-administration.

5. GENETIC ASSOCIATIONS OF THE EWcp Ucn1 SYSTEM WITH EXCESSIVE ALCOHOL CONSUMPTION

Early studies noted that selective breeding for increased continuous-access alcohol consumption and increased alcohol-induced hypothermia often resulted in higher levels of Ucn1 immunoreactivity in the EWcp (Bachtell et al., 2002, 2003; Fonareva et al., 2009; Ryabinin & Weitemier, 2006). Since continuous alcohol consumption results in signs of dependence (Smith, Hostetler, Heinricher, & Ryabinin, 2016), it can be considered as a translational model of alcoholism. Therefore, these data suggest that Ucn1 contributes to predisposition to alcohol dependence. The relationship between high alcohol drinking and high EWcp-Ucn1 peptide levels was also documented between inbred strains that differ significantly in alcohol consumption (C57BL/6J; B6 vs DBA/2J; D2 mice) (Giardino et al., 2012).

Since these original findings focused primarily on expression of Ucn1 protein levels, we next conducted quantitative real-time PCR profiling studies in B6 vs D2 mouse strains. Results of these studies revealed that EWcp mRNA levels of Ucn1 (as well as several other EWcp-enriched neuropeptides) are also genetically associated with higher levels of alcohol consumption (Giardino et al., 2012). Interestingly, selective breeding for binge-like nondependent alcohol consumption in the so-called drinking-in-the-dark (DID) procedure is associated with decreased Ucn1 peptide and mRNA expression (A.E. Ryabinin, unpublished), suggesting that Ucn1 can counteract occasional episodes of drinking.

The caveats of selective breeding include the possibility of genetic drift and developmental compensations. However, results described later using alternative strategies agree with the results of genetic breeding. Genetic comparisons of mice selectively bred for alcohol-induced hyperlocomotion found no significant differences in EWcp-Ucn1 expression (Bachtell et al., 2002), suggesting that this trait may be genetically segregated from the traits controlling the relationship between EWcp-Ucn1 activity and increased alcohol drinking.

6. EFFECTS OF EXCESSIVE ALCOHOL CONSUMPTION ON THE Ucn1 SYSTEM AND EWcp

To further pursue the genetic associations between alcohol drinking and Ucn1, additional studies sought to determine whether excessive alcohol consumption itself resulted in changes in levels of Ucn1 peptide, Ucn1 gene expression, and CRFR binding.

Exposure of genetically heterogeneous mice first to continuous access and then to limited access procedure did not lead to significant changes in Ucn1 immunoreactivity or number of Ucn1-positive cells in EWcp. However, it resulted in a significant reduction of the number of Ucn1 fibers in one of its target areas, the lateral septum (Bachtell et al., 2003). Repeated injections of ethanol (2 g/kg, IP) in B6 mice and D2 mice also did not change the number of Ucn1 cells of Ucn1 immunoreactivity in EWcp, but decreased the number of Ucn1 fibers in the lateral septum and dorsal raphe. This effect of repeated ethanol injection was accompanied by a marginal increase in CRFR2 binding in the lateral septum and a strong increase in CRFR2 binding in the dorsal raphe (Weitemier & Ryabinin, 2005a). These finding indicated that alcohol exposure not only changes activity of the Ucn1 system but also results in long-term adaptations within this system.

To further elucidate these effects, recent studies performed quantitative real-time PCR profiling of microdissected EWcp samples from male B6 mice exposed to the intermittent (every other day) alcohol drinking procedure (Giardino et al., 2017). Analysis of EWcp gene expression at 24 h following cessation of the long-term intermittent-access drinking paradigm identified significant increases in the expression of six genes encoding neuropeptide-related components, including Ucn1, in the alcohol-experienced mice, compared to matched water-only controls. These results point to important long-term adaptations in regulation of EWcp activity of neuropeptide-related genes as a result of chronic binge-like alcohol drinking, providing further evidence for a key role for the EWcp-Ucn1 system in driving excessive levels of alcohol consumption.

7. EFFECTS OF GENETIC MANIPULATIONS OF EWcp Ucn1 SYSTEM ON ALCOHOL CONSUMPTION

While these multiple converging lines of evidence suggested that Ucn1 expression can predispose to excessive alcohol consumption, and excessive alcohol consumption can elevate EWcp-Ucn1 gene expression, it was critical to determine whether alcohol consumption is dependent on the Ucn1 system itself.

To address this possibility, we first tested single-gene knockout (KO) mice-lacking Ucn1 in several models of alcohol drinking and alcohol-related behavior. Compared to wild-type (WT) littermate controls, Ucn1 KO mice displayed normal levels of binge-like drinking in the short-term 4-day DID procedure (Kaur et al., 2012), but had significantly reduced levels of alcohol intake and preference in models of long-term heavy continuous-access drinking (10%–40%, two-bottle choice), as well as long-term intermittent-access drinking (3%–20%, two-bottle choice; Giardino et al., 2017). These results suggest that Ucn1 is essential for excessive alcohol consumption that involves progressive escalation across time, but that Ucn1 is dispensable for short-term periods of exposure to alcohol drinking.

Genetic KO of Ucn1 also attenuated ethanol-induced conditioned place preference (Giardino, Cocking, Kaur, Cunningham, & Ryabinin, 2011). This effect contrasted with lack of effects of genetic deletion of Ucn1 on ethanol-induced conditioned place aversion indicating that Ucn1 contributes to excessive alcohol consumption in part through its ability to regulate alcohol-induced reward. We further tested Ucn1 KO and WT mice for alcohol-induced sedation, locomotion, and behavioral inhibition, identifying only marginal sex-specific genotype-dependent effects (Giardino et al., 2017). The latter findings also served as a confirmation that Ucn1 contributes to regulation of alcohol intake by affecting rewarding properties of alcohol. The ethanol-induced conditioned place preference was also sensitive to deletion of CRFR2 suggesting that Ucn1 regulates alcohol reward through CRFR2 receptors (Giardino et al., 2011).

To achieve further specificity for manipulating Ucn1 directly within EWcp neurons, we adopted a combinatorial approach in which we asked whether genetic KO of Ucn1 altered the impact of electrolytic EWcp lesion on alcohol drinking. Indeed, early studies in WT B6 mice found that EWcp lesions could disrupt continuous-access alcohol consumption (Bachtell et al., 2004; Weitemier & Ryabinin, 2005b). Therefore, we hypothesized interactive effects between EWcp lesion and Ucn1 genetic KO, in which the effect of EWcp manipulation on alcohol drinking would be dependent on the expression of Ucn1.

In a continuous-access 3%–10% alcohol two-bottle choice drinking paradigm, we found that alcohol preference was significantly disrupted by EWcp lesion only in WT mice, but not in Ucn1 KO littermates (Giardino et al., 2011). Furthermore, we found that alcohol preference was significantly disrupted by Ucn1 KO only in sham-operated, but not EWcp-lesioned mice. Together, these results provided further evidence suggesting that Ucn1 is a particularly important neuropeptide driving the EWcp's role in promoting alcohol consumption, and that global KO of Ucn1 impacts alcohol drinking via its specific absence from EWcp neurons (rather than from neurons in the lateral superior olive or supraoptic nucleus).

8. THE CRITICAL ROLE OF THE EWcp Ucn1 SYSTEM IN EXCESSIVE ALCOHOL CONSUMPTION

Closer examination of specific effects of genetic manipulations of the Ucn1 system in studies described earlier revealed a curious pattern. Thus, if alcohol was available to the animals in the two-bottle choice procedure at a constant concentration (i.e., 10%), then deletion on the Ucn1 gene did not affect its consumption. However, if alcohol was available to the animals at increasing concentrations, then the deletion decreased alcohol consumption of the subsequent higher concentration (Giardino et al., 2011, 2017). Thus, when alcohol was available at first at 3% and later at 6% and 10%, then only consumption and preference of 6% and 10% solutions (but not 3%) was sensitive to Ucn1 deletion. And when alcohol was available first at 10%, and later at 20% and 40%, then only consumption and preference of 20% and 40% (but not 10%) was affected to the Ucn1 deletion. This observation suggested that Ucn1 could be selectively regulating escalation of alcohol consumption.

To directly address this possibility, we took advantage of the intermittent-access procedure, known to result in escalated alcohol consumption (Hwa et al., 2011; Loi et al., 2010; Melendez, 2011; Wise, 1973). Importantly, this procedure can result in increased handling induced convulsions during withdrawal, a sign of severe alcohol dependence (Hwa et al., 2011). To be able to discern specific drinking patterns regulated by Ucn1, these studies were done using the lickometer approach precisely monitoring the drinking behaviors. In agreement with the hypothesis, Ucn1 KO mice demonstrated significantly lower licking rates on days when the drinking escalated in WT animals (Giardino et al., 2017). Analysis of licking patterns showed that the genetic effect was due to decreased approaches to the drinking spout at the beginning of the circadian dark, the peak time of consummatory behavior in mice and that the effect increased with subsequent behavioral sessions. The latter finding suggested that Ucn1 contributes to plasticity mechanisms, presumably associated with learning the rewarding consequences of higher levels of alcohol intoxication, as in the conditioned place preference procedure (Giardino et al., 2011).

To completely eliminate any possibility of contribution of developmental compensations in Ucn1 KO to observed effects and to achieve even further specificity in genetically manipulating the EWcp-Ucn1 system, we implemented RNA interference knockdown of Ucn1. The knockdown was achieved specifically within the EWcp by lentiviral-encoded short hairpin RNAs targeted against Ucn1 (Giardino et al., 2017). Compared to matched viral controls, infection of EWcp-Ucn1 neurons reduced peptide immunoreactivity levels at ~40%–50%. Mice were then tested in the intermittent-access procedure. The knockdown of Ucn1 expression in EWcp was accompanied by significant reductions in escalated intake and preference. These data conclusively demonstrate that Ucn1 is the specific gene residing within the EWcp that is required for excessive, escalated consumption of alcohol. In addition to alcohol drinking studies, we also compared baseline anxiety-like behaviors between control and EWcp-Ucn1 viral knockdown mice, identifying no significant group differences (Giardino et al., 2017). These data indicate that EWcp-Ucn1 RNA interference does not alter locomotor activity and anxiety levels, and is therefore behaviorally specific for disrupting excessive alcohol consumption.

9. CONTRASTING THE ROLES OF Ucn1 VS CRF IN CONTRIBUTION TO EXCESSIVE ALCOHOL CONSUMPTION

The findings of ethanol-specific involvement with Ucn1 in escalated drinking place this peptide in a unique position in the development of excessive drinking and contrast its role with role of CRF in this behavior (Fig. 1).

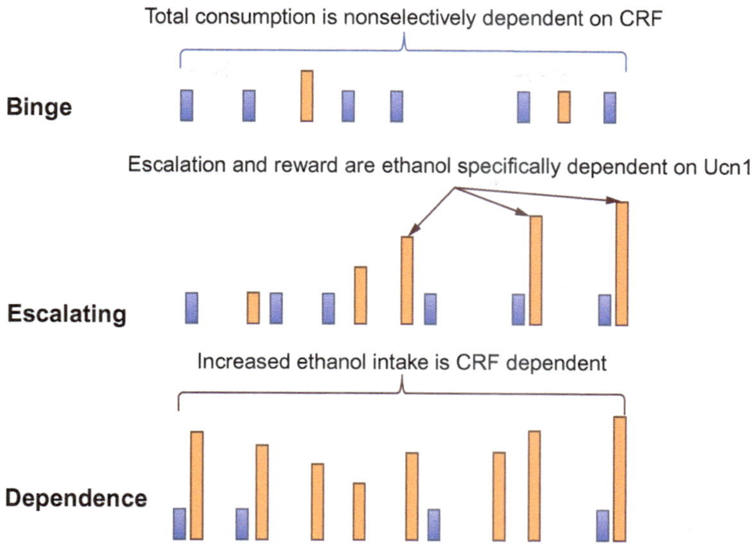

Fig. 1 Patterns of excessive alcohol consumption and their dependence on the CRF system. Consumption of alcoholic beverages is represented by *beige bars*, consumption of other fluids by *blue bars*. The nonspecific dependence of consumption of total fluid on specific peptides is indicated by the *blue bracket*. The dependence of alcohol consumption on specific peptides is indicated by *red arrows* or *bracket*.

The studies using the DID model of binge drinking show that a CRFR1-mediated tone is involved in consummatory behaviors and recently identify some of the potential neurocircuitry involved in these behaviors (Kaur et al., 2012; Lowery & Thiele, 2010; Rinker et al., 2017). However, the contribution of CRF to these behaviors is not specific to ethanol, as CRFR1 KO and CRFR1 pharmacological blockade decrease accompanying water and food intake. Paradoxically, CRFR1 KO and CRFR1 antagonists increase (rather than decrease) consumption of sucrose, indicating that using this rewarding solution as a "control solution" for testing the specificity of the manipulations may be inappropriate (Giardino & Ryabinin, 2013). Since the DID procedure involves testing mice at the time when they transition into their behaviorally active circadian phase, one possibility is that CRF is involved in basal state of arousal, and therefore, inhibition of CRF's action leads to nonspecific decrease in fluid and food consumption.

On the other side of the spectrum of CRF's effects is its contribution to alcohol drinking in models of severe alcohol dependence (Chu, Koob, Cole, Zorrilla, & Roberts, 2007; Heilig & Koob, 2007; Koob, 2010; Roberto et al., 2010). This contribution has been extensively reviewed. In particular, the enhanced alcohol drinking observed after intermittent exposure to ethanol vapors is dependent on well-described CRF plasticity in the CeA (Herman et al., 2013; Roberto et al., 2010). Both binge-like DID and dependence-enhanced drinking have also been shown to be negatively influenced by endogenous stimulation of CRFR2 receptors (Albrecht-Souza et al., 2015; Funk & Koob, 2007; Lowery et al., 2010). However, there are no studies confirming that these behaviors are regulated by endogenous CRFR2-specific peptides (Ucn2 and Ucn3).

Complementary to CRF's role in dependence-induced drinking, studies described in Section 8 of this review indicate a critical role of Ucn1 in the escalation of alcohol consumption (Fig. 1). The critical involvement of Ucn1 in this process has been shown in both KO and shRNA knockdown experiments (Giardino et al., 2017). At least in part the contribution of Ucn1 to escalated drinking is thought to be mediated through modulation of rewarding aspects of alcohol intoxication. This notion is based on studies testing ethanol-induced conditioned place preference procedure in Ucn1 KO mice (Giardino et al., 2011). However, this effect could also be aided by Ucn1's potential role in symptoms of withdrawal. The latter mechanism is suggested by the observation of increased EWcp Ucn1 expression during withdrawal from escalated alcohol consumption in C57BL/6J mice (Giardino et al., 2017). Regardless of the mechanism, escalation of alcohol drinking in humans occurs during the transition from normative social drinking to compulsive excessive alcohol consumption in dependent subject. As such, Ucn1 is likely to play a critical role in the transition to alcoholism.

Ucn1 is unique among the CRF-like peptides in that it has high affinity to both CRFR1 and CRFR2. While first attempts at use of CRF antagonists in clinical trials have not been successful, one possibility for lack of effects is that only one target of Ucn1 (i.e., CRFR1) has been affected by this treatment. Nonspecific CRF antagonists (affecting both CRFR1 and CRFR2 receptors), better replicating inhibition of Ucn1, have been shown to be effective across many animal models of excessive drinking and dependence (Finn et al., 2007; Funk, O'Dell, Crawford, & Koob, 2006; Le, Funk, Coen, Li, & Shaham, 2013; Valdez et al., 2002). However, the currently available antagonists are peptide-based and not blood–brain penetrant. Development of less-selective small molecule-based antagonists of both CRFR1 and CRFR2 could potentially aid in inhibiting actions of Ucn1, and thereby help curb the escalated drinking underlying alcohol use disorder.

ACKNOWLEDGMENTS

While writing this review authors were supported by NIH Grants RO1 AA019793 (A.E.R.), R21 AA025548 (A.E.R.), and F32 AA022832 (W.J.G.).

REFERENCES

Albrecht-Souza, L., Hwa, L. S., Han, X., Zhang, E. Y., DeBold, J. F., & Miczek, K. A. (2015). Corticotropin releasing factor binding protein and CRF2 receptors in the ventral tegmental area: Modulation of ethanol binge drinking in C57BL/6J mice. *Alcoholism, Clinical and Experimental Research*, *39*(9), 1609–1618. http://dx.doi.org/10.1111/acer.12825.

Anacker, A. M., Loftis, J. M., Kaur, S., & Ryabinin, A. E. (2011). Prairie voles as a novel model of socially facilitated excessive drinking. *Addiction Biology*, *16*(1), 92–107. http://dx.doi.org/10.1111/j.1369-1600.2010.00234.x.

Bachtell, R. K., Tsivkovskaia, N. O., & Ryabinin, A. E. (2002). Strain differences in urocortin expression in the Edinger-Westphal nucleus and its relation to alcohol-induced hypothermia. *Neuroscience*, *113*(2), 421–434.

Bachtell, R. K., Weitemier, A. Z., Galvan-Rosas, A., Tsivkovskaia, N. O., Risinger, F. O., Phillips, T. J., et al. (2003). The Edinger-Westphal-lateral septum urocortin pathway and its relationship to alcohol consumption. *The Journal of Neuroscience*, *23*(6), 2477–2487.

Bachtell, R. K., Weitemier, A. Z., & Ryabinin, A. E. (2004). Lesions of the Edinger-Westphal nucleus in C57BL/6J mice disrupt ethanol-induced hypothermia and ethanol consumption. *The European Journal of Neuroscience*, *20*(6), 1613–1623. http://dx.doi.org/10.1111/j.1460-9568.2004.03594.x.

Bale, T. L., & Vale, W. W. (2004). CRF and CRF receptors: Role in stress responsivity and other behaviors. *Annual Review of Pharmacology and Toxicology*, *44*, 525–557. http://dx.doi.org/10.1146/annurev.pharmtox.44.101802.121410.

Bamberger, C. M., & Bamberger, A. M. (2000). The peripheral CRH/urocortin system. *Annals of the New York Academy of Sciences*, *917*, 290–296.

Bittencourt, J. C., Vaughan, J., Arias, C., Rissman, R. A., Vale, W. W., & Sawchenko, P. E. (1999). Urocortin expression in rat brain: Evidence against a pervasive relationship of urocortin-containing projections with targets bearing type 2 CRF receptors. *The Journal of Comparative Neurology*, *415*(3), 285–312. http://dx.doi.org/10.1002/(SICI)1096-9861(19991220)415:3<285::AID-CNE1>3.0.CO;2-0.

Boorse, G. C., & Denver, R. J. (2006). Widespread tissue distribution and diverse functions of corticotropin-releasing factor and related peptides. *General and Comparative Endocrinology*, *146*(1), 9–18. http://dx.doi.org/10.1016/j.ygcen.2005.11.014.

Cavalcante, J. C., Sita, L. V., Mascaro, M. B., Bittencourt, J. C., & Elias, C. F. (2006). Distribution of urocortin 3 neurons innervating the ventral premammillary nucleus in the rat brain. *Brain Research*, *1089*(1), 116–125. http://dx.doi.org/10.1016/j.brainres.2006.03.043.

Chen, A. M., Perrin, M. H., Digruccio, M. R., Vaughan, J. M., Brar, B. K., Arias, C. M., et al. (2005). A soluble mouse brain splice variant of type 2alpha corticotropin-releasing factor (CRF) receptor binds ligands and modulates their activity. *Proceedings of the National Academy of Sciences of the United States of America*, *102*(7), 2620–2625. http://dx.doi.org/10.1073/pnas.0409583102.

Chu, K., Koob, G. F., Cole, M., Zorrilla, E. P., & Roberts, A. J. (2007). Dependence-induced increases in ethanol self-administration in mice are blocked by the CRF1 receptor antagonist antalarmin and by CRF1 receptor knockout. *Pharmacology, Biochemistry, and Behavior*, *86*(4), 813–821. http://dx.doi.org/10.1016/j.pbb.2007.03.009.

Deussing, J. M., Breu, J., Kuhne, C., Kallnik, M., Bunck, M., Glasl, L., et al. (2010). Urocortin 3 modulates social discrimination abilities via corticotropin-releasing hormone receptor type 2. *The Journal of Neuroscience*, *30*(27), 9103–9116. http://dx.doi.org/10.1523/JNEUROSCI.1049-10.2010.

Deussing, J. M., Ryabinin, A. E., Zuniga, A., & Kozicz, T. (2016). Action of CRF/urocortin peptides. In *Molecular and cellular mechanisms: Vol. 3. Hormones, brain and behavior*, Amsterdam: Elsevier.

Dos Santos Junior, E. D., Da Silva, A. V., Da Silva, K. R., Haemmerle, C. A., Batagello, D. S., Da Silva, J. M., et al. (2015). The centrally projecting Edinger-Westphal nucleus-I: Efferents in the rat brain. *Journal of Chemical Neuroanatomy*, *68*, 22–38. http://dx.doi.org/10.1016/j.jchemneu.2015.07.002.

Finn, D. A., Snelling, C., Fretwell, A. M., Tanchuck, M. A., Underwood, L., Cole, M., et al. (2007). Increased drinking during withdrawal from intermittent ethanol exposure is blocked by the CRF receptor antagonist D-Phe-CRF(12-41). *Alcoholism, Clinical and Experimental Research*, *31*(6), 939–949. http://dx.doi.org/10.1111/j.1530-0277.2007.00379.x.

Fonareva, I., Spangler, E., Cannella, N., Sabino, V., Cottone, P., Ciccocioppo, R., et al. (2009). Increased perioculomotor urocortin 1 immunoreactivity in genetically selected alcohol preferring rats. *Alcoholism, Clinical and Experimental Research*, *33*(11), 1956–1965. http://dx.doi.org/10.1111/j.1530-0277.2009.01033.x.

Funk, C. K., & Koob, G. F. (2007). A CRF(2) agonist administered into the central nucleus of the amygdala decreases ethanol self-administration in ethanol-dependent rats. *Brain Research*, *1155*, 172–178. http://dx.doi.org/10.1016/j.brainres.2007.04.009.

Funk, C. K., O'Dell, L. E., Crawford, E. F., & Koob, G. F. (2006). Corticotropin-releasing factor within the central nucleus of the amygdala mediates enhanced ethanol self-administration in withdrawn, ethanol-dependent rats. *The Journal of Neuroscience*, *26*(44), 11324–11332. http://dx.doi.org/10.1523/JNEUROSCI.3096-06.2006.

Giardino, W. J., Cocking, D. L., Kaur, S., Cunningham, C. L., & Ryabinin, A. E. (2011). Urocortin-1 within the centrally-projecting Edinger-Westphal nucleus is critical for ethanol preference. *PloS One*, *6*(10), e26997. http://dx.doi.org/10.1371/journal.pone.0026997.

Giardino, W. J., Cote, D. M., Li, J., & Ryabinin, A. E. (2012). Characterization of genetic differences within the centrally projecting Edinger-Westphal nucleus of C57BL/6J and DBA/2J mice by expression profiling. *Frontiers in Neuroanatomy*, *6*, 5. http://dx.doi.org/10.3389/fnana.2012.00005.

Giardino, W. J., Rodriguez, E. D., Smith, M. L., Ford, M. M., Galili, D., Mitchell, S. H., et al. (2017). Control of chronic excessive alcohol drinking by genetic manipulation of the Edinger-Westphal nucleus urocortin-1 neuropeptide system. *Translational Psychiatry*, 7(1), e1021. http://dx.doi.org/10.1038/tp.2016.293.

Giardino, W. J., & Ryabinin, A. E. (2011). Corticotropin-releasing factor: Innocent until proven guilty. *Nature Reviews. Neuroscience*, *13*(1), 70. http://dx.doi.org/10.1038/nrn3110-c1. author reply 70.

Giardino, W. J., & Ryabinin, A. E. (2013). CRF1 receptor signaling regulates food and fluid intake in the drinking-in-the-dark model of binge alcohol consumption. *Alcoholism, Clinical and Experimental Research*, *37*(7), 1161–1170. http://dx.doi.org/10.1111/acer.12076.

Heilig, M., & Koob, G. F. (2007). A key role for corticotropin-releasing factor in alcohol dependence. *Trends in Neurosciences*, *30*(8), 399–406. http://dx.doi.org/10.1016/j.tins.2007.06.006.

Henckens, M. J., Deussing, J. M., & Chen, A. (2016). Region-specific roles of the corticotropin-releasing factor-urocortin system in stress. *Nature Reviews. Neuroscience*, *17*(10), 636–651. http://dx.doi.org/10.1038/nrn.2016.94.

Herman, M. A., Kallupi, M., Luu, G., Oleata, C. S., Heilig, M., Koob, G. F., et al. (2013). Enhanced GABAergic transmission in the central nucleus of the amygdala of genetically selected Marchigian Sardinian rats: Alcohol and CRF effects. *Neuropharmacology*, *67*, 337–348. http://dx.doi.org/10.1016/j.neuropharm.2012.11.026.

Hwa, L. S., Chu, A., Levinson, S. A., Kayyali, T. M., DeBold, J. F., & Miczek, K. A. (2011). Persistent escalation of alcohol drinking in C57BL/6J mice with intermittent access to 20% ethanol. *Alcoholism, Clinical and Experimental Research*, *35*(11), 1938–1947. http://dx.doi.org/10.1111/j.1530-0277.2011.01545.x.

Kaur, S., Li, J., Stenzel-Poore, M. P., & Ryabinin, A. E. (2012). Corticotropin-releasing factor acting on corticotropin-releasing factor receptor type 1 is critical for binge alcohol drinking in mice. *Alcoholism, Clinical and Experimental Research*, *36*(2), 369–376. http://dx.doi.org/10.1111/j.1530-0277.2011.01610.x.

Kaur, S., & Ryabinin, A. E. (2010). Ghrelin receptor antagonism decreases alcohol consumption and activation of perioculomotor urocortin-containing neurons. *Alcoholism, Clinical and Experimental Research*, *34*(9), 1525–1534. http://dx.doi.org/10.1111/j.1530-0277.2010.01237.x.

Koob, G. F. (2008). A role for brain stress systems in addiction. *Neuron*, *59*(1), 11–34. http://dx.doi.org/10.1016/j.neuron.2008.06.012.

Koob, G. F. (2010). The role of CRF and CRF-related peptides in the dark side of addiction. *Brain Research*, *1314*, 3–14. http://dx.doi.org/10.1016/j.brainres.2009.11.008.

Koob, G. F., & Le Moal, M. (2001). Drug addiction, dysregulation of reward, and allostasis. *Neuropsychopharmacology*, *24*(2), 97–129. http://dx.doi.org/10.1016/S0893-133X(00)00195-0.

Kozicz, T., Bittencourt, J. C., May, P. J., Reiner, A., Gamlin, P. D., Palkovits, M., et al. (2011). The Edinger-Westphal nucleus: A historical, structural, and functional perspective on a dichotomous terminology. *The Journal of Comparative Neurology*, *519*(8), 1413–1434. http://dx.doi.org/10.1002/cne.22580.

Le, A. D., Funk, D., Coen, K., Li, Z., & Shaham, Y. (2013). Role of corticotropin-releasing factor in the median raphe nucleus in yohimbine-induced reinstatement of alcohol seeking in rats. *Addiction Biology*, *18*(3), 448–451. http://dx.doi.org/10.1111/j.1369-1600.2011.00374.x.

Lewis, K., Li, C., Perrin, M. H., Blount, A., Kunitake, K., Donaldson, C., et al. (2001). Identification of urocortin III, an additional member of the corticotropin-releasing factor (CRF) family with high affinity for the CRF2 receptor. *Proceedings of the National Academy of Sciences of the United States of America*, *98*(13), 7570–7575. http://dx.doi.org/10.1073/pnas.121165198.

Li, C., Vaughan, J., Sawchenko, P. E., & Vale, W. W. (2002). Urocortin III-immunoreactive projections in rat brain: Partial overlap with sites of type 2 corticotrophin-releasing factor receptor expression. *The Journal of Neuroscience*, *22*(3), 991–1001.

Loi, B., Lobina, C., Maccioni, P., Fantini, N., Carai, M. A., Gessa, G. L., et al. (2010). Increase in alcohol intake, reduced flexibility of alcohol drinking, and evidence of signs of alcohol intoxication in Sardinian alcohol-preferring rats exposed to intermittent access to 20% alcohol. *Alcoholism, Clinical and Experimental Research*, *34*(12), 2147–2154. http://dx.doi.org/10.1111/j.1530-0277.2010.01311.x.

Lovejoy, D. A., & de Lannoy, L. (2013). Evolution and phylogeny of the corticotropin-releasing factor (CRF) family of peptides: Expansion and specialization in the vertebrates. *Journal of Chemical Neuroanatomy*, *54*, 50–56. http://dx.doi.org/10.1016/j.jchemneu.2013.09.006.

Lowery, E. G., Spanos, M., Navarro, M., Lyons, A. M., Hodge, C. W., & Thiele, T. E. (2010). CRF-1 antagonist and CRF-2 agonist decrease binge-like ethanol drinking in C57BL/6J mice independent of the HPA axis. *Neuropsychopharmacology*, *35*(6), 1241–1252. http://dx.doi.org/10.1038/npp.2009.209.

Lowery, E. G., & Thiele, T. E. (2010). Pre-clinical evidence that corticotropin-releasing factor (CRF) receptor antagonists are promising targets for pharmacological treatment of alcoholism. *CNS & Neurological Disorders Drug Targets*, *9*(1), 77–86. http://dx.doi.org/10.2174/187152710790966605.

May, P. J., Reiner, A. J., & Ryabinin, A. E. (2008). Comparison of the distributions of urocortin-containing and cholinergic neurons in the perioculomotor midbrain of the cat and macaque. *The Journal of Comparative Neurology*, *507*(3), 1300–1316. http://dx.doi.org/10.1002/cne.21514.

Melendez, R. I. (2011). Intermittent (every-other-day) drinking induces rapid escalation of ethanol intake and preference in adolescent and adult C57BL/6J mice. *Alcoholism, Clinical and Experimental Research*, *35*(4), 652–658. http://dx.doi.org/10.1111/j.1530-0277.2010.01383.x.

Molander, A., Vengeliene, V., Heilig, M., Wurst, W., Deussing, J. M., & Spanagel, R. (2012). Brain-specific inactivation of the Crhr1 gene inhibits post-dependent and stress-induced alcohol intake, but does not affect relapse-like drinking. *Neuropsychopharmacology*, *37*(4), 1047–1056. http://dx.doi.org/10.1038/npp.2011.297.

Park, H. J., Ran, Y., Jung, J. I., Holmes, O., Price, A. R., Smithson, L., et al. (2015). The stress response neuropeptide CRF increases amyloid-beta production by regulating gamma-secretase activity. *The EMBO Journal*, *34*(12), 1674–1686. http://dx.doi.org/10.15252/embj.201488795.

Reyes, T. M., Lewis, K., Perrin, M. H., Kunitake, K. S., Vaughan, J., Arias, C. A., et al. (2001). Urocortin II: A member of the corticotropin-releasing factor (CRF) neuropeptide family that is selectively bound by type 2 CRF receptors. *Proceedings of the National Academy of Sciences of the United States of America*, *98*(5), 2843–2848. http://dx.doi.org/10.1073/pnas.051626398.

Rinker, J. A., Marshall, S. A., Mazzone, C. M., Lowery-Gionta, E. G., Gulati, V., Pleil, K. E., et al. (2017). Extended amygdala to ventral tegmental area Corticotropin-releasing factor circuit controls binge ethanol intake. *Biological Psychiatry*, *81*, 930–940. http://dx.doi.org/10.1016/j.biopsych.2016.02.029.

Roberto, M., Cruz, M. T., Gilpin, N. W., Sabino, V., Schweitzer, P., Bajo, M., et al. (2010). Corticotropin releasing factor-induced amygdala gamma-aminobutyric acid release plays a key role in alcohol dependence. *Biological Psychiatry*, *67*(9), 831–839. http://dx.doi.org/10.1016/j.biopsych.2009.11.007.

Ryabinin, A. E., Tsivkovskaia, N. O., & Ryabinin, S. A. (2005). Urocortin 1-containing neurons in the human Edinger-Westphal nucleus. *Neuroscience*, *134*(4), 1317–1323. http://dx.doi.org/10.1016/j.neuroscience.2005.05.042.

Ryabinin, A. E., Tsoory, M. M., Kozicz, T., Thiele, T. E., Neufeld-Cohen, A., Chen, A., et al. (2012). Urocortins: CRF's siblings and their potential role in anxiety, depression and alcohol drinking behavior. *Alcohol*, *46*(4), 349–357. http://dx.doi.org/10.1016/j.alcohol.2011.10.007.

Ryabinin, A. E., & Weitemier, A. Z. (2006). The urocortin 1 neurocircuit: Ethanol-sensitivity and potential involvement in alcohol consumption. *Brain Research Reviews*, *52*(2), 368–380. http://dx.doi.org/10.1016/j.brainresrev.2006.04.007.

Schank, J. R., Ryabinin, A. E., Giardino, W. J., Ciccocioppo, R., & Heilig, M. (2012). Stress-related neuropeptides and addictive behaviors: Beyond the usual suspects. *Neuron*, *76*(1), 192–208. http://dx.doi.org/10.1016/j.neuron.2012.09.026.

Schwandt, M. L., Cortes, C. R., Kwako, L. E., George, D. T., Momenan, R., Sinha, R., et al. (2016). The CRF1 antagonist verucerfont in anxious alcohol-dependent women: Translation of neuroendocrine, but not of anti-craving effects. *Neuropsychopharmacology*, *41*(12), 2818–2829. http://dx.doi.org/10.1038/npp.2016.61.

Sillaber, I., Rammes, G., Zimmermann, S., Mahal, B., Zieglgansberger, W., Wurst, W., et al. (2002). Enhanced and delayed stress-induced alcohol drinking in mice lacking functional

CRH1 receptors. *Science*, *296*(5569), 931–933. http://dx.doi.org/10.1126/science.1069836 296/5569/931.

Smith, M. L., Hostetler, C. M., Heinricher, M. M., & Ryabinin, A. E. (2016a). Social transfer of pain in mice. *Science Advances*, *2*(10), e1600855. http://dx.doi.org/10.1126/sciadv.1600855.

Smith, M. L., Li, J., Cote, D. M., & Ryabinin, A. E. (2016b). Effects of isoflurane and ethanol administration on c-Fos immunoreactivity in mice. *Neuroscience*, *316*, 337–343. http://dx.doi.org/10.1016/j.neuroscience.2015.12.047.

Spangler, E., Cote, D. M., Anacker, A. M., Mark, G. P., & Ryabinin, A. E. (2009). Differential sensitivity of the perioculomotor urocortin-containing neurons to ethanol, psychostimulants and stress in mice and rats. *Neuroscience*, *160*(1), 115–125. http://dx.doi.org/10.1016/j.neuroscience.2009.02.030.

Tanaka, Y., Makino, S., Noguchi, T., Tamura, K., Kaneda, T., & Hashimoto, K. (2003). Effect of stress and adrenalectomy on urocortin II mRNA expression in the hypothalamic paraventricular nucleus of the rat. *Neuroendocrinology*, *78*(1), 1–11. http://dx.doi.org/10.1159/000071700.

Valdez, G. R., Roberts, A. J., Chan, K., Davis, H., Brennan, M., Zorrilla, E. P., et al. (2002). Increased ethanol self-administration and anxiety-like behavior during acute ethanol withdrawal and protracted abstinence: Regulation by corticotropin-releasing factor. *Alcoholism, Clinical and Experimental Research*, *26*(10), 1494–1501. http://dx.doi.org/10.1097/01.ALC.0000033120.51856.F0.

Van Pett, K., Viau, V., Bittencourt, J. C., Chan, R. K., Li, H. Y., Arias, C., et al. (2000). Distribution of mRNAs encoding CRF receptors in brain and pituitary of rat and mouse. *The Journal of Comparative Neurology*, *428*(2), 191–212. http://dx.doi.org/10.1002/1096-9861(20001211)428:2<191::AID-CNE1>3.0.CO;2-U.

Weitemier, A. Z., & Ryabinin, A. E. (2005a). Brain region-specific regulation of urocortin 1 innervation and corticotropin-releasing factor receptor type 2 binding by ethanol exposure. *Alcoholism, Clinical and Experimental Research*, *29*(9), 1610–1620.

Weitemier, A. Z., & Ryabinin, A. E. (2005b). Lesions of the Edinger-Westphal nucleus alter food and water consumption. *Behavioral Neuroscience*, *119*(5), 1235–1243. http://dx.doi.org/10.1037/0735-7044.119.5.1235.

Weitemier, A. Z., Tsivkovskaia, N. O., & Ryabinin, A. E. (2005). Urocortin 1 distribution in mouse brain is strain-dependent. *Neuroscience*, *132*(3), 729–740. http://dx.doi.org/10.1016/j.neuroscience.2004.12.047.

Weitemier, A. Z., Woerner, A., Backstrom, P., Hyytia, P., & Ryabinin, A. E. (2001). Expression of c-Fos in Alko alcohol rats responding for ethanol in an operant paradigm. *Alcoholism, Clinical and Experimental Research*, *25*(5), 704–710.

Wise, R. A. (1973). Voluntary ethanol intake in rats following exposure to ethanol on various schedules. *Psychopharmacologia*, *29*(3), 203–210.

Kolibri 360 GmbH
Ferdinand-Jühlke-Straße 7
99095 Erfurt, Deutschland
produktsicherheit@kolibri360.de